Developments in Primatology:
Progress and Prospects

Series Editor: Louise Barrett

For further volumes:
http://www.springer.com/series/5852

Alexander Weiss · James E. King
Lindsay Murray
Editors

Personality and Temperament in Nonhuman Primates

Editors
Alexander Weiss
Department of Psychology
School of Philosophy
Psychology and Language Sciences
The University of Edinburgh
Scotland, Edinburgh, UK
alex.weiss@ed.ac.uk

James E. King
Department of Psychology
University of Arizona
Tucson, AZ, USA
kingj@u.arizona.edu

Lindsay Murray
Department of Psychology
University of Chester
Chester, UK
l.murray@chester.ac.uk

Please note that additional material for this book can be downloaded from http:///extras.springer.com

ISBN 978-1-4614-0175-9 e-ISBN 978-1-4614-0176-6
DOI 10.1007/978-1-4614-0176-6
Springer New York Dordrecht Heidelberg London

Library of Congress Control Number: 2011933598

Printed on acid-free paper

Springer is part of Springer Science+Business Media (www.springer.com)

To my wife Emily; siblings Katherine, Elizabeth, and Chris; and to my parents David and Gisela

Alexander Weiss

To Penny

James E. King

Thanks and love to Raymond and my favourite little primates, Ethan, Alyssa and Aaron, for all your support and inspiration during this journey

Lindsay Murray

Preface

This volume evolved from a symposium held at the American Society of Primatologists conference in Portland, Oregon, USA in August 2005. The symposium was organized by Lindsay Murray, Alex Weiss, and Sam Gosling with Jim King as the discussant. The symposium's purpose was to present an overview of the current status of research on personality in nonhuman primates. Sometime after the symposium Russ Tuttle approached two of the present editors (Alex Weiss and Jim King) about the idea of putting together this edited volume. We eagerly agreed and invited Lindsay Murray to participate in the project.

Primate personality research has progressed much since the 1930s and 1940s when primatologists began to notice large individual differences in the behavioral and emotional dispositions of monkeys and apes. Interest in the area waned and then briefly resurfaced when Jane Goodall, during her initial researches at Gombe National Park, described striking personality differences among the chimpanzees. These sometimes dramatic manifestations of distinct personalities should have been celebrated by the scientific community as much as the existence of tool making among chimpanzees. Instead the personality differences were largely ignored, no doubt because they were subjective and anthropomorphic at a time when primate behavior was synonymous with specific observable behaviors. Personality assessments were just as objectionable as Goodall's ascribing names to individual chimpanzees.

Fortunately, the attempts by behaviorists and ethologists to discourage Goodall from violating taboos against description of individual differences were spectacularly unsuccessful. Moreover, her interest in individual differences inspired other researchers to study nonhuman primate personality, though the more cautious sometimes have preferred identifying these phenomena as temperament or individuality. This behaviorally oriented approach to personality was exemplified by Chamove, Eysenck, and Harlow's early study of behavioral traits in rhesus monkeys that revealed dimensions similar to Eysenck's three human dimensions: Psychoticism, Extraversion, and Neuroticism. However, most notable were studies in 1978 on the personalities of chimpanzees in Gombe assessed with the Emotions Profile Index by Peter Buirski and the studies of Joan Stevenson-Hinde's group with the Madingley Questionnaire. These studies provided unambiguous evidence that subjective ratings of personality traits from different raters converged and, in fact, were stable over time and related to behavior, thus meeting criteria set by early skeptics of personality trait theory.

These early studies have had long-lasting impacts. Notably, the Madingley scale developed by Joan Stevenson-Hinde has been adapted for use in several species. Moreover, this scale is still being used and adapted by several researchers, including some in this volume, who conduct research in a diverse range of areas. Thus, the present volume would not be complete without Joan Stevenson-Hinde's and Camilla Hinde's chapter describing the development of this questionnaire and the early history of the field.

A fundamental issue in primate personality is the relationship between behavioral observations and subjective rater judgments in personality measurement. The chapter by Freeman and Gosling presents a detailed comparison of these two approaches. A further question is whether personality measures should be initially approached through purely behavioral measures or through personality ratings based on subjective rater judgments. Jana Uher, in her chapter makes a case for the former strategy within a broader context of methodological issues in measurement of primate personality.

Lindsay Murray's chapter also emphasized the importance of overall context in making personality measurements. In this chapter she describes a series of studies which highlight the contextual variables that mediate the relationships between personality and behavior in captive great apes. These findings and those of other chapters stress the need to take such contextual factors into account even when validating personality ratings of small captive groups.

Similarly, the chapter by Bard, Gaspar, and Vick on individual differences among chimpanzees in facial expression might, at first, appear to be an unusual contribution to a book on primate personality. However, as documented in the chapter, chimpanzees' facial expressions are clearly related to their emotional state and most likely to stable personality traits. We suspect that individual consistency in patterns of facial expression may offer new insights into personality differences not currently measured by other behaviors or by rating scales, particularly for traits related to emotional and agonistic states. The subtle differences in facial expression may offer some far more sensitive personality measures than more conventional personality measures.

As this chapter further indicates, measurement of facial expressions and categorization of those expressions into a meaningful taxonomy is enormously complex. This complexity has probably impeded research into the relationship between facial expression and personality. However, the Facial Action Coding System (FACS) devised by the authors and described in their chapter may now be a basis for more studies on this new behavioral dimension and personality.

As research in primate personality develops, questions about the causal links between personality variables and the outcomes should come under increasing scrutiny. The chapter by Capitanio describes how the rated sociability of rhesus monkeys is positively correlated with the vigor of their immune response. He also describes evidence of the physiological variables that mediate the link between personality and the immune response.

As with other endeavors that seek to understand complex systems, progress can be made by understanding the simplest components of the system. In this respect, we believe that the study of personality in nonhuman primates is well-positioned to aid researchers in understanding the relationship of personality to other variables. For example, studies of nonhuman primates have been used to better understand the proximate genetic and environmental causes of personality or temperament. In this vein, the chapters by Dee Higley, Steve Suomi, and Andrew Chaffin as well as that by Lynn Fairbanks and Matthew Jorgensen demonstrate how studies of primates have been used to examine the effects of genes, the environment, and gene by environment interactions which give rise to individual differences underlying personality or temperamental traits related to psychopathology, aggression, and alcoholism in humans. In addition, Stephanie Anestis' chapter reveals that nonhuman primate research may help researchers untangle the relationship between traits such as personality and hormones.

Finally, evolutionary psychologists including David Buss and Daniel Nettle have hypothesized and conducted research on the evolutionary bases of human personality dimensions and variation. We believe a major contribution of all of the chapters in this volume is to inform this area of research. In particular, King and Weiss describe their work applying the comparative method to methods and measures adapted from human personality research improve our understanding of the evolutionary bases of personality dimensions such as Conscientiousness. The final chapter describes a new approach to studying personality, viz., the study of behavioral syndromes. This approach is rooted in behavioral ecology and we believe that it, while not yet being fully implemented in studies of nonhuman primates, has great promise in developing an understanding of how stable personality differences arose and are maintained by physiological characteristics of individuals (e.g., body size or metabolic rate), the nature of the social and physical environments, and the constitution of the population.

In conclusion, we believe that the present volume offers much to those interested in studying the personality of nonhuman primates and other species (including humans). We also hope that this volume dispels concerns about the field's rigor with respect to methods and subject matter. Finally, we think that this volume demonstrates the richness and diversity of findings and inspires new researchers to take the sorts of bold steps needed to further our understanding of diversity in primates, whether they are nonhuman or human.

Edinburgh, UK Alexander Weiss
Tucson, AZ James E. King
Chester, UK Lindsay Murray

Contents

Contributors

Stephanie F. Anestis Department of Anthropology, Yale University, New Haven, CT 06520-8277, USA
stephanie.anestis@yale.edu

Kim A. Bard Psychology Department, University of Portsmouth, King Henry Building, Portsmouth, PO1 2DY, UK
Kim.Bard@port.ac.uk

John P. Capitanio California National Primate Research Center, University of California, One Shields Avenue, Davis, CA 95616, USA
jpcapitanio@ucdavis.edu

Andrew C. Chaffin Department of Psychology, Brigham Young University, 1042 SWKT, Provo, UT 84602, USA
andrew_chaffin@byu.edu

Lynn A. Fairbanks Department of Psychiatry and Biobehavioral Sciences, Semel Institute, University of California, 760 Westwood Plaza, Los Angeles, CA 90095, USA
lfairbanks@mednet.ucla.edu

Hani Freeman Department of Psychology, Georgia State University, PO Box 5010, Atlanta, GA 30302-5010, USA
hani.freeman@gmail.com

Augusta D. Gaspar ISCTE-IUL, Department of Psychology, CIS-IUL, Centre for Psychological Research and Intervention, Av. das Forças Armadas, 1649-026 Lisbon, Portugal
augusta.gaspar@iscte.pt

Samuel D. Gosling Department of Psychology, University of Texas, 1 University Station A8000, Austin, TX 78712, USA
samg@mail.utexas.edu

J. Dee Higley Department of Psychology, Brigham Young University, 1042 SWKT, Provo, UT 84602, USA
james_higley@byu.edu

Camilla A. Hinde Department of Zoology, The Edward Grey Institute, University of Oxford, South Parks Road, Oxford, OX1 3PS, UK
camilla.hinde@zoo.ox.ac.uk

Matthew J. Jorgensen Department of Pathology, Section on Comparative Medicine, Wake Forest University School of Medicine, Winston-Salem, NC 27157, USA
mjorgens@wfubmc.edu

James E. King Department of Psychology, University of Arizona, Tucson, AZ 85721, USA
kingj@u.arizona.edu

Lindsay Murray Department of Psychology, University of Chester, Parkgate Road, Chester, CH2 1DH, UK
l.murray@chester.ac.uk

Steven J. Schapiro Department of Veterinary Sciences, Michale E. Keeling Center for Comparative Medicine and Research, University of Texas M.D. Anderson Cancer Center, 650 Cool Water Drive, Bastrop, TX 78602, USA
sschapir@mdanderson.org

Andrew Sih Department of Environmental Science and Policy, University of California, One Shields Avenue, Davis, CA 95616, USA
asih@ucdavis.edu

Joan Stevenson-Hinde Department of Zoology, Sub-Department of Animal Behaviour, University of Cambridge, Madingley, Cambridge, CB23 8AA, UK
jgs11@cam.ac.uk

Stephen J. Suomi Laboratory of Comparative Ethology, National Institute of Child Health and Human Development, 6105 Rockledge Drive Suite 8030, MSC 7971, Bethesda, MD 20892–7971, USA
suomis@lce.nichd.nih.gov

Jana Uher Comparative Differential and Personality Psychology, Freie Universität Berlin, Habelschwerdter Allee 45, Berlin 14195, Germany
uher@primate-personality.net

Sarah-Jane Vick Department of Psychology, University of Stirling, Stirling FK9 4LA, UK
sarah-jane.vick@stir.ac.uk

Alexander Weiss Department of Psychology, School of Philosophy, Psychology and Language Sciences, The University of Edinburgh, 7 George Square, Scotland, Edinburgh, EH8 9JZ, UK
alex.weiss@ed.ac.uk

Part I
Beginnings of Personality Measurement in Nonhuman Primates

Chapter 1
Individual Characteristics: Weaving Psychological and Ethological Approaches

Joan Stevenson-Hinde and Camilla A. Hinde

Abstract The chapter begins by outlining the development of the questionnaire for assessing individual characteristics of rhesus monkeys in the Madingley colony. The steps taken in preparing items, obtaining reliability and validity, and condensing items are presented, along with examples of how scores arising from principal component analyses can be put to use. In our own research, the scores were referred to simply as "individual characteristics," in an attempt to avoid going beyond the data and implying either heritability or human-like qualities. *Fearfulness* is then explored as it is a particularly strong dimension, not only for our monkeys, but also *across* many species, from fish to humans. This suggests that fearful behavior may have been selected for during the course of evolution, with a presumed function of protection from harm. *Within* a species, individual differences in fearfulness may be a result of differing selection pressures, and recent research on how individual variation could be adaptive is discussed.

1.1 A Personal Note (by JSH)

In the late 1950s and early 1960s, at least at Mount Holyoke College and Brown University where I studied, the Psychology Departments' strengths were in the experimental realm: perception, psychophysics, physiological psychology, and operant conditioning. But in a graduate seminar on social learning in naturally occurring contexts, I came across the seminal work of William Thorpe on song learning in chaffinches. Here was learning of complex patterns that could be quantitatively assessed via sonograms. If Skinner was right, reinforcement should be involved. This led to a postdoctoral fellowship in Professor Thorpe's thriving laboratory at

J. Stevenson-Hinde (✉)
Department of Zoology, Sub-Department of Animal Behaviour,
University of Cambridge, Madingley, Cambridge, CB23 8AA, UK
e-mail: jgs11@cam.ac.uk

A. Weiss et al. (eds.), *Personality and Temperament in Nonhuman Primates*,
Developments in Primatology: Progress and Prospects, DOI 10.1007/978-1-4614-0176-6_1,

Madingley, for a visit that was supposed to last 1 year. In fact, this lasted longer than a year, and raised some basic, persistent issues. In using hand-reared or autumn-caught chaffinches with song as a reinforcer, I came to appreciate that, unlike laboratory-reared pigeons working for food, other factors had to be taken into account, such as early experience, hormonal levels, and even individual differences (see Stevenson-Hinde and Roper 1975). We also documented individual differences in the calls of adult sandwich terns and common terns as they returned from fishing to feed their young (Hutchison et al. 1968). The chicks in turn responded to their own parents' calls but not to those of near neighbors (Stevenson et al. 1970).

Also at Madingley was the rhesus monkey colony set up by Robert Hinde in 1959 with the encouragement of John Bowlby, to study the effects of maternal separation on the mother–infant relationship and the developing infant (Hinde 1977). The colony consisted of six indoor/outdoor enclosures, each with an adult male plus several adult females and their offspring. Within each group, the frequency and duration of a variety of social interactions were recorded daily. My interest in this research grew, and in 1972 (the year of our own first-born), I began to assess individual differences in young monkeys by coding discrete behaviors over a battery of behavioral tests administered outside the colony. However, another aspect of individual differences emerged. The observers who had been recording behavior in the colony over the years already had clear impressions of individuals. Each monkey in the colony had a name, such as Yuri or Miranda, and I recall walking past the Madingley coffee room and overhearing an in-depth discussion about someone I assumed to be human. I thought to myself, "They should not be talking in public like that, and I should not be listening!," when I realized the discussion concerned one of our monkeys. Given my background in psychology, I decided to capitalize on these observers' impressions by developing a "personality" questionnaire. Just as parents complete questionnaires on their children, observers could complete them on the monkeys. For example, in the field of childhood temperament (e.g., Thomas and Chess 1977), behaviorally defined items are rated by parents, thereby tapping into their long hours of observations over a range of social contexts. That is not to say that parents are not without biases, but their ratings are nevertheless reliable and valid (Simpson and Stevenson-Hinde 1985; Vaughn et al. 1992; Stevenson-Hinde and Glover 1996).

1.2 An Empirical Approach Toward Developing a "Personality" Questionnaire

Personality assessments for primates did exist in the 1970s, but they did not use rating scales. One involved directly recorded social behavior of rhesus monkeys who had been separated from their mothers at birth, with three principal factors emerging: fearful, hostile, and affectionate (Chamove et al. 1972). The other did not involve factor analysis, but rather imposed a structure based on Plutchik's theory of emotions. Although developed with humans, the theory was said to be based on "evolutionary considerations" and focused on "the reality and ubiquity of conflict of emotions in man and animals" (Plutchik 1970, p. 22). Pairwise comparisons of behaviorally defined adjectives were made for olive baboons (Buirski et al. 1973) and chimpanzees (Buirski et al. 1978). The outcome for each subject was eight scores, representing the relative strength of each of the "eight basic primary emotions." Thus, the method carried the questionable assumption that this is an adequate framework for human emotions, and further "that the same components (emotional dispositions) apply to both humans and lower animals" (Buirski et al. 1978). Fortunately, such an assumption is no longer made, but rather item ratings are used to empirically test analogies to factors found in humans, most notably the Five-Factor Model of personality (King and Figueredo 1997; Gosling and John 1999; Gosling 2001; Nettle 2006; Weiss et al. 2006). My training in Experimental Psychology, including Skinnerian antipathy toward traditional learning theorists such as Hull and Tolman, led me toward an empirical approach. This fitted nicely with the British and Dutch approach to Ethology which I found at Madingley. Thus, our first step was to make use of observers' knowledge of the monkeys within the social context in which they were living, by asking observers to provide descriptions of individuals. From these, and with reference to the adjectives used within Sheldon's *Scale for Temperament* (1942, p. 26), we created a list of 33 adjectives, each with a behavioral definition. Ratings were to be made on a 7-point scale, adapted from Sheldon (1942, p. 28). Pilot studies indicated which items did not give a good spread of ratings and which definitions needed clarifying. In this way, the list of items was tailored to fit our colony as well as our research objectives (Stevenson-Hinde and Zunz 1978).

The resulting list was accompanied by the following instructions:

Preliminary instructions. Although the ratings must be made independently, observers should first reach agreement on:

- The behavioral definition of each item with respect to their subjects
- The period of time the ratings should cover (e.g., the past 2 weeks)
- The period of time to complete the ratings (e.g., the next few days)

Instructions for rating. Consider each item according to its definition, and independently of any other item. Rate all subjects over one item before proceeding to the next item. (This is intended to prevent raters from unconsciously adjusting their ratings to give a coherent overall view of each subject, rather than focusing on the behavior specified for each item.)

Assign the following ratings according to a normal distribution over all the animals, giving about 30% of the subjects a rating of 4:

1. Extreme antithesis to the behavior.
2. The item is weakly represented, although traces are present.
3. The item is distinctly present, but falls a little below average.
4. The individual falls just about halfway between the two extremes or slightly above the general average.
5. The item is strong, although not outstanding.
6. The item is very strong and conspicuous, approaching the extreme.
7. Extreme manifestation of the behavior (Stevenson-Hinde et al. 1980a, Appendix).

Reliability. The observers, who were accustomed to identifying and recording preselected items of behavior, felt a little uncomfortable at being asked to be more than a mere recording instrument. A defense of *subjective* rating scales vs. *objective* coding is nicely set out in a recent article by Vazire et al. (2004). In making subjective ratings, the observer becomes an *active* instrument – filtering, cumulating, weighing, and integrating data, including rare but relevant behavior. This confers many advantages (Block 1977; Stevenson-Hinde 1983) that may outweigh any decrease in objectivity. However, because of the subjectivity involved, at least two observers should make ratings independently, so that reliability may be assessed and then the *mean rating* for each item used in further analyses. For our colony in 1977, inter-observer reliabilities ranged from a low of Pearson $r=0.24$ (for protective) to 0.75 (for subordinate), with a mean of .56 over all 25 items and with 21 items significant at $p<0.001$, one-tailed ($N=48$). Since we always used *mean ratings* of two observers, agreement is in fact higher than this. That is, if our mean ratings were correlated with the mean ratings of two equivalent observers, then the agreement estimated by the Spearman–Brown prophecy formula would be 0.72 for the 25 items (Stevenson-Hinde et al. 1980a).

Validity. Despite the advantages of rating scales, it is not possible to validate them properly against recorded frequencies or duration of particular behavior patterns. Nevertheless, six of our final 21 questionnaire items bore some correspondence with direct observations of behavior. Spearman correlations between those six items and corresponding direct observations were all significant ($p<0.001$, one-tailed, $N=48$). However, the correlations ranged from 0.73 (for *Effective* cf. the sum of occurrences of displacement of others and avoidance by others) down to 0.45 (for *Excitable* cf. the sum of displays and threats directed outside the pen). The lower correlations could reflect a lack of overlap in meaning between the ratings and direct observations rather than poor validity (Stevenson-Hinde et al. 1980a).

1.2.1 Condensing Items

In order to condense items, we relied on principal component analyses. Over *each* of four successive years, the following three main components emerged and accounted for over 65% of the variance: (C1) Confident to Fearful; (C2) Excitable to Slow; and (C3) Sociable to Solitary. The following items loaded either highly positively or highly negatively on one of the main components:

C1

(+) *Confident*: Behaves in a positive, assured manner; not restrained or tentative
(+) *Effective*: Gets own way; can control others
(+) *Aggressive*: Causes harm or potential harm
(−) *Apprehensive*: Is anxious about everything; fears and avoids any kind of risk
(−) *Subordinate*: Gives in readily to others; submits easily
(−) *Fearful*: Makes fear grins; retreats from others or from outside disturbances

C2

(+) *Excitable*: Overreacts to change
(+) *Active*: Moves about a lot
(−) *Equable*: Reacts to others in an even, calm way; not easily disturbed
(−) *Slow*: Moves and sits in a relaxed manner; moves slowly and deliberately, not easily hurried

C3

(+) *Sociable*: Seeks companionship of others
(+) *Opportunistic*: Seizes a chance as soon as it arises
(−) *Solitary*: Spends time alone

Since specific item loadings are subject to sampling error, they will vary arbitrarily from year to year. Therefore in calculating component scores, we recommend using unit weightings for those items that load either highly positively or highly negatively. That is, over the above items, a *score for each monkey on each component* was calculated by multiplying the standardized rating by either +1 or −1 (according to component loadings), summing over the items, and then dividing by the number of items in each component. Thus, each component score had the properties of a standardized score, with a mean of zero over all individuals rated that year and a standard deviation of just less than one (Stevenson-Hinde et al. 1980a).

1.2.2 Consistency

Our monkeys were rated each November – a quiet time after summer births but before mating. Regarding consistency in component scores over time, say for Confident to Fearful scores, significant correlations occurred from age 1 to 2, 2 to 3, and 3 to 4 years. With adult males, consistency over four successive years was 0.65 (Kendall correlation, $p<0.001$, $n=5$) and adult female consistency was even higher at 0.90 ($p<0.001$, $n=11$). Not included in the latter group were the seven primiparous females assessed over the year from pre- to postbirth. Their Confident-Fearful consistency was nonsignificant and far lower (0.46) than that of the other adult females (Stevenson-Hinde et al. 1980a).

1.2.3 Specific Issues

Component scores were also used to address specific issues, including effects of adverse experience such as separation or being orphaned (Stevenson-Hinde et al. 1980a); predictability to test situations outside the colony (Stevenson-Hinde et al. 1980b, c); or how mothers' rather stable characteristics relate to those of her developing infant. Interestingly, mothers and their year-old daughters were alike in that their scores were significantly correlated for each of the three main components: Confident, Excitable, and Sociable. With sons, M-I Sociable scores were significantly correlated. The other significant correlation was negative, with Excitable mothers having less Confident sons (Spearman $r = -0.76$, $p < 0.01$, two-tailed, $n = 14$; Stevenson-Hinde et al. 1980a).

Direct observations of mother–infant interactions in the colony over the first year of life suggested how this pattern of correlations might have arisen. For example, while Confident mothers tended not to reject or leave their daughters, they rejected their sons early on (at 8 weeks). On the other hand for both sons and daughters, mothers' Excitable scores were positively correlated with mothers approaching as well as leaving, and with restricting at the age of weaning. Reciprocally, infants approached Excitable mothers relatively frequently and year-old sons played a relatively high role in maintaining proximity with mother. "In terms of attachment theory, Confident mothers appear to provide a *secure base* for their daughters, but reject their sons when very young. Excitable mothers appear to behave inappropriately to both sons and daughters, producing infants who may be *insecurely attached*" (Stevenson-Hinde and Simpson 1981).

In natural situations a daughter will remain in her natal group, whereas a son will leave at 3.5–5 years and have to rely on his own characteristics. Indeed, *sons'* component scores were more predictive of their behavior in test situations outside the colony than were daughters' scores (Stevenson-Hinde et al. 1980b). We have seen that sons' Confidence is related to being raised by low-Excitable, relaxed mothers. Overall, "the findings suggest that a son may bring to a new situation some characteristics related to those of his mother, which may also be stable across situations and across time" (Stevenson-Hinde and Simpson 1981). Moreover, subsequent research using our methods on adult male rhesus produced dimensions similar to ours, with "predictive power that is both long-term and cross-situational" (Capitanio 1999).

1.3 Fearfulness: Function and Causation

We have seen that the strongest dimension to emerge from the Madingley colony was *Confident–Fearful*. Fearful behavior in unfamiliar or challenging situations has been observed *across diverse taxa* – crustaceans, cephalopods, fish, birds, and mammals (e.g., Wilson et al. 1994; Gosling and John 1999; Dingemanse et al. 2004), including humans (Kagan et al. 1988; John 1990; Stevenson-Hinde and Marshall 1999). This suggests that such behavior may have been selected for by one route or another during the course of evolution. That is, those individuals who exhibited fearful behavior to an appropriate degree would have been more apt to survive and leave offspring – i.e., to have increased their inclusive fitness – compared with those who did not. Thus, as John Bowlby suggested many years ago for humans, a propensity for fearful behavior may have been guided by natural selection, with the function of protection from harm (Bowlby 1969).

Furthermore, variation in fearfulness occurs across individuals *within* a species, as reflected by the *Confident to Fearful* dimension for our rhesus monkeys. Based on mammalian studies, neurobiological models for individual differences in fearfulness indicate that fearful individuals may differ from others at several levels – in peripheral sensory receptor systems; in the early relaying of sensory information through the thalamus; in the processing of sensory data in the cortex or in subcortical structures such as the amygdala; in the efferent autonomic, neuroendocrine, and motor responses; or in the feedback from these responses (Marshall and Stevenson-Hinde 2001). Particularly involved is the central nucleus of the amygdala: Damage to the amygdala interferes with fear-related behavior, stimulation of the central nucleus of the amygdala activates neural circuitry underlying startle responses, stimulation heightens attention toward fearful events, and neurons within the amygdala are reactive to fearful signals. "Thus data from many avenues strongly suggest that the amygdala and its associated neural circuitry appraise fearful signals and orchestrate behavioral and autonomic responses to these events" (Schulkin and Rosen 1999). Neurohormonal studies of fearfulness in birds and other taxa may enable examination of how a common behavioral outcome may be achieved via different underlying mechanisms.

Back when we discovered that the main principal component emerging year after year involved a *Confident to Fearful* dimension, Jerry Kagan at Harvard, who studied *behavioral inhibition* in children, took this result with monkeys to indicate the *inherent biological basis* of fearfulness. On the other hand, we focussed on the influence of antecedent rhesus *mother–infant interactions* on the traits of their developing infants (Stevenson-Hinde and Simpson 1981). Of course both views are compatible, with inherent biological predispositions developing within the context of early close relationships. Across many species, gene × environment interactions involving fearfulness are being specified throughout the course of development (e.g., Sih et al. 2004b; Suomi 2004).

1.4 Selection for Individual Differences

From an evolutionary perspective, why should differences in fearfulness as well as other individual differences persist? Behavioral Ecologists have studied fearful behavior across a range of species, including shy and bold behavior in pumpkinseed fish (Wilson et al. 1994), fast and slow exploratory behavior in great tits (Verbeek et al. 1994), and responses to novel environment in rats (Cavigelli and McClintock 2003). Individual differences had been assumed to be "noise" around presumably adaptive norms. However, the fact that many individual differences in behavior are consistent, heritable, and correlated either with each other or with life history variables suggests they may have adaptive significance (Wilson 1998). The interesting question for Behavioral Ecologists is how are these differences maintained? If they are adaptive, why doesn't selection drive behavior toward one optimum? Is it not best to show behavioral plasticity around this optimum to cope with different situations rather than a fixed strategy?

Both theoretical (John 1990; Sih et al. 2004a; Wolf et al. 2007) and empirical (Sih et al. 2004b; Dingemanse and Réale 2005; van Oers et al. 2005; Smith and Blumstein 2008) research is beginning to unravel these questions. A pioneering study of great tits in the Netherlands allows fitness consequences of personality types to be measured in the wild (Dingemanse et al. 2002), combined with the study of selection lines for these heritable personality traits in captivity (Drent et al. 2003; van Oers et al. 2004). Wild-caught great tits were identified as "fast" or "slow" explorers by housing them overnight and then allowing them individual access to a room containing five artificial "trees." The number of flights and hops within the first 2 min was used as an index of personality type, which was shown to be repeatable and heritable (Dingemanse and Réale 2005). Males with "fast" personality types were found to produce more offspring than "slow" males in good years, when competition for territories was high, and females with "fast" personalities produced more offspring than "slow" females in poor years when they were presumably better at competing for food (Dingemanse et al. 2004). This work shows that fluctuating environmental pressures could lead to fluctuations in competition and variation in selection pressure between years, maintaining genetic variation in personalities (Dingemanse et al. 2004).

In addition to fluctuating environments, variation could be maintained through frequency dependence, where the fitness consequence of one strategy depends on the frequency of others in the population (Wilson et al. 1994; Dall et al. 2004; Bell 2007). For example, a sneaking strategy could be selected for if it were unusual (Bell 2007). A recent model suggests that personality may reflect an underlying life history trade-off, which is linked to life history strategies such as whether to reproduce early, or to be more risk averse and delay reproduction (Wolf et al. 2007). Further work should explore reproductive strategies between personality types to see if this is the case. Our knowledge of the fitness consequences of variation in personality traits is limited and the field is still developing. Even less is known about the possible link between personality traits and aspects of sexual selection (but see van Oers et al. 2008).

The question still arises; why not be flexible, instead of having relatively fixed behavioral types? It could be that there is a cost to plasticity, and more specialized behavioral types perform better under certain situations (see Dall et al. 2004). Wilson et al. (1994) have speculated about how natural selection may produce *phenotypically inflexible* genotypes as well as *phenotypically plastic* genotypes. Referring to shyness and boldness in pumpkinseed fish, they argue that in a constant environment the inflexible *shy* or *bold* individuals should replace the plastic form. However, "If the opportunities for risk-prone and risk-averse individuals are temporally variable … natural selection will promote a mixture of innate and facultative forms, whose relative proportions will depend on the magnitude of temporal variation" (p. 445). Alternative explanations include a reaction norm perspective (van Oers et al. 2005; Penke et al. 2007) whereby personality genotypes may differ not only in how *phenotypically plastic* they are, but also in their behavioral expression in a range of environments. If this phenotypic plasticity shows a nonparallel pattern, reaction norms may cross (genotype × environment interaction). Absolute behavioral differences between personality types may thereby be context specific depending on the environment.

In this relatively young field that integrates personality theory with behavioral ecology, main issues include plasticity and genotype × environment interactions, the link with life-history trade-offs, and sexual selection. On a more integrative level, questions arise as to how general the personality construct is when comparing populations under different selection pressures, or even when comparing species. The field would benefit greatly from a clear conceptual framework with clear hypotheses on these different matters. Integration across disciplines is essential.

1.5 A Note on Terminology

In the field of Ethology or Behavioral Ecology, unlike Psychology, individual differences are typically assessed by direct observations of the frequency or duration of specific behaviors which are then organized in some way. For example, the concept of a *behavior system* refers to the organization of behavior patterns which share a common causation and subserve a particular biological function (e.g., Bowlby 1969; Baerends 1976). Emphasis was placed on the *context* in which the behavior occurred, such as an unfamiliar or challenging situation for fear behavior or separation from mother for attachment behavior. Changes in behavior *within an individual* were conceptualized in terms of the activation or deactivation of particular systems. Thus, activation of a *fear behavior system* would lead to activation of an *attachment behavior system* and deactivation of an *exploratory behavior system* (e.g., Greenberg and Marvin 1982). Furthermore, *individual differences* could be conceptualized in terms of the threshold of activation of any particular system, so that a shy child could be said to have a *fear behavior system* with a low threshold of arousal (e.g., Stevenson-Hinde 1991).

With the more recent concept of a *behavioral syndrome*, the organization of behavior relies not on postulated common causation, but rather on correlations. A *behavioral syndrome* is "a suite of correlated behaviors reflecting between-individual consistency in behavior across multiple (two or more) situations (see Sih 2011). A population or species can exhibit a behavioral syndrome. Within the syndrome, individuals have a behavioral type (e.g., more aggressive vs. less aggressive behavioral types)" (Sih et al. 2004a). Thus, a *behavioral syndrome* implies a set of correlations (see also Dall et al. 2004; Sih 2011), without the conceptual overtones of a *behavior system*.

In studies of individual differences in infants and young children, the term *temperament* (Thomas and Chess 1977) is preferred over *personality*, with its overtones of adult-like motivations and attitudes. However, since *temperament* implies some degree of heritability, our rhesus monkey scores were referred to simply as *individual characteristics*, without prejudging heritability. Although the term "personality" is eye-catching when applied to any species, it carries the risk of carrying "excess baggage" in terms of implying more than is warranted by the data.

1.6 Full Circle

In the mid-1970s, Robert Hinde and I (JSH) turned to studying *human* offspring. We explored the relation between mother–child interactions at home and peer–peer interactions in playgroup. Two characteristics stood out: the *shyness* of some of the children, and the *security* gained by having a sensitively responsive mother (Ainsworth et al. 1978; Bowlby 1988). Perhaps both constructs came into focus thanks to earlier observations with rhesus monkey mothers and infants, together with the centrality of the Confident to Fearful scores year after year. Thus we came full circle, from Robert Hinde starting the rhesus monkey colony in order to help John Bowlby establish attachment theory, to our using attachment theory to understand the development of individual characteristics of rhesus monkeys (Stevenson-Hinde and Simpson 1981) and later young children (Stevenson-Hinde and Marshall 1999; Stevenson-Hinde 2005). As this chapter and indeed this volume suggest, interchanges are continuing between psychological and ethological approaches, shedding light on the development, causation, function, and evolution of consistent individual characteristics.

Acknowledgments We would like to thank Kees van Oers for his invaluable input, as well as Sam Gosling and Robert Hinde for their constructive comments. We would also like to thank Alex Weiss for his helpful editorial suggestions.

References

Ainsworth MDS, Blehar MC, Waters E et al. (1978) Patterns of attachment. Erlbaum, Hillsdale, NJ

Baerends GP (1976) The functional organization of behaviour. Anim Behav 24:726–738

Bell AM (2007) Animal personalities. Nature 447:539–540

Block J (1977) Advancing the psychology of personality: Paradigmatic shift or improving the quality of research. In: Magnusson D, Endler NS (eds) Personality at the crossroads. Wiley, New York

Bowlby J (1969) Attachment and loss, Vol. I: Attachment. Hogarth Press, London

Bowlby J (1988) A secure base: Clinical applications of attachment theory. Routledge, London

Buirski P, Kellerman H, Plutchik R et al. (1973) A field study of emotions, dominance, and social behavior in a group of baboons (*Papio anubis*). Primates 14:67–78

Buirski P, Plutchik R, Kellerman H (1978) Sex differences, dominance, and personality in the chimpanzee. Anim Behav 26:123–129

Capitanio JP (1999) Personality dimensions in adult male rhesus macaques: Prediction of behaviors across time and situation. Am J Primatol 47:299–320

Cavigelli SA, McClintock MK (2003) Fear of novelty in infant rats predicts adult corticosterone dynamics and an early death. Proc Natl Acad Sci USA 100:16131–16136

Chamove AS, Eysenck HJ, Harlow HF (1972) Personality in monkeys: Factor analysis of rhesus social behaviour. Q J Exp Psychol 24:496–504

Dall SRX, Houston AI, McNamara JM (2004) The behavioural ecology of personality: Consistent individual differences from an adaptive perspective. Ecol Lett 7:734–739

Dingemanse NJ, Both C, Drent PJ et al. (2004) Fitness consequences of avian personalities in a fluctuating environment. Proc Biol Sci 271:847–852

Dingemanse NJ, Both C, Drent PJ et al. (2002) Repeatability and heritability of exploratory behaviour in great tits from the wild. Anim Behav 64:929–938
Dingemanse NJ, Réale D (2005) Natural selection and animal personality. Behaviour 142:1165–1190
Drent PJ, van Oers K, van Noordwijk AJ (2003) Realized heritability of personalities in the great tit (*Parus major*). Proc Biol Sci 270:45–51
Gosling SD (2001) From mice to men: What can we learn about personality from animal research? Psychol Bull 127:45–86
Gosling SD, John OP (1999) Personality dimensions in nonhuman animals: A cross-species review. Curr Dir Psychol Sci 8:69–75
Greenberg MT, Marvin RS (1982) Reactions of preschool children to an adult stranger: A behavioral systems approach. Child Dev 53:481–490
Hinde RA (1977) Mother-infant separation and the nature of inter-individual relationships: Experiments with rhesus monkeys. Proc Biol Sci 196:29–50
Hutchison RE, Stevenson J, Thorpe WH (1968) The basis for individual recognition by voice in the Sandwich Tern (*Sterna sandvicensis*). Behaviour 32:150–157
John OP (1990) The "Big Five" factor taxonomy: Dimensions of personality in the natural language and in questionnaires. In: Pervin LA (ed) Handbook of personality: Theory and research. Guilford, New York
Kagan J, Reznick JS, Snidman N (1988) Biological bases of childhood shyness. Science 240: 167–171
King JE, Figueredo AJ (1997) The Five-Factor Model plus Dominance in Chimpanzee Personality. J Res Pers 31:257–271
Marshall PJ, Stevenson-Hinde J (2001) Behavioral inhibition: Physiological correlates. In: Crozier WR, Alden LE (eds) International handbook of social anxiety. Wiley, Chichester, UK
Nettle D (2006) The evolution of personality variation in humans and other animals. Am Psychol 61:622–631
Penke L, Denissen JJA, Miller GF (2007) The evolutionary genetics of personality. Eur J Pers 21:549–587
Plutchik R (1970) Emotions, evolution, and adaptive processes. In: Arnold M (ed) Feelings and emotions. Academic Press, New York
Schulkin J, Rosen JB (1999) Neuroendocrine regulation of fear and anxiety. In: Schmidt LA, Schulkin J (eds) Extreme fear, shyness, and social phobia. Oxford University Press, Oxford
Sheldon WH (1942) The varieties of temperament: A psychology of constitutional differences. Harper & Brothers, New York
Sih A, Bell A, Johnson JC (2004a) Behavioral syndromes: An ecological and evolutionary overview. Trends Ecol Evol 19:372–378
Sih A, Bell AM, Johnson JC et al. (2004b) Behavioral syndromes: An integrative review. Q Rev Biol 79:241–277
Sih A (2011) Behavioral syndromes: A behavioral ecologist's view on the evolutionary and ecological implications of animal personality. In: Weiss A, King JE, Murray L (eds) Personality and temperament in nonhuman primates. Springer, New York
Simpson AE, Stevenson-Hinde J (1985) Temperamental characteristics of three- to four-year-old boys and girls and child-family interactions. J Child Psychol Psychiatry 26:43–53
Smith BR, Blumstein DT (2008) Fitness consequences of personality: A meta-analysis. Behav Ecol 19:448–455
Stevenson-Hinde J (1983) Individual characteristics of mothers and their infants. In: Hinde RA (ed) Primate social relationships. Blackwell, Oxford
Stevenson-Hinde J (1991) Temperament and attachment: An eclectic approach. In: Bateson P (ed) Development and integration of behaviour. Cambridge University Press, Cambridge
Stevenson-Hinde J (2005) On the interplay between attachment, temperament, and maternal style. In: Grossmann KE, Grossmann K, Waters E (eds) Attachment from infancy to adulthood. Guilford, New York
Stevenson-Hinde J, Glover A (1996) Shy girls and boys: A new look. J Child Psychol Psychiatry 37:181–187

Stevenson-Hinde J, Marshall PJ (1999) Behavioral inhibition, heart period, and respiratory sinus arrhythmia: An attachment perspective. Child Dev 70:805–816

Stevenson-Hinde J, Roper R (1975) Individual differences in reinforcing effects of song. Anim Behav 23:729–734

Stevenson-Hinde J, Simpson MJ (1981) Mothers' characteristics, interactions, and infants' characteristics. Child Dev 52:1246–1254

Stevenson-Hinde J, Stillwell-Barnes R, Zunz M (1980a) Subjective assessment of rhesus monkeys over four successive years. Primates 21:66–82

Stevenson-Hinde J, Stillwell-Barnes R, Zunz M (1980b) Individual differences in young rhesus monkeys: Consistency and change. Primates 21:498–509

Stevenson-Hinde J, Zunz M (1978) Subjective assessment of individual rhesus monkeys. Primates 19:473–482

Stevenson-Hinde J, Zunz M, Stillwell-Barnes R (1980c) Behaviour of one-year-old rhesus monkeys in a strange situation. Anim Behav 28:266–277

Stevenson JG, Hutchison RE, Hutchison JB et al. (1970) Individual recognition by auditory cues in the common tern (*Sterna hirundo*). Nature 226:562–563

Suomi SJ (2004) How gene-environment interactions shape biobehavioral development: Lessons from studies with rhesus monkeys. Res Hum Dev 1:205–222

Thomas A, Chess S (1977) Temperamant and development. Brunner/Mazel, New York

van Oers K, de Jong G, van Noordwijk AJ et al. (2005) Contribution of genetics to the study of animal personalities: A review of case studies. Behaviour 142:1185–1206

van Oers K, Drent PJ, de Goede P et al. (2004) Realised heritability and repeatability of risk-taking behaviour in relation to avian personalities. Proc Biol Sci 271:65–73

van Oers K, Drent PJ, Dingemanse NJ et al. (2008) Personality is associated with extrapair paternity in great tits, *Parus major*. Anim Behav 76:555–563

Vaughn BE, Stevenson-Hinde J, Waters E et al. (1992) Attachment security and temperament in infancy and early childhood: Some conceptual clarifications. Dev Psychol 28:463–473

Vazire S, Gosling SD, Dickey AS et al. (2004) Assessing animal personality: Behavioral codings or trait ratings? In Robins, RW, Fraley, RC, Krueger, RF (eds) Handbook of research methods in personality psychology. Guilford Press, New York

Verbeek MEM, Drent PJ, Wiepkema PR (1994) Consistent individual differences in early exploratory behavior of male great tits. Anim Behav 48:1113–1121

Weiss A, King JE, Perkins L (2006) Personality and subjective well-being in orangutans (*Pongo pygmaeus* and *Pongo abelii*). J Pers Soc Psychol 90:501–511

Wilson DS (1998) Adaptive individual differences within single populations. Philos Trans R Soc Lond B Biol Sci 353:199–205

Wilson DS, Clark AB, Coleman K et al. (1994) Shyness and boldness in humans and other animals. Trends Ecol Evol 9:442–446

Wolf M, van Doorn GS, Leimar O et al. (2007) Life-history trade-offs favour the evolution of animal personalities. Nature 447:581–585

Part II
Some Basic Issues in Personality Measurement in Nonhuman Primates

Chapter 2
Comparison of Methods for Assessing Personality in Nonhuman Primates

Hani Freeman, Samuel D. Gosling, and Steven J. Schapiro

Abstract A review of the field of primate personality revealed that two main methods have been used to study primate personality: behavioral codings and observer trait ratings. Both of these methods can be broken down by the conditions in which they are used – naturalistic observation, testing context, or cumulative observation. The most commonly used method-condition combination to assess primate personality has been the behavioral coding/naturalistic observation combination. This combination was used in more than half (60%) of the studies reviewed. Strengths and weaknesses of each of the method-condition combinations are discussed. A review of empirical studies comparing the behavioral-coding and trait-rating methods revealed that the trait-rating method seems to be better at assessing primate personality. However, only three scales have been used for 66% of the primate personality studies using a trait-rating method. Two procedures are suggested in order to develop new ratings scales for studying primate personality. Recommendations are made for the best method/condition combination to use and how to improve the use of each combination when assessing primate personality.

2.1 Introduction

Empirical research on nonhuman primate personality began in earnest in the 1930s. In 1938, for example, Crawford conducted a study of chimpanzees, showing that a reliable rating scale could be developed to assess personality. In this groundbreaking study, personality was assessed by having individuals who were familiar with the animals rate them on a list of 22 items. Soon afterwards, Robert Yerkes (1940) assessed personality as a factor in dominance behavior in chimpanzees, but instead of using a

H. Freeman (✉)
Department of Psychology, Georgia State University, PO Box 5010,
Atlanta, GA 30302-5010, USA
e-mail: hani.freeman@gmail.com

A. Weiss et al. (eds.), *Personality and Temperament in Nonhuman Primates*,
Developments in Primatology: Progress and Prospects, DOI 10.1007/978-1-4614-0176-6_2,

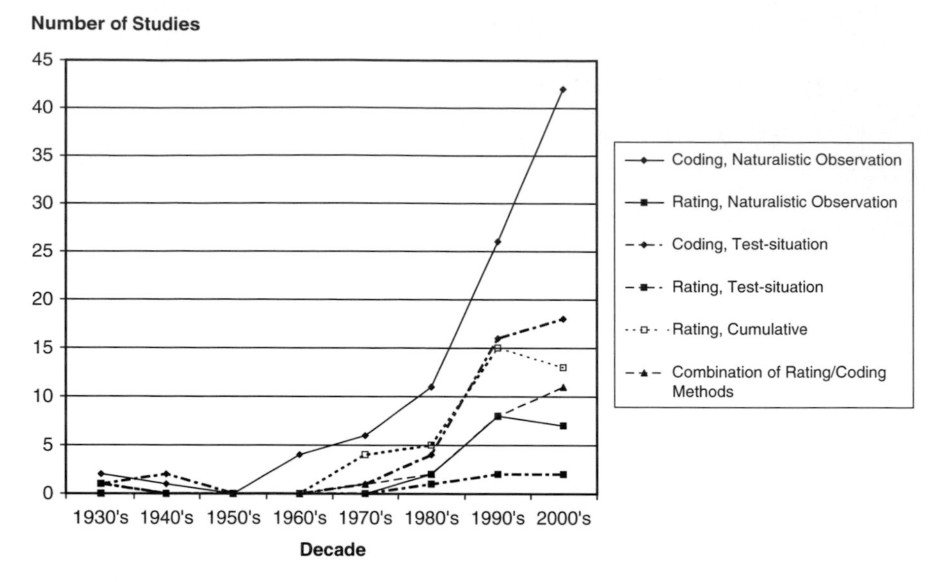

Fig. 2.1 Number of primate personality articles published each decade using each method of personality assessment

rating method, Yerkes assessed personality by recording the frequency and duration of different behaviors in a natural setting (Yerkes 1940). Thus, even in the early days of primate research, diverse methods were being used to assess personality.

Despite this important early work, studies of primate personality were still infrequent. As shown in Fig. 2.1, it was not until the 1960s and 1970s that primate personality research began to gain momentum (Stevenson-Hinde and Hinde 2011). The increased attention continued to be marked by methodological diversity, a trend that has persisted to the present day. However, only a few studies have directly compared assessment methods in terms of their accuracy and efficiency in capturing an animal's personality (Capitanio 1999; Gosling 2001; Itoh 2002; Gosling et al. 2003; Rouff et al. 2005; Vazire et al. 2007).

In this chapter, we will examine and evaluate different methods to assess primate personality. Our analyses will draw on an in-depth review of the literature. We shall discuss the strengths and weaknesses of each method, paying particular attention to the studies that have directly compared assessment methods. We shall conclude by offering some concrete steps that researchers may wish to consider as they measure personality in their own research.

2.2 Literature Review

The review of methods presented in this chapter draws on a much more extensive review of the primate personality literature (Freeman and Gosling 2010). The present review included full-length research articles published in journals or books.

In some cases the latest research is reported as an abstract published in conference proceedings and may never become a published article. Therefore, we included in our review abstracts not published in full-length elsewhere, if they reported sufficient substantive details to contribute to our review.

We based our literature search on keyword searches (or their equivalent) in AnthroPlus, Biosis, PrimateLit, Psychinfo, and Web of Science databases. We searched for all articles containing the keywords "primate and temperament" or "primate and personality." In addition, some relevant articles mentioned the species by name without mentioning the term "primate" in the search fields. Therefore, we repeated the searches replacing "primate" with terms that would allow us to capture all primate species and subspecies (e.g., "Monkey," "Lemur," "Gibbon," "Bushbaby"). The only primate species excluded from our search was humans. The resulting pool of reports was supplemented with articles cited in the papers identified in the keyword search and from nominations by experts in the field. Articles not directly relevant to the present review were eliminated from the pool (for details see Freeman and Gosling 2010).

The 154 studies identified in our review served as the basis for our analyses of the different methods used to assess primate personality and their strengths and weaknesses. In addition, we consider several arguments drawn from the human personality literature concerning the reliability and efficiency of the different methods.

2.3 Types of Assessment Methods

Two main methods of data collection have been used in primate-personality studies: behavioral codings and observer trait ratings. These two methods also characterize almost all research on nonprimate personality as well (Gosling 2001).

Behavioral coding (used in 87% of studies) involves observing animals and recording their behavior, typically in terms of bout frequency and/or duration. For example, one behavioral-coding study investigated the relationship between behavioral style, dominance rank, and cortisol levels in chimpanzees (Anestis 2005, 2011); the study used an all-occurrences sampling technique (i.e., one in which behavior was recorded anytime one chimpanzee interacted with another) to record social interactions of various kinds (e.g., agonistic, affiliative, etc.).

Trait rating (used in 40% of studies) involves having people who are familiar with the subjects rate them on a set of predefined traits or adjectives. Typically a rating scale is used in which high numbers indicate the trait is strongly characteristic of the target animal and low numbers indicate the trait is strongly uncharacteristic of the animal. For example, King and Figueredo (1997) assessed personality in 100 chimpanzees across 12 US zoos using rating data from an instrument that contained 43 adjectives, such as "playful" and "affectionate." These two methods (behavioral coding and trait rating) can further be broken down by the conditions in which they are used – naturalistic observation, testing contexts, or cumulative observation. Naturalistic observation (used in 71% of studies) entails coding or rating the animals over a specific period of time based on their ordinary daily behavior. French's (1981) study of Japanese macaques and individual differences in play behavior would fall into this category because the method of data collection was based on sessions in which the animals were observed in their home cages.

Testing contexts (used in 29% of studies) involve rating or coding the behavior of target animals in response to a particular stimulus or experiment designed to elicit differing reactions from animals with different personalities; Fairbanks' (2001) study of social impulsivity in vervet monkeys would fall into this category because the behaviors of the animals were recorded in response to a change in their typical daily routine, i.e., a new individual being shown to, but not introduced to, the group (Fairbanks and Jorgensen 2011).

The third condition, cumulative observation, (used in 25% of studies) applies only to the rating method and refers to the studies in which ratings are based on the knowledge and experience that each rater has accumulated since he or she has known the target animal; Martin's (2005) study of chimpanzees would fall into this category because the observers rated the animals based on their memories of experiences accumulated over the course of their acquaintance, which ranged from 1 to over 20 years.

The various combinations of assessment methods and assessment conditions can be combined into a five-cell grid shown in Table 2.1. Note, the cell representing the combination of the behavioral-coding method/cumulative-experience condition is absent because implementing such a method would be beyond the capacity of human memory; in practice, the missing cell would be equivalent to the cell that combines the rating method/cumulative experience. Also note that the percentages add up to more than 100 because some studies used more than one assessment method (14%).

2.4 How Often Are Each of These Methods Used?

In terms of the five-cell grid depicted in Table 2.1, the most commonly used method-condition combination was behavioral codings/naturalistic observation. This combination was used in more than half (60%) of the studies reviewed. The next most commonly used method-condition combination was behavioral codings/testing contexts, which was used in 27% of studies. This was closely followed by trait-rating/cumulative observation method-condition, which was used in 25% of the studies. Many fewer studies used the remaining combinations. Trait rating/naturalistic observation was used in only 11% of the studies and the trait-rating/testing combination was used in a mere 4% of the studies. To better understand the relative prevalence of these method-condition combinations, we next consider the strengths and weaknesses of the different methods. This analysis will help guide researchers in choosing the method or methods best suited to their research questions.

Table 2.1 Method of collecting/recording data

	Rating (40% of articles)	Coding (87% of articles)
	Pros – Faster than coding – Takes variability due to noise into account – Takes cross-situational consistency into account Cons – Requires subjective judgment by observer – Comparisons between animals can be hard to interpret – Raters may weigh salient events more	Pros – Requires less subjective judgment by observer – Easier to make direct comparisons between animals Cons – Time consuming – Hard to account for cross-situational consistency – Hard to account for variability due to noise
Naturalistic behavior (71% of articles) Pros – Convenient – Less vulnerable (than cumulative observation) to rater differences in familiarity Cons – Based on only observable animals and behaviors – Observations based on behaviors in only one context	Example Capitanio (1999). Trained observers rated 42 adult male rhesus macaques on adjectives (e.g., sociable, aggressive) along a 7-point scale. Ratings were completed after 4 weeks of observations of the animals in their natal groups	Example Kaplan et al. (2002). Twenty-five adult male cynomolgus macaques were coded using a combination of all-occurrences and scan sampling to record behaviors (e.g., fights wins/losses, grooming) from a predetermined ethogram. The behaviors were recorded while the subjects were in their home cage
Test context (29% of articles) Pros – High level of control over situation – Increased data collection on specific behaviors – Less vulnerable (than cumulative observation) to rater differences in familiarity Cons – Requires manipulating the subjects – Based on specific behaviors during a limited time	Example Bard and Gardner (1996). Twenty-nine infant chimpanzees were rated on 30 items (e.g., responsiveness to objects, cooperativeness) along a 9-point scale, following tests of infant behavior	Example Hebb (1949). Thirty captive adult chimpanzees were coded using all-occurrence sampling to record behaviors (e.g., grooming, gesturing with fingers through mesh) from a predetermined ethogram. The behaviors were recorded while the subjects were involved in various test situations

(continued)

Table 2.1 (continued)

Cumulative experience (25% of articles)	Example
Pros – Fastest of all methods – Based on years of observations of the subjects (i.e., takes into account many behaviors and contexts) Cons – Not based on direct data collection	Stevenson-Hinde and Zunz (1978). Trained observers with a minimum of 200 h experience, rated 46 rhesus macaques on a list of 33 adjectives (e.g., confident, eccentric) along a 7-point scale

Note: The cell representing the combination of the behavioral-coding method and the cumulative-experience context is absent because implementing such a method would be beyond the capacity of human memory; in practice, the missing cell would be equivalent to the cell that combines the rating method and cumulative experience

2.5 Range of the Present Analysis

Before beginning our discussion about the strengths and weaknesses of the different methods of primate personality assessment we want to be clear about the range of applications covered by our analysis. Our discussion specifically focuses on rating and coding methods in the context of measuring personality. Many, or most, of the points we make in this context may have absolutely no bearing on the use of these methods in other contexts, such as observing behaviors in experimental contexts.

2.6 Strengths and Weaknesses of the Different Methods in Personality Assessment

A summary of the pros and cons of the different method-condition combinations along with examples of each is given in Table 2.1. Several designs have been used because different methods reflect different solutions to the trade-offs that arise when measuring behavior. For example, in an effort to combat the apparent subjectivity of trait ratings, behavioral codings may be used, which rely on direct observations such as recording the frequency of agonistic encounters. However, the added objectivity of behavioral codings comes with its own costs, such as the fact that codings take much longer to collect than do ratings. In addition to methodological considerations, some methods are preferred over others purely because they have been traditionally associated with a particular research tradition. Here we outline the strengths and weaknesses of the different methods so that researchers can make informed decisions about the design most suited to the circumstances of their own research. It should be noted that the present authors' backgrounds span a variety of disciplines.

2.7 Trait Ratings

2.7.1 Advantages of Trait Ratings

As shown in the cell labeled *Trait Rating* in Table 2.1, the first advantage of trait ratings is that they allow data to be collected more quickly and efficiently than is possible with coding systems. Typically, a judge uses a rating scale to indicate how well an adjective or behavioral description reflects the behavior of an animal; the rating can be based on recent observations, testing situations, or past experiences with the subject. By summarizing in just one rating many behaviors performed over time and across situations, ratings are an exceptionally efficient way of capturing behavioral information and distilling it down to a single number in a matter of moments.

The second advantage of ratings as a method for measuring personality is that they implicitly control for variability in behavior due to nonpersonality factors (Vazire et al. 2007). For researchers trying to assess the aspects of behavior that are consistent across time and situations (i.e., personality), this variance represents noise. Some variability in behavior is due to changes in an animal's situation or environment, including seasonal effects, daily fluctuations, or changes in the social or physical environment (e.g., presence of a predator). For example, an individual might be active at one moment but inactive at another due to situational changes (e.g., feeding time). Observers can reduce the effects of this kind of variability by discounting situational influences on behavior when making their ratings. As Martin and Bateson (1993) noted,

> the human rater has played an active role in filtering, accumulating, weighting, and integrating information over a considerable period of time (p 81).

Another portion of variability in behavior can be attributed to random variance. One way to reduce the effects of random variance is to aggregate measures across time so that nonsystematic sources of variance will cancel out. Trait ratings inherently benefit from aggregation because when observers rate an animal they implicitly summarize that animal's behavior across all the days, months, or years they have known it. Thus, data collected from trait ratings are essentially already aggregated across all the times the observer has observed the target animal. This quality of trait ratings drastically improves their reliability.

Another way to reduce the effects of random variance is to aggregate measures across raters so that irrelevant random differences among raters will tend to cancel out each other. Strictly speaking, this advantage could be applied to any of the methods discussed here, e.g., there is nothing to stop behavioral codings being aggregated across coders. However, in practice this step of aggregating across observers has been most widely used for the rating method, almost certainly because the relative efficiency of ratings makes it much easier to collect assessments from multiple raters than from multiple coders.

2.7.2 *Disadvantages of Trait Ratings*

One important disadvantage of trait ratings is that they rely on potentially idiosyncratic trait definitions. Any rating is affected by how the researcher has defined each trait and how the rater interprets that meaning as he or she makes a rating. At both of these stages their subjectivity may enter the process – two different researchers might have different ideas of what constitutes sociability and even if they do agree, two raters using the same scales may apply them differently. By relying on researchers' definitions, ratings of broad traits (e.g., sociability) may make assumptions that have not been verified empirically. For example, behaviors that co-occur in one species also co-occur in another species.

An additional problem with rating scales is that comparisons between animals can be hard to interpret; absolute counts (e.g., number of agonistic encounters, duration of exploration) can readily be compared across time periods, animals, experimental conditions, and labs, but it is harder to interpret the meaning of such differences using rating values. If a common set of raters rate the same individuals, the simple mean ratings of individuals can be compared, but means cannot be compared across species or different settings. For example, direct between-zoo comparisons are not meaningful if different zoos have different animals and different raters. The reason is that ratings are group relative and group means are therefore clustered around the center of the rating scale. Of course, behavior scores may have similar limitations for between-setting comparisons as a result of differences in group density, environmental differences, and other factors.

A further problem with relying on human judgment is that human raters may not appropriately weigh the information they observe when they combine elements of behavior into a single rating. For example, they may put undue emphasis on salient events (e.g., fights) and fail to take into account the degree to which behaviors overlap. A large body of literature in clinical psychology has shown that humans are relatively poor at integrating information in a statistically optimal manner (Meehl 1954).

Another problem with trait ratings is that they rely on the availability of raters experienced in the behavioral repertoire of the target species in various contexts. Raters who are experienced in a wide range of behavior may not always be available. In these situations, it would not be appropriate to use the trait-rating method.

2.8 Behavioral Codings

2.8.1 *Advantages of Behavioral Codings*

As shown in the cell labeled *Behavioral Codings* in Table 2.1, behavioral codings have an advantage over trait ratings in that they involve direct data collection of the duration or frequency of specific behaviors. This method may eliminate much of the subjectivity associated with ratings and is ideal for certain types of behavioral observation, such as trying to identify the true frequency of behaviors or proportion of time the animal is displaying the behavior.

Another advantage of behavioral codings is that they allow for easy comparison between animals because they are based on frequencies of behavior. For example, if one animal is recorded as playing 50 times and another is recorded as playing 100 times in a year, then a strong case can be made that the second animal is more playful than the first.

2.8.2 Disadvantages of Behavioral Codings

There are several disadvantages to behavioral codings when used *as a method for assessing personality*. However, many of these disadvantages are not serious when the coding method is used in nonpersonality contexts (e.g., in experiments). Indeed the rich history of research on animal behavior stands as a testament to the value of coding methods for many research purposes.

One disadvantage is that collecting behavioral codings is often very time consuming compared to collecting trait ratings. Behavioral codings typically require the focal animal to be observed anywhere from 5 to 30 min at a time. This duration is not very long, but the observations are almost always repeated, between 6 and 100 times (more often for the shorter periods of time) in order to capture variations of behavior during different times of day and temperature, resulting in a large amount of total time dedicated to each animal. When conducting behavioral codings, researchers must decide which method to use. Possible methods include all-occurrences, instantaneous, and zero-one sampling. All-occurrences involves recording the frequency and duration all instances of a behavior in a specified observation period. This method has the advantage of capturing data on frequency and duration. The disadvantage to all-occurrences sampling is that it can be difficult to record many different behaviors occurring all at once. Instantaneous sampling involves recording whether a particular behavior is occurring at a specific instance. Zero-one sampling involves recording whether a behavior is occurring or occurred at all during a specific time interval. Both instantaneous and zero-one sampling reduce the time spent focused on recording behaviors. Zero-one sampling captures frequent short duration as well as behaviors that are expressed infrequently for long durations; however, neither frequency nor duration can be accurately estimated from zero-one sampling. In the case of instantaneous sampling, scores give unbiased estimates of the total proportion of time a subject is displaying any recorded behavior. All three methods are not good at capturing very rare behaviors and instantaneous sampling can easily miss behaviors that last a short time (Martin and Bateson 1993). Also, researchers wanting to use behavioral-coding methods face the problem of identifying multiple individuals who are willing to spend hundreds of hours necessary to undergo the requisite training and then code behavior (Vazire et al. 2007). Although raters also need to have spent many hours observing the animals in different situations and contexts, the observations do not have to be done systematically, in the same way as in behavioral coding. Instead, they could be done in connection with other projects or tasks that the rater was completing at the time (e.g., as a member of the animal care staff).

In addition to being time consuming, behavioral codings do not typically control for as much of the variability in behavior due to nonpersonality factors as trait ratings do. As mentioned earlier in relation to trait ratings, this variability in behavior due to nonpersonality factors can include situational factors and random noise.

Another problem with behavioral-coding systems as they are often used is that the ethograms often implement a rule in which any behavior is assigned to one and only one category. This can be problematic when situations arise where the researcher wants to take into account two behaviors that are occurring simultaneously. For example, if an animal is playing, an observer might want to take into account, not only the fact that play was occurring, but also the type of locomotion (e.g., stationary, walk, run) that was occurring during play to get an idea of energy level. If an ethogram only allows the researcher to assign a behavior to one category, the researcher will not be able to take both the play and locomotion into account for that occurrence. A single behavioral sequence (e.g., rambunctious play) can involve multiple behaviors and could reflect multiple personality traits so a system that assigns behaviors to only one category may not be a valid reflection of all the relevant traits. It should be emphasized that this issue is more a problem with the way that behavior codings are practiced than anything intrinsic to the method itself because there is no reason why behavioral-coding systems need to insist that behaviors are assigned to only a single coding category. One practical reason why behaviors have often been assigned to just a single category in ethograms arises from limitations of observational software programs. These programs are sometimes designed in a way that makes it tricky to assign a behavior to more than one category. However, there are some simple solutions to the issues with using a specific observational software program, Observer version 5.0, which we will discuss later. In addition, a new version, Observer XT has recently been released and may address some of these issues.

Another drawback of behavioral-coding methods is their susceptibility to the effects of random noise because they capture only a narrow snapshot of all the behaviors in which the animal engages throughout the day. Many behaviors relevant to personality only reveal their consistency of performance over time periods too long to be captured by coding systems. For example, even if an animal has 15 fights a year, it is unlikely that any of these events will be recorded in the narrow windows afforded by coding sessions.

In addition, there are some personality traits that would be difficult to translate into a feasible list of behaviors to code. For example, the trait unpredictable would be difficult to operationalize in terms of specific behaviors.

In summary, our analyses of the two methods of assessment suggest that using trait ratings to assess primate personality provides several advantages over behavioral codings. We next consider the strengths and weaknesses of the three types of experimental designs in which these methods are used (i.e., naturalistic observation, testing context, and cumulative observation).

2.9 Naturalistic Observation

2.9.1 Advantages of Naturalistic Observation

As shown in the cell labeled *Naturalistic Observation* in Table 2.1, one advantage of naturalistic observation over a testing context is convenience. Naturalistic observation does not require the rater/observer to manipulate the subjects at all.

In studies based on naturalistic observation, each rater spends a standard amount of time observing the subjects before rating them. Therefore, naturalistic observation has the benefit of controlling for differences in familiarity among the observer/raters.

2.9.2 Disadvantages of Naturalistic Observation

One disadvantage of naturalistic observation is that the observations are limited to animals that can be seen. If a subject happens to be out of view at any time during the observation period, then its behavior will not be recorded or taken into consideration. In a testing context it is possible, depending on the particular circumstance, that a subject cannot be seen, but unobservability is less likely to be an issue in testing contexts where an effort is usually made to conduct the tests where the subject can be seen.

Another disadvantage of naturalistic observation is that observations are obtained in only one specific context. For example, if the animals are simply observed in their home cage with the same individuals they are typically near, there is no measure of what the animal is like outside the home cage. These other contexts could elicit behavior relevant to understanding the individual's personality (Capitanio 1999; Fairbanks and Jorgensen 2011).

2.10 Testing Context

2.10.1 Advantages of Testing Context

As shown in Table 2.1, testing contexts have several advantages. One advantage is that all behaviors performed by the subject during an observation session are seen and recorded because the subjects are often tested individually or, if the testing is done in a group setting, an effort is made to test each subject in the group.

Another advantage of the testing context is that it can examine the effects of context-specific behavior (e.g., novelty seeking) and take the context into account. Testing situations permit subjects to be coded/rated in places other than their home cage and often with different social groupings (e.g., isolated or with only one or two other individuals).

Testing contexts also give researchers control over the situation in which the rating/coding is occurring. For example, a subject can be placed in different social groupings or away from distractions. This contrasts with studies using naturalistic observation where subjects cannot be manipulated at all and where animals cannot be tested separately from other group members or out of their home cages.

Another advantage of testing contexts is the ability to elicit specific types of behavior. For example, Watson and Ward (1996) used the testing context to assess the relationship between temperament and problem solving. Use of this method to create problem-solving opportunities elicited more instances of observed problem solving than would have occurred if the researchers had relied on naturalistic observation.

2.10.2 Disadvantages of Testing Context

There are also disadvantages to assessing personality in testing contexts. One issue is that manipulating subjects can be time consuming or in some situations (e.g., when only one cage is available) impractical.

Another problem with using testing contexts is that they typically focus on only a few specific behaviors during a limited period of time, allowing only a relatively restricted view of the species' behavioral repertoire.

2.11 Cumulative Observation

2.11.1 Advantages of Cumulative Observation

Cumulative observations are recorded exclusively with trait ratings and enjoy several advantages over both naturalistic observations and testing contexts. One benefit of cumulative observation is that raters are familiar with all of the animals they are rating because they have observed and/or interacted with each of the subjects at many different times over an extended period of time, ranging from several months to many years. The raters' familiarity with the animals is an advantage because it suggests that they have typically been able to observe the animals in different contexts and over different periods of time, rather than having just a few interactions on which to base their impressions.

Of the three contexts of assessment, ratings based on cumulative observations require the shortest amount of time to complete. The rater does not have to perform any additional observations or tests before rating the subjects because ratings are based on previous experiences. In turn, the relative ease of obtaining trait ratings reduces the likelihood of errors due to lack of training, misunderstandings, or fatigue on the part of the observer (Vazire et al. 2007).

2.11.2 Disadvantages of Cumulative Observation

One practical limitation of cumulative observation methods is that it may be difficult to identify multiple observers who are sufficiently familiar with the target animals to provide expert ratings.

Another issue in cumulative-observation studies is that raters may vary in their experiences of the contexts and amount of time in which they have observed each subject. This can be problematic for some of the same reasons associated with the behavior-coding methodology. If a person has worked with an animal in only one context at intermittent times, then he or she may not have accumulated a broad impression of the subject's personality. However, if another individual has worked with the animal during regular intervals of time, across many contexts, then he or she might have a more general and accurate sense of the subject's personality.

Another danger with cumulative observations is that the raters almost certainly have talked about the animals with each other, thereby compromising the independence of the observations. Of course, in some instances these conversations may communicate valid information, e.g., reporting that an animal was involved in a fight.

2.12 Empirical Studies Comparing Behavioral-Coding and Trait-Rating Data

The arguments summarized earlier suggest several theoretical strengths and weaknesses of the different methods. How do these strengths and weaknesses play out in practice? Six empirical papers have focused on directly comparing behavioral-coding with trait-rating methods. Although other papers have used both behavioral-coding and trait-rating methods to assess personality, this was not the focus of the studies. If both methods are valid they should converge; assessments of the same animals using both methods should correlate with each other when there is a prediction that they will converge or when the adjective and behavioral meanings are the same, e.g., aggressive the adjective and aggressive behaviors. If they do not converge when predicted to do so, then at least one of them is not valid. Examples of high correlations found in the six studies are summarized in Tables 2.2 and 2.3. The studies in Table 2.2 incorporated the naturalistic observation condition and those in Table 2.3 incorporated the testing context procedure.

Each of the six studies includes two kinds of assessment method (behavioral codings and trait ratings). For example, the study by Vazire et al. (2007) assessed the personality of 52 chimpanzees using both trait ratings based on cumulative observations and behavioral codings based on naturalistic observation. The studies differ slightly in the level at which the behavioral codings were measured. Five studies coded behavior as aggregates of individual behaviors. For example, the term "aggressive behavior" might include individual behaviors such as biting and attacking. These aggregated measures are listed in the top halves of Tables 2.2 and 2.3 under

Table 2.2 Examples of high correlations in empirical studies comparing behavioral codings and trait ratings through naturalistic observations

Species	*r*	Dimension	Item label	Study
Aggregated behaviors				
Chimpanzee	0.53	Emotionality	Agonistic	Pederson et al. (2005)
	−0.37	Playfulness	Solitary behavior	Vazire et al. (2007)
Gorilla	−0.54	Dominant	Displacement – receiving	Kuhar et al. (2006)
	0.55	Extroverted	Contact aggression-initiating	
Vervet monkey	0.82	Curious	Play	McGuire et al. (1994)
Stumptailed macaque	−0.46	Excitable	Nonsocial behavior	Mondragon-Ceballos and Santillan-Doherty (1994)
Individual behaviors				
Chimpanzee	−0.48	Extraversion	Idle	Pederson et al. (2005)
	0.59	Playfulness	Social play	Vazire et al. (2007)
Rhesus macaque	0.54	Confident	Groom receive	Capitanio (1999)
Stumptailed macaque	−0.83	Sociable	Resting	Mondragon-Ceballos and Santillan-Doherty (1994)

Table 2.3 Examples of high correlations in empirical studies comparing behavioral codings and trait ratings in a testing situation

Species	*r*	Dimension	Item label	Study
Aggregated behaviors				
Rhesus macaque	0.41	Confident	Affiliative	Capitanio (1999) video playbacks
Individual behaviors				
Rhesus macaque	0.39	Excitable	Back of the cage	Capitanio (1999) responsiveness behaviors
	0.70	Sociable	Approach receive	Capitanio (1999) group behaviors
	0.67	Sociable	Lipsmack initiate to nonsocial video	Capitanio (1999) video playbacks
	0.78	Excitable	Fear grimace receive	Capitanio (1999) social dyad

the heading "aggregate behaviors." Four studies coded, in addition to or instead of aggregate behavior terms, specific behaviors such as "aggress receive." These behaviors are listed in the bottom halves of Tables 2.2 and 2.3 under the heading "individual behaviors." Three of the studies used the behavioral codings to code both aggregate and individual behaviors and the other three used the behavioral codings to code either aggregate or individual behaviors.

The first column of Table 2.1 shows the species, and the second through fourth columns show examples of high correlations between ratings and behavior in each of the studies. The last column lists the relevant citation.

Although Table 2.2 gives evidence of some high correlations between ratings and behavior, indicating that convergence is possible, the cross-method convergence correlations are generally quite low indicating that behavioral codings and trait ratings often result in different assessments of the same individuals' personalities. For example, contrary to expectations, the correlation between scratching behavior and emotionality was –0.32 (Pederson et al. 2005). These findings suggest that codings and ratings are not always measuring the same thing. The aggregated codings were substantially less reliable than the aggregated ratings suggesting that the codings were less likely to be measuring any valid construct (including personality) than the ratings (Vazire et al. 2007). The low reliability of the codings reflected the fact that behavioral counts from one coding period correlated poorly with the behavioral counts generated by the same animal in another coding period. The low correlations between the two coding periods may have been a result of taking observations at different times of day, so the animal had a different energy level and as a result behaved differently during the two coding periods. It is not known if poor coding reliabilities could explain the low convergence between trait ratings and behavioral codings in the other studies evaluating validity because reliability was not reported for behavior codings in these studies (McGuire et al. 1994; Mondragon-Ceballos and Santillan-Doherty 1994; Capitanio 1999; Pederson et al. 2005; Kuhar et al. 2006). An alternative explanation for the low instances of convergence between trait ratings and behavioral codings is that only a small fraction of behaviors on an ethogram might be actually captured by a behavioral-coding strategy through naturalistic observation performed in the subject's home cage (Capitanio 1999). Instead, higher validity can be achieved between trait ratings and behavioral codings, when the codings are obtained in a testing situation (Capitanio 1999). This improved validity may occur because the testing situations can be designed to target specific behaviors that might be seen at a much lower frequency under conditions of naturalistic observation.

Table 2.3 shows examples of high correlations when assessing the cross-method convergence correlations obtained under conditions of a testing situation as opposed to naturalistic observation summarized in Table 2.2. The same aggregate/individual trait separation scheme is used in Table 2.3 as was used in Table 2.2. Studies using aggregate terms are shown in the top half of the table and those using individual terms are shown in the bottom half. It should be noted that Table 2.3 is made up of four different studies performed by Capitanio. In most cases, the testing conditions (Table 2.3) did yield stronger cross-method convergence correlations than those obtained in naturalistic observations (Table 2.2). However, comparisons of this sort could not be made in the other five studies because they collected behavior codings exclusively with naturalistic observations so no testing-situation data were available. In addition, it is worth noting that the cross-method convergence correlations obtained in the testing situation were in some cases lower than the convergence correlations obtained in other studies where the behavioral codings were obtained in naturalistic observations. This finding suggests that although collecting behavioral codings in a testing situation might improve the convergence between trait rating and behavioral codings to some degree, it still may not fully account for the poor convergence between the two methods.

The data presented in the six empirical papers that compared trait ratings to behavioral-coding methods suggest that, in some cases, the two methods are not producing the same results. Both Vazire et al. (2007) and Capitanio (1999) have suggested possible explanations for why there may be instances of low convergence between the trait ratings and behavioral-coding methods (i.e., low reliability of the codings, behavioral observations completed through naturalistic observation as opposed to testing situation). Although these potential explanations have been tested by only one paper each, they both point to shortcomings of the behavioral-coding methodology, rather than the trait-rating method for assessing personality. Further studies are needed to fully understand what is driving the disparity between the behavioral codings and the trait ratings.

Based on the evidence presented on the strengths and weaknesses of each of the method combinations and from the six studies directly comparing behavioral-coding with trait-rating methods, we propose that the method combination of trait rating/cumulative observation is the most efficient and reliable method for assessing primate personality.

2.13 "Etic" and "Emic" Approaches to Rating Scale Development

There are two broad strategies to developing rating scales, which are somewhat analogous to the etic and emic approaches used in cross-cultural research.

2.13.1 "Etic" Approach

In human cross-cultural research, the etic approach is one in which personality scales from one culture are imported and applied to another culture (Berry 1999; Gosling and John 1999; Weiss et al. 2006). An etic approach in animals involves using scales that have been developed and used for assessment in one species to inform the development and implementation of a scale for another species. For example, taking a scale that was developed for use in rhesus monkeys and using it to measure personality in gorillas would be an etic approach.

One benefit of using an etic approach is that the scale has usually been shown to be reliable and valid in previous research using that scale. In addition, using a previously used scale allows researchers to easily perform cross-species or cross-colony comparisons because of the commonalities in the adjective names and definitions used in all the samples using those scales.

The main problem with using an etic approach is that the adjectives used to describe one species may not be the best ones for describing another species. As a result, researchers could include adjectives that do not fit the behavior of a new species. An etic approach might also lead researchers to neglect or omit adjectives that reflect important traits in the new species.

2.13.2 "Emic" Approach

In human cross-cultural research the emic approach is one in which personality scales for a culture are developed within that culture (Berry 1999; Gosling and John 1999; Weiss et al. 2006). An emic approach to rating-scale development in animals involves developing a list of traits based on systematic observations made by trained researchers familiar with the specific species' behavior. To make sure the adjective set is representative of the full range of the species' behavior, adjectives would be included based on previous research on the behavior of the species. For example, an emic approach to developing a scale for use in rhesus monkeys would draw both from observations and previous research on rhesus monkeys.

An emic approach to scale development allows a researcher to include an exhaustive list of trait adjectives that are representative of the species examined. Doing so increases the chances of comprehensively capturing the behaviors expressed in that species.

The problem with using an emic approach to scale development is that cross-species comparisons are hindered by cross-study differences in the instruments used.

In sum, there are strengths and weaknesses to both the etic and emic approaches. As we shall see below, we recommend using a combined etic and emic approach. That is, using some items from previous instruments to permit cross-species and cross-study comparisons, and also generating items specifically for use in the study species to ensure the full breadth of the species' behavioral repertoire is captured.

2.14 Most Commonly Used Rating Scales for Assessing Primate Personality

The present review revealed that nearly all personality rating research has been done using just a few instruments. In fact two thirds of the 59 studies used just three scales.

2.14.1 Emotions Profile Index

Buirski et al.'s (1973) Emotions Profile Index (EPI) was used in 10% of the primate rating studies. It was developed using an etic approach and was based on methods designed to assess human personality (Plutchik 1965; Kellerman and Plutchik 1968). The primate version of the scale was first described and assessed in Buirski et al. (1973), where it was used in a field study of 7 olive baboons using 3 raters.

2.14.2 Trait Descriptive Adjective Set

King and Figuerdo's (1997) scale was used in 17% of the primate rating studies. It was also designed using an etic approach. It was based on adjectives taken from Goldberg's (1990) adjective list originally developed for use in humans. The scale was first described and assessed in King and Figuerdo (1997), where it was used to rate 100 chimpanzees at 12 zoological parks using 53 raters.

2.14.3 Stevenson-Hinde and Zunz Instrument

Stevenson-Hinde and Zunz's (1978) scale was used in 40% of the primate rating studies. It was designed using aspects of an emic process, in which adjectives were generated by three people familiar with the 45 rhesus macaques and their behavior; it did not include a thorough literature search for additional behaviors that might not have been captured initially.

The development and widespread use of these scales has probably played a large role in the acceptance of trait rating as a reliable and valid method for assessing primate personality. A number of benefits accrue from having similar instruments used in multiple studies, most notably, the ability to perform cross-study comparisons. For example, broad comparisons can be made across all 23 studies that have used the Stevenson-Hinde and Zunz (1978) instrument.

However, the reliance on just a few instruments also raises a number of dangers. First, there is a danger that an instrument will be used in a species other than that for which it was specifically developed. For example, although the Stevenson-Hinde and Zunz (1978) instrument was developed for use in rhesus macaques it has also been used in gorillas (Gold and Maple 1994) and tufted capuchins (Byrne and Suomi 2002). Second, if an instrument is created and widely adopted before there has been time to replicate the personality findings in multiple labs then subsequent research could be overly influenced by the idiosyncrasies of the particular group of subjects on which the instrument was based or the perspectives of the particular group of researchers who developed it.

2.15 The Need for New Scale Development

The three scales most commonly used to assess primate personality have helped establish the legitimacy of the trait-rating method. However, it is important that existing scales be augmented by scales tailored to unique features of other primate species. Continued development and improvement of existing scales will also bolster the foundations upon which subsequent primate personality research is built. Below, we outline how such scale development could proceed. We draw heavily on Uher (2008a, b) and Gosling (1998).

Uher (2011) discusses a behavioral repertoire approach to developing a list of traits that can be used to assess personality. A full example of this approach can be seen in recent publications as applied to the great ape species (Uher and Asendorpf 2008; Uher et al. 2008). The behavioral repertoire approach, which can be broken down into a four-step method, was initially designed to develop a list of traits that could be operationalized using behavior codings, behavior ratings, and adjective ratings. However, for this chapter it will just be discussed for the purposes of describing the development of new adjective ratings scales. Gosling (1998) developed a three-step method, originally used to establish a reliable and valid personality assessment scale in spotted hyenas

The choice of whether to use the four- or three-step approach depends on how much previous behavioral research has been conducted on the species of interest and on the availability of humans with necessary experience observing the species. If the amount of previous behavioral research is limited or few experienced researchers are available, the four-step emic approach to adjective trait-rating scale development should be used. If there is an extensive body of previous behavioral research and/or the experienced individuals are available to assist with the research, then the three-step approach is recommended.

2.15.1 Four-Step Approach

This approach is designed to maximize comprehensiveness and explicitly acknowledges the importance of basing the rating categories on behavioral data acquired across a variety of contexts and occasions (Uher 2008a, b).

Step 1: Begin by doing a thorough review of the literature to discover all behavioral aspects of a species' life, both in the wild and captivity. This should encompass only observable and measurable behaviors. The list of behaviors should be noted along with the different situations in which they are found to occur.

Step 2: Merge the behavioral domains and situations into theoretical trait constructs. For example, aggressive behavior towards a novel object may be related to the trait aggressiveness.

Step 3: Operationalize the trait constructs into behavioral rating (behavior-descriptive verbs) and adjective rating lists and collect data based on these lists. For example, in Uher and Asendorpf (2008) as a behavioral rating, curiosity was operationalized as "animal often touches new objects at great length." In Uher and Asendorpf (2008), as an adjective rating, curiosity was operationalized as "animal is very curious."

Step 4: Analyze data for stable interindividual differences and perform a multivariate analysis on the data to understand relationships among different traits.

The emic approach endorsed by Uher (2008a, b) is a comprehensive approach to match the behavior of a specific species with the personality traits being assessed. However, in tailoring an instrument to a particular species, the ability to compare the findings across studies is compromised. Thus, a natural tension exists between the demands of comprehensiveness and comparability (Gosling 1998, 2001). In capturing the idiosyncrasies of a particular species, researchers may be forced to use traits that are not applicable to other species. A balance needs to be reached in which a basic set of standardized descriptors (gleaned from other research and operationally defined in species-appropriate terms) is supplemented by important species-specific descriptors (using procedures along the lines of those proposed by Uher 2008a, b). It should be noted that the emic and etic approaches are not the only approaches to the development of rating scales. A more thorough discussion of various approaches and their benefits and drawbacks can be seen in recent papers by Uher (2008a, b).

2.15.2 Three-Step Approach

This approach is designed to strike a balance between comparability and comprehensiveness in generating trait-rating items (Gosling 1998).

Step 1: Generate a comprehensive list of behavioral traits from three sources: "previous research on animal personality, previous research on human personality, and expert nominations" (p. 108 Gosling 1998).
Step 2: Eliminate redundancy from the list of behavioral traits suggested in step 1, preferably with the help of experts.
Step 3: Define each of the adjectives in terms of species-specific behavior.

2.16 Recommendations

2.16.1 What Is the Best Method for Assessing Primate Personality?

Based on our analysis of the advantages and limitations of each of the methods and conditions, as well as our review of the empirical comparisons between trait-rating and behavioral-coding methods, we conclude that the trait-rating/cumulative observation method-condition is the most reliable and practical method-condition combination.

However, the recommendation of a trait-rating/cumulative observation method-condition does not negate the usefulness of a combination of behavioral-coding and trait-rating methods, when the time and resources are available to include them. In fact, over the past two decades (see Fig. 2.1) a trend has emerged where a combination of rating and coding methods is increasingly being used. A combination of methods offers several advantages over studies that use just one personality assessment method. One benefit to using both methods is that a researcher obtains information about both a context-relevant behavior (from behavioral coding) as well as a broader view of the animal's personality (from trait rating). In addition, multimethod designs allow better assessment of reliability and validity (McGuire et al. 1994; Capitanio 1999; Pederson et al. 2005; Kuhar et al. 2006).

2.16.2 What Improvements Can Be Made to Methods of Assessing Primate Personality?

Several suggestions can be made to improve the current methods of primate personality assessment. Although we suggest the cumulative observation/trait-rating method for use in most assessment contexts, there may be situations in which another method-condition combination must be used. In such cases, there are several ways in which current methods can be improved.

2.16.2.1 All Methods

Both raters and coders should be well experienced in working with or observing the animals they are assessing. They should also be thoroughly trained on the assessment method. In addition, interrater reliability should always be obtained.

2.16.2.2 Behavioral Coding of Naturalistic Observation or Testing Situation conditions

There are circumstances where behavioral codings are useful in personality assessment. For example, an important feature of the emic approach is gathering a broad array of species-specific behaviors from which ratings can be derived. Behavioral coding of naturalistic observation and testing situations can help accomplish this task in order to identify behavioral domains and situations that can serve as the foundation for the development of behavior rating and adjective-based rating scales (Uher 2008a, b).

Another recommendation is to use a behavioral ethogram that allows behaviors to be assigned to more than one category. As mentioned earlier, this is often difficult when using behavioral observation software programs, because of their design. However, there are several ways to work around this issue in some programs, including Observer 5.0 (Noldus 1991). One strategy is to identify behaviors that are likely to be seen along with other behaviors (e.g., screaming). Then identify the different behaviors to which this one behavior is likely to be linked (e.g., running, sitting, climbing). Then, list screaming as a modifier for each of these possible scenarios. This technique can be effective if a researcher is able to accurately predict the behaviors that might co-occur, but it does not allow for situations where the co-occurrence of behaviors has not been anticipated. An alternative method is to turn on the comments feature of the ethogram in Observer. This step will allow the addition of a comment anytime the researcher records an observation. Then, if two behaviors occur simultaneously, the observer can enter the code for one behavior and just add the other behavior in the comments section. This procedure may add more work during the data-analysis phase because the comments will have to be added to the coded behaviors but it will allow for a more accurate representation of the behavior.

2.16.2.3 Testing Situation

If the goal of a study is to investigate a specific aspect of personality (e.g., novelty seeking, social impulsivity) a behavioral coding/testing situation combination will allow the researcher to capture more of the specific behaviors than is possible with a general rating scale assessment. However, in order to insure the validity of the behavioral coding/testing situation results, a trait-rating method should also be used. The combination of these two methods (behavioral coding/testing situation and trait-rating methods) allows researchers to assess whether the behavioral coding/ testing situation is accurately capturing the specific aspects of personality that are the focus of the study. For example, if a behavioral coding/testing situation is used to assess different aspects of impulsivity, there should be a strong correlation between the frequency of impulsive behaviors recorded and the rating of impulsivity of the animals being assessed.

2.17 Conclusions

In this chapter, we examined and evaluated the different methods used to assess primate personality. Our analyses drew on an in-depth review of the literature. We discussed the strengths and weaknesses of each method, paying particular to attention to the studies that have directly compared assessment methods. We concluded that in most cases the trait-rating/cumulative observation method combination is the best method for primate personality assessment. However, more research is needed to compare empirically the different methods of personality assessment. We also suggested some improvements for primate personality assessment methods that can be implemented for any chosen method. The suggested steps and improvements provide a blueprint for a solid foundation for future research on primate personality.

Acknowledgments We are grateful to Stephanie Anestis, John Capitanio, Frans de Waal, Lynn Fairbanks, James Dee Higley, James King, Lindsay Murray, Jon Sefcek, and Alex Weiss for their help with this task.

References

Anestis SF (2005) Behavioral style, dominance rank, and urinary cortisol in young chimpanzees (*Pan troglodytes*). Behaviour 142:1245–1268

Anestis SF (2011) Primate personality and behavioral endocrinology. In: Weiss A, King JE, Murray L (eds) Personality and temperament in nonhuman primates. Springer, New York

Bard KA, Gardner KH (1996) Influences on development in infant chimpanzees: Enculturation, temperament, and cognition. In: AE Russon, KA Bard (eds) Reaching into thought: The minds of the great apes. Cambridge University Press, Cambridge

Berry JW (1999) Emics and etics: A symbiotic conception. Cult Psychol 5:165–171

Buirski P, Kellerman H, Plutchik R et al. (1973) A field study of emotions, dominance, and social behavior in a group of baboons (*Papio anubis*). Primates 14:67–78

Byrne G, Suomi SJ (2002) Cortisol reactivity and its relation to homecage behavior and personality ratings in tufted capuchin (*Cebus paella*) juveniles from birth to six years of age. Psychoneuroendocrinology 27:139–154

Capitanio JP (1999) Personality dimensions in adult male rhesus macaques: Prediction of behaviors across time and situation. Am J Primatol 47:299–320

Crawford MP (1938) A behavior rating scale for young chimpanzees. J Comp Psychol 26:79–91

Cronbach LJ, Meehl PE (1955) Construct validity in psychological tests. Psychol Bull 52:281–302

Fairbanks LA (2001) Individual differences in response to a stranger: Social impulsivity as a dimension of temperament in vervet monkeys (*Cercopithecus aethiops sabaeus*). J Comp Psychol 115:22–28

Fairbanks LA, Jorgensen MJ (2011) Objective behavioral tests of temperament in nonhuman primates. In: Weiss A, King JE, Murray L (eds) Personality and temperament in nonhuman primates. Springer, New York

Freeman H, Gosling SD (2010) Personality in nonhuman primates: A review and evaluation of past research. Am J Primatol 71:1–19

French JA (1981) Individual differences in play in *Macaca fuscata*: The role of maternal status and personality. Int J Primatol 2:237–246

Gold KC, Maple TL (1994). Personality assessment in the gorilla and its utility as a management tool. Zoo Biol 13:509–522

Goldberg LR (1990). An alternative 'description of personality': The Big-Five factor structure. J Pers Soc Psychol 59:1216–1229

Gosling SD (1998) Personality dimensions in spotted hyenas (*Crocuta crocuta*). J Comp Psychol 112:107–118

Gosling SD (2001) From mice to men: What can we learn about personality from animal research. Psychol Bull 127:45–86

Gosling SD, John OP (1999) Personality dimensions in nonhuman animals: A cross-species review. Curr Dir Psychol Sci 8:69–74

Gosling SD, Kwan VSY, John O P (2003). A dog's got personality: A cross-species comparative approach to evaluating personality judgments. J Pers Soc Psychol 85:1161–1169

Hebb DO (1949) Temperament in chimpanzees: I. Method of analysis. J Comp Physiol Psychol 42:192–206

Itoh K (2002) Personality research with non-human primates: theoretical formulation and methods. Primates 43:249–261

Kaplan JR, Manuck SB, Fontenot MB et al. (2002) Central nervous system monoamine correlates of social dominance in cynomologus monkeys (*Macaca fascicularis*). Neuropsychopharmacology 26:431–443

Kellerman H, Plutchik R (1968) Emotion-trait interrelations and the measurement of personality. Psychol Rep 23:1107–1114

King JE, Figueredo AJ (1997) The Five-Factor Model plus Dominance in chimpanzee personality. J Res Pers 31:257–271

Kuhar CW, Stoinski TS, Lukas KE, Maple TL (2006) Gorilla Behavior Index revisted: Age, housing and behavior. Appl Anim Behav Sci 96:315–326

Martin JE (2005) The influence of rearing on personality ratings of captive chimpanzees (*Pan troglodytes*). Appl Anim Behav Sci 90:167–181

Martin P, Bateson P (1993) Measuring Behaviour. Cambridge University Press, Cambridge

McGuire MT, Raleigh MJ, Pollack DB (1994) Personality features in vervet monkeys: the effects of sex, age, social status, and group composition. Am J Primatol 33:1–13

Meehl PE (1954) Clinical vs. statistical prediction: A theoretical analysis and review of the evidence. University of Minneapolis Press, Minneapolis

Mondragon-Ceballos R, Santillan-Doherty AM (1994) The relationship between personality and age, sex and rank in captive stumptail macaques. In: Roeder JJ, Thierry B, Anderson JR, Herrenschmidt N (eds) Current primatology, Vol. II: Social development, learning and behavior. University of Louis Pasteur, Strasbourg

Noldus LPJJ (1991) The Observer: A software system for collection and analysis of observational data. Behav Res Methods Instrum Comput 23:415–429

Pederson AK, King JE, Landau VI (2005) Chimpanzee (*Pan troglodytes*) personality predicts behavior. J Res Pers 39:534–549

Plutchik R (1965) What is an emotion? J Psychol 61:295–303

Rouff JH, Sussman RW, Strube MJ (2005) Personality traits in captive lion-tailed macques (*Macaca silenus*). Am J Primatol 67:177–198

Stevenson-Hinde J, Zunz M (1978) Subjective assessment of individual rhesus monkeys. Primates 19:473–482

Stevenson-Hinde J, Hinde CA (2011) Individual characteristics: Weaving psychological and ethological approaches. In: Weiss A, King JE, Murray L (eds) Personality and temperament in nonhuman primates. Springer, New York

Uher J (2008a) Comparative personality research: Methodological approaches. Eur J Pers 22:427–455

Uher J (2008b) Three methodological core issues of comparative personality research. Eur J Pers 22:475–496

Uher J, Asendorpf JB (2008) Personality assessment in the Great Apes: Comparing ecologically valid behavior measures, behavior ratings, and adjective ratings. J Res Pers 42:821–838

Uher J, Asendorpf JB, Call J (2008) Personality in the behaviour of the great apes: Temporal stability, cross-situatational consistency and coherence in response. Anim Behav 75:99–11

Uher J (2011) Personality in nonhuman primates: What can we learn from personality psychology. In: Weiss A, King JE, Murray L (eds) Personality and temperament in nonhuman primates. Springer, New York

Vazire S, Gosling SD, Dickey AS et al. (2007) Measuring personality in nonhuman animals. In: Robins RW, Fraley RC, Krueger RF (eds), Handbook of research methods in personality psychology. Guilford Press, New York

Watson S, Ward JP (1996) Temperament and problem solving in the small-eared bushbaby (*Otolemur garnetti*). J Comp Psychol 110:377–385

Weiss A, King JE, Perkins L (2006) Personality and subjective well-being in orangutans (*Pongo pygmaeus* and *Pongo abelii*). J Pers Soc Psychol 90:501–511

Yerkes RM (1940) Social behavior of chimpanzees: Dominance between mates, in relation to sexual status. J Comp Psychol 30:147–186

Chapter 3
Personality in Nonhuman Primates: What Can We Learn from Human Personality Psychology?

Jana Uher

Abstract Primate personality research encounters a number of puzzling methodological challenges. Individuals are unique and comparable at the same time. They are characterized by relatively stable individual-specific behavioral patterns that often show only moderate consistency across situations. Personality is assumed to be temporally stable, yet equally incorporates long-term change and development. These are all déjà vus from human personality psychology. In this chapter, I present classical theories of personality psychology and discuss their suitability for nonhuman species. Using examples from nonhuman primates, I explain basic theoretical concepts, methodological approaches, and methods of measurement of empirical personality research. I place special emphasis on theoretical concepts and methodologies for comparisons of personality variation among populations, such as among species.

3.1 Introduction

All species consist of individuals. These individuals share many characteristics in genome, morphology, physiology, biochemistry, and behavior that define their species membership. But despite this essential similarity, individuals are in no sense uniform; beyond age and sex differences, individuals also differ in their specific genotypic and phenotypic characteristics. The behavioral phenotypes of individuals and their variation within populations are covered by theoretical concepts of personality differences (Stern 1911; Uher 2008a, 2011).

J. Uher (✉)
Comparative Differential and Personality Psychology, Freie Universität Berlin, Habelschwerdter Allee 45, Berlin 14195, Germany
e-mail: uher@primate-personality.net

A. Weiss et al. (eds.), *Personality and Temperament in Nonhuman Primates*, Developments in Primatology: Progress and Prospects, DOI 10.1007/978-1-4614-0176-6_3,

The scientific study of personality differences in human (Galton 1869; Stern 1911; Allport 1937) and nonhuman primates (Crawford 1938; Yerkes 1939; Hebb 1949) started about 100 years ago. Whereas human personality research has evolved into a discipline of its own within psychology, nonhuman primate personality research developed only incompletely within several heterogeneous disciplines. Yet many challenges are structurally similar if not identical to those in human research, allowing primate researchers to profit greatly from the theoretical, methodological, and statistical advances made in psychology.

In this chapter, I introduce theoretical concepts, methodological approaches, and methods of measurement from human personality psychology, and discuss their suitability for empirical studies in nonhuman primates. My special concern is to show how established theoretical concepts and methodologies for cross-cultural comparisons of human personality variation generalize to cross-population comparisons of personality variation within and across species. Using examples from nonhuman primates, I discuss theoretical foundations and typical methodological challenges that provide the necessary background for empirical research. How can we compare individuals when they are all unique? What role do situations play in studying individuals? How can personality variation be compared among different species? How can we decide what is important to study within a species? And what methods can we use to measure the personality of nonhuman primate individuals? This chapter explores theoretical concepts and suitable methodological tools for these and other puzzling issues in empirical research on primate personality.

3.2 Theoretical Concepts for Primate Personality Research

To explain personality differences in human primates, psychologists have developed various classical schools of thinking. They differ in basic ideas of man, theoretical concepts, investigative methods, and explanatory approaches (Buss 1991; Funder 2007; Cervone and Pervin 2008). Perhaps most commonly known outside psychology are the psychoanalytic approaches grounded in Freud's theories that assume infantile psychodynamics determine an individual's personality. Humanistic psychologists try to explain the individual through its unique conscious experience of the world driven by its free will and the striving for personal growth and for an understanding of the meaning of life.

To oppose the mainly introspective methods of these approaches, which hinder empirical investigation, behaviorists tried to explain an individual's personality as a result of its learning history. They assumed individuals are born as *tabulae rasae*, as "blank slates" with no innate content, whose development is largely determined by acquired stimulus-response connections. Cognitive psychologists filled the behaviorists' black box with structures of information processing that explain personality differences with variations in the architecture and processing parameters of the individuals' cognitive systems. Social constructivist psychologists view personality as created through interactions and negotiations with others. Developmental psychologists focus on continuous dynamic transactions between developing individuals and their changing environments to investigate processes of individual personality development.

In search of the biological basis of personality, biological psychologists and neuropsychologists study processes in the neural, hormonal, and immune systems that underlie observable individual differences. Behavior genetic approaches estimate average contributions of genes and environment to behavioral differences on the basis of twin and adoption studies. They are increasingly refined by molecular genetic approaches that study the transaction between specific genes and specific environments over the course of life. A promising approach models systematic transactions between intertwined genetic and environmental influences (Johnson 2007). Finally, evolutionary psychologists understand personality differences as proximate mechanisms that have evolved in adaptation to environmental conditions.

All these classical schools of thought with their different philosophical, theoretical, and methodological principles have contributed to our understanding of human personality. Clearly, some of them, such as psychoanalytic or humanistic schools that rely on introspective methods, are not suitable for empirical research in nonhuman primates. Behaviorism was too one-sided because it neglected genetic influences. Yet many others have broad intersections with nonhuman research, in particular those focusing on information-processing, genetics, neurobiology, ontogeny, and evolution. They try to unravel mechanisms and processes governing observable behavioral differences – provided we already have a sketchy road map of what kind of individual differences a species exhibits. It was probably not by chance that one of the oldest and most influential schools in personality psychology, trait psychology, focuses on this essential first task of measuring and cataloguing individual differences. Stern (1911) laid the methodological foundations of empirical and statistical approaches that form the basis of much of today's personality psychology. They also provide an excellent foundation for empirical research on nonhuman personality.

3.2.1 Variable-oriented and Individual-oriented Perspectives on Individuals

Primate individuals exhibit individual-specific behavioral patterns that are commonly construed as their personality. Individual-specificity implies that these patterns are relatively stable within each individual over time, and that the individuals vary in the degree to which they exhibit certain behavioral patterns (Uher 2011). Their empirical interindividual variation across the composite of the population can be described with theoretical dimensions of personality differences (Stern 1911). Theoretical conceptions of such behavioral patterns are also called personality traits, personality constructs, or trait constructs; accordingly, the dimensions that describe their interindividual variation are also called trait dimensions or personality dimensions.

In explanatory models of personality, individual-specific patterns of behavior are interpreted as reflecting the individuals' psychobiological organization that determines their unique adaptations to their environments (Allport 1937). Personality traits are thus conceived as reflecting behavior-regulating mechanisms that can have genetic, physiological, cognitive, motivational, and behavioral components (Buss 1991; Mischel et al. 2003; Funder 2007).

To make these behavioral patterns accessible for empirical research, Stern (1911) introduced the differential perspective to psychology that was the groundbreaking shift in viewpoint from the average individual to differences among individuals. He laid the methodological foundations of empirical personality research by conceiving two complementary methodological perspectives.

The first perspective focuses on the measurement variables. Variable-oriented analyses address the individuals' relative positions along shared trait dimensions. First, they analyze the statistical distributions of trait scores in specified populations. In many populations, many trait scores, such as human extraversion, are normally distributed. Most individuals' scores center around the mean of the dimension, and only a few individuals are on its extremes. If a trait's variability is limited on one side of the dimension, the distribution pattern can be skewed. On aggressiveness, for example, most humans score rather low, and only a few are high scoring. Furthermore, variable-oriented analyses address the covariation of individual trait score distributions among various trait dimensions in a population which I explain further below in the section about personality taxonomy. That is, variable-oriented analyses characterize the population.The second perspective focuses on individuals. Individual-oriented analyses address the individual's unique configuration of its relative positions across multiple trait dimensions that have been identified in its population with variable-oriented analyses. This allows us to quantify the individuals' uniqueness based on their empirical comparability along shared dimensions. Quantification of an individual's personality thus depends on the personality variation of the other individuals to which it is compared and that are called the reference population.

Individual-oriented analyses rely mostly on standardized scores that depict the individual's relative scores in comparison to those of other individuals in its sample. Absolute score profiles, in contrast, are confounded with the mean profile of the sample. For example, since all individuals generally score higher on locomotion than on social play, absolute scores may fail to reveal that some individuals may score higher than others in social play and lower than others in locomotion (as in Suomi et al. 1996). The pattern of an individual's relative trait scores can be illustrated as a profile across trait dimensions; the shape of this trait profile characterizes the individual (Stern 1911; Cairns et al. 1998; Mervielde and Asendorpf 2000).

Standardized personality profiles can be illustrated with behavioral data from great apes. In a methodological study, Uher et al. (2008) repeatedly observed 20 great apes (five each of bonobos, chimpanzees, gorillas, and orangutans) in 14 different laboratory test situations and two different group situations. They studied 19 different personality trait constructs that they measured with 76 behavior variables, most of which could be obtained from all four species. The data were analyzed systematically from both variable-oriented and individual-oriented perspectives.

Figure 3.1 shows z-scored trait profiles from two individuals in that study. A z-score is a measure of deviation from the sample's mean that is standardized

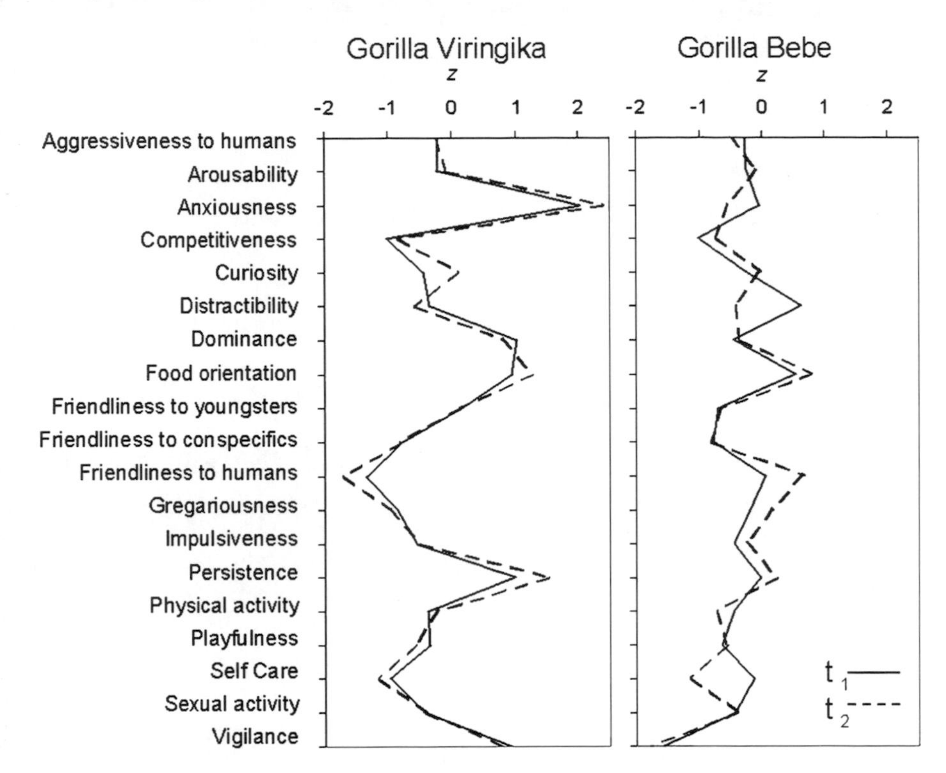

Fig. 3.1 Personality profiles of two individuals based on ethological measures of behavior obtained in a series of 14 laboratory tests and group observations in two different group situations. All trait scores represent behavioral measures that were aggregated over several occasions of measurement. For details on ethological behavior measurement see Uher et al. (2008). The *z*-standardized trait scores depict the individuals' positions on each trait dimension in relation to those of the other individuals of the sample. The sample's mean score is thereby 0, and the standard deviation is 1. The data were aggregated and the aggregate scores were standardized across individuals separately within two nonoverlapping test periods; t_1 is the first test period, t_2 second is the test period 3–6 weeks later

such that the sample mean is 0, and the sample standard deviation is 1. Standardization allows three kinds of direct comparisons: (1) an individual's score can be placed within the trait distribution of the population, (2) its scores can be compared across different traits, and (3) different individuals can be compared on the same trait.

Viringika, for example, is high scoring on anxiousness; her trait score is about two standard deviations above the sample mean. Her scores on dominance, food orientation, and persistence are also about one standard deviation above average. But her scores on competitiveness, friendliness to humans, and self-care are one standard deviation below average. Across traits, these deviations are the most pronounced in her profile, whereas her scores on other traits are rather average. Comparing Viringika's trait profile with that of Bebe, one can see that both are more food oriented than the sample average, but that Viringika is even more "greedy" than Bebe. These females score equally low on competitiveness.

Their scores on dominance, in contrast, are quite different; Bebe tends to be submissive, whereas Viringika is quite dominant. How do we know that these behavioral scores can be used to infer these individuals' personality?

3.2.2 Temporal Consistency of Interindividual Behavioral Differences

The everyday connotation of the word "trait" already implies characterization by lasting attributes. Personality traits imply characterization by relatively stable individual-specific behavioral tendencies. The stability criterion is important since interindividual variation can also derive from momentary behavioral fluctuations that are unrelated to the individuals' lasting behavioral tendencies. Thus, measuring personality differences in the flood of individual behavior requires repeated observations and evidence of temporal consistency. A basic criterion of personality measurement is therefore test–retest reliability. Individuals having low scores on a personality trait should retain their position relative to other individuals in retest assessments at least over intermediate time periods. Variable-oriented test–retest reliability means that the individuals' rank orders on that dimension should correlate over time. Individual-oriented test–retest reliability means that the individuals should retain their individual-specific behavioral patterns; their individual profile shapes across multiple trait dimensions should correlate over time (Cairns et al. 1998).

The fluctuating nature of behavior often hinders establishment of test–retest reliability and entails particular methodological difficulties (Hebb 1949; Stevenson-Hinde et al. 1980; Suomi et al. 1996). A strategy to reduce the impact of random variation and measurement error, and to increase the reliability of personality measurement is aggregation at least over multiple occasions, if not over different trait-related behaviors and situations (Rushton et al. 1983).

These methodological principles can be demonstrated with results from Uher et al. (2008). In that study, the individuals were observed repeatedly in the same test and group situations over a period of about 2–3 weeks. After a break of about a fortnight, all individuals were again observed repeatedly in the same behaviors and situations in a second 2- to 3-week period. Overall, each individual was observed for more than 67 h within a 50-day period. Due to the intense and repeated observations in this design, the behavioral raw data could be aggregated within each of the two nonoverlapping periods and analyzed for temporal reliability between them. Mean variable-oriented temporal reliability of the 76 behavioral variables was high ($r=0.78$) as was temporal reliability of 19 trait indices each composed of several different trait-related behavioral variables ($r=0.77$). This shows that the data were sufficiently aggregated and that reliable personality measures were obtained. Personality differences can thus be measured in great ape behavior as reliably as those in human behavior – provided the data are aggregated sufficiently (Uher et al. 2008).

Insufficient behavior observation can result in unreliable personality measures that compromise comparisons and coherence with measures obtained by other methods, such as ratings. For example, focal samples of 15 min are obviously insufficient to measure personality in chimpanzees reliably outside controlled laboratory settings. Comparisons of behavior observations across such extremely short time periods yielded only low to zero reliability scores (Vazire et al. 2007) that are best interpreted as reflecting estimates of the daily fluctuations of behavior rather than of the individuals' personality. Comparisons of such unreliable measures with rating measures based on the raters' mental aggregations of everyday observations over 7 years (as in Vazire et al. 2007) are therefore necessarily compromised. Based on these unreliable measures it cannot be concluded that ethological behavior measures would be per se unreliable or even inferior measures of personality (as assumed by Vazire et al. 2007; Gosling 2008).

Reliable measures of personality, whether ethological behavior measures or rating measures, can only be obtained with sufficient aggregation across repeated observations. When this principle is considered adequately, behavioral personality measures were shown to be as reliable as those obtained with ratings in nonhuman primates (Uher and Asendorpf 2008). Similarly, raters must have sufficient observational experiences with the target individuals; ratings provided by raters who hardly know the individuals will be meaningless.

Most primate studies focus on variable-oriented analyses. Yet, temporal stability at the population level may mask changes occurring at the individual level. Even high rank-order stability does not mean that each individual retains the same relative position over time. Instead, a few individuals may change, while the majority of individuals remain the same. Such differences in individual trajectories are essentially a question of stability, gradual change, and long-term development of individual personality that can only be studied with individual-oriented analyses. These analyses can reveal information beyond those shown in variable-oriented analyses and can therefore contribute meaningfully to personality studies (Block 1971; Magnusson 1988; Mervielde and Asendorpf 2000). To show more of their potential, I will discuss individual-oriented analyses often in this chapter.

Figure 3.1 illustrates the test–retest reliability of individual trait profiles. Profiles indicated by continuous lines are based on aggregated measures obtained in the first observation period of the Uher et al. (2008) study; those indicated by broken lines derive from the second period 3–6 weeks later. Given that these profiles were measured and standardized independently in two nonoverlapping observation periods, their shapes are remarkably similar. These findings show that these behavioral profiles are reliable measures of the individuals' personality. Among individuals, test–retest reliability scores varied from $r=0.49$ to 0.94 for profiles across all 76 single behavioral variables and from $r=0.38$ to 0.97 for profiles across 19 composed trait indices. Temporal correlations were significant ($p<0.05$) in all single behavior profiles, and in 90% of the composed trait profiles. This shows that stability and change are also manifested at the individual level. It suggests that some individuals are more consistent in their behavior, whereas others may be guided more by environmental influences. Hence, behavioral consistency itself seems to be interindividually different (Caspi and Roberts 1999; Funder 2007).

3.2.3 The Role of Situations in Personality Research

Early personality theorists had assumed not only temporal consistency but also substantial consistency across situations to be central to personality. When Hartshorne and May (1928) reported low consistency in the behavior of 850 school children across different situations, the concept of personality seemed to be challenged fundamentally. It culminated in Mischel's (1968) finding that cross-situational consistency in behavior rarely exceeds the "magic" correlation of $r=0.30$. Individual behavior appeared to be highly situation specific rather than individual specific and cross-situationally consistent. These puzzling findings provoked the person-situation controversy that lasted four decades in psychology (Mischel 1968; Funder and Colvin 1991; Fleeson 2004; Funder 2006).

When looking for consistency across situations, primate researchers came across exactly the same findings. In their famous series of studies on personality differences in rhesus macaques (*Macaca mulatta*), Stevenson-Hinde et al. (1980) reported that reliable behavior measures were lacking significant correlations across situations. But, instead of reflecting a "failure to look at appropriate measures rather than a characteristic of the … [individuals] themselves" (p 508), these findings mirror the core issues of cross-situational consistency. They show that careful methodological considerations are needed to avoid misinterpretations of empirical findings.

First, cross-situational consistency is no mere illusion. Moderate correlations show that individuals do display some consistency in their behavior across situations; it is just less than initially expected, and behavioral correlations across situations are lower than their correlations over time. This also shows that situations exert significant impacts on individual behavior because individuals respond to them differently (Mischel and Peake 1982). Individual-oriented analyses can reveal whether such differences are individual specific; they quantify and illustrate the individuals' unique patterns of responsiveness to different situations in behavior profiles across situations. Such a situation-behavior profile depicts the individual's scores on the same trait dimension measured in different situations (Mischel et al. 2002).

If individuals' trait scores are standardized within each situation, the profile informs about situational influences that are specific to the target individual. For example, most chimpanzees react more fearfully to snakes than to petrol cans (Goodall 1986). Unless their responses are standardized within situations, most chimpanzees will therefore exhibit higher fear scores for snakes than for petrol cans. Their individual situation-behavior profiles would be confounded with that of the average chimpanzee. After standardization, chimpanzees that are generally more fearful toward everything will have positive z-scores. Those individuals that are less snake-fearful than the average chimpanzee will have negative z-scores for snakes. Independent of that, some chimpanzees will show large positive z-scores for snakes as compared to their z-scores for other situations; these chimpanzees are more specifically snake-fearful.

Such differences in situational responsiveness are individual-specific if they are temporally consistent (Mischel and Shoda 1995). Individual-oriented test–retest reliability analysis of situation-behavior profiles is illustrated in the Uher et al. (2008) study that obtained behavioral data for the same traits in various situations. Aggressiveness, for example, was measured in four laboratory-based test situations involving familiar keepers, observers entering neighboring cages, friendly masked humans offering food, and playbacks of radio news records. The average correlation of aggressiveness scores across these situations on the sample level was $r=0.25$; yet, the individuals' aggressiveness-situation profiles correlated on average $r=0.77$ over time (3–6 weeks), ranging from an outlier with an almost inversed profile of $r=-0.49$ to 0.99 ($N=16$; Uher et al. 2008). This means that the individuals differed substantially in how strongly they responded with aggression to these four situations, yet within each individual, aggressiveness patterns across these situations were fairly stable.

Test–retest reliable situation-behavior profiles reflect consistent interactional patterns between situations and the individuals' responses over time. Individuals may not only respond to situations in particular ways, they may also actively choose particular environments that are suited to their personality; they may evoke certain reactions from the environment, in particular, from their social environment; and they may also actively shape their environments. Such interactions may be the mechanisms behind the increasing matches between certain personalities and certain environments, and thus behind continuity in personality development (Magnusson 1988; Matthews et al. 2003).

Situations, conceived as complex constellations of stimuli, vary in how they permit personality differences to emerge. Two qualitatively different aspects, situational strength and trait-relevance, are distinguished. Situational strength denotes how compelling a situation is for the individuals' behavior. For example, variations in aggressiveness might emerge most clearly in situations that typically elicit low to moderate aggression. Situations that permit easy emergence of personality differences are referred to as weak situations (Mischel 1977). Strong situations, by contrast, may mask interindividual variability because they force behavior into specific channels by either inhibiting the behavior substantially or by evoking heightened responses from all individuals (Tett and Guterman 2000). For example, most captive primates react strongly to veterinarians with blow guns in front of their cages, making interindividual differences less pronounced.

The second aspect of situations is trait relevance, which refers to the type of information to which the individuals are responding. That a behavior cannot be observed does not necessarily mean that the individual has a low trait score, or that the assumed trait construct is a mere theoretical hypothesis without any empirical relations to observable behavior. Situations have to activate relevant behavior. For example, aggressions are responses to stimuli indicating that aggressive behavior might be functional. Individuals that are more sensitive to them and that react more quickly or more intensely with aggression than others are assumed to be more aggressive (Tett and Guterman 2000; Capitanio 2004).

3.2.4 Individual Response Specificity

Typically, trait constructs are inferred from different behavioral responses. For example, human shyness is inferred from long pauses in speech, hesitant speaking, gaze aversion, or restricted gestures (Asendorpf 1988). Chimpanzee arousability in prefeeding contexts can be inferred from rocking, grinning, vocalizing, or pacing (Uher et al. 2008). Since these responses are assumed to indicate the same trait, they should be correlated. Surprisingly, they are not; both studies report low to zero correlations among the different behavioral indicators of these two personality traits.

We can gain some understanding of this puzzling finding with individual-oriented analyses. Analogous to cross-situational consistency, correlations among behavioral trait indicators can be low on the sample level because they lack validity for the trait in question, or just vary randomly. Yet they can also be low due to stable individual response specificity. It imposes methodological difficulties since restricting personality measurement to single behaviors can result in misclassifying those individuals who primarily exhibit behaviors that are not measured. In fact, traits can often be inferred from a variety of responses that are not necessarily shown by all individuals (Asendorpf 1988; Marwitz and Stemmler 1998; Uher et al. 2008).

Individual response specificity can be analyzed and illustrated in individual response profiles that depict their scores across different behavioral indicators of the same trait. Behavior measures are standardized within the sample because absolute behavior scores would confound interindividual differences with sample-level differences. For example, while all chimpanzees may generally show more rocking than pacing in prefeeding situations, some individuals may show, in comparison to others, more pacing than rocking. Standardized behavior scores thus inform about individual-specific patterns of behavioral trait indicators. They also allow comparisons of different types of behavior measures such as durations, latencies, and frequencies that can be neither directly compared nor simply averaged since they may be distributed differently.

To capture such interindividual differences in response specificity, the Uher et al. (2008) study measured most traits with multiple behaviors. Figure 3.2 illustrates individual arousability profiles across different arousal responses (rocking, grinning, vocalizing, or pacing) of four chimpanzees prior to their noon feedings. The z-scores indicate the individuals' relative positions on each response variable and allow direct comparisons. One can see, for example, that Frodo showed pleasure grins much more often than the others; he scored three standard deviations above the sample's mean. Robert and Fraukje were rocking much more often than Dorien or Frodo; they scored two standard deviations higher than the others. Their particular profile shapes illustrate the typical arousal responses of these individuals. For Fraukje, it was most characteristic to rock, vocalize, and change position when awaiting the feeding, whereas she hardly ever paced. Comparison of these four response profiles also shows that measuring arousability only with rocking would misclassify Frodo.

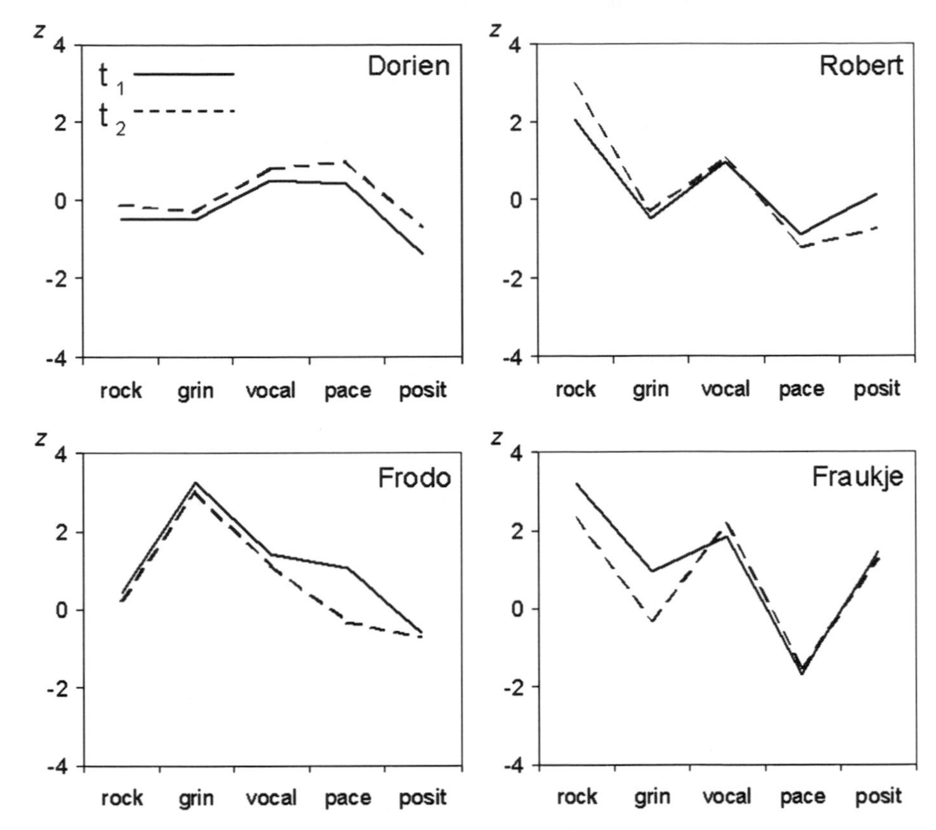

Fig. 3.2 Response profiles depicting individual response specificity. Five arousal-indicating behaviors observed prior to the noon feeding were analyzed: rock = rocking, grin = pleasure grin, vocal = vocalizing, pace = pacing, posit = changing position, defined as rising from the waiting position, and sitting down again or staying within 1.5 m from the original place within 10 s. The *z*-scores depict the individuals' positions on each response in relation to the other individuals of the sample; the sample's mean score is thereby zero. t_1 is the first test period, t_2 is the second, nonoverlapping test period 3–6 weeks later. Within each test period separately, the data were aggregated across multiple occasions of observation; the aggregated scores were then standardized across individuals

Empirical test–retest reliability reveals whether such response profiles are individual-specific. On the sample level, variable-oriented correlations of the individuals' rank orders among the different behavioral indicators of some of the traits studied by Uher et al. (2008) were on average $r=0.16$. This reflects that individuals can vary in which and how many of multiple trait-related behaviors they show. Yet, on the individual level, individual response profiles consisting of the individuals' relative scores on different trait-related behaviors correlated on average $r=0.66$ over 3–6 weeks, indicating temporally consistent individual response specificity. Figure 3.2 illustrates these findings. The shapes of the individual response profiles of the four chimpanzees measured in the first observation period are very similar to those measured in the second nonoverlapping observation period, as indicated with continuous vs. broken lines.

3.2.5 Personality Types

Although individual profiles are distinct and unique, there may be groups of individuals showing similarities in their profile shapes. The response profiles of Fraukje and Robert in Fig. 3.2, for example, have strikingly similar shapes. This may indicate a shared response profile *type*. Similarly, there may be individuals sharing similar situation-behavior profiles or similar personality profiles. Such personality types can be identified statistically with cluster or Q-factor analysis; they represent prototypes of similar individuals (Asendorpf and van Aken 1999).

Extreme scores on a single trait dimension are also sometimes referred to as types, for example, the extravert type. These "univariate" types are special cases of the configurational "multivariate" types that are based on multiple traits. To my knowledge, personality types have not yet been analyzed empirically in behavior-based studies of nonhuman primate personality. Yet some rating-based studies in chimpanzees identified distinct personality types that were defined by characteristic trait score patterns, such as those labeled as "socially confident" (Murray 2002) or "deferent apprehensive" types (Martin 2005).

3.2.6 Personality Taxonomy

Typically, certain personality traits go together in a population. Their covariation can be subsumed statistically within broader, higher order trait constructs underlying this shared covariation, thus making the less complex trait constructs subtrait constructs of the emergent, more complex trait constructs. Such patterns of variable-oriented trait correlations can be analyzed with multilevel, cluster, or factor analysis. They can be organized in hierarchical trait taxonomies. At the top of such hierarchies are a few abstract trait constructs, often called personality factors, which summarize the shared variance of the correlated lower order traits they comprise (see King and Weiss 2011). Preferably, linear, relatively independent factors are extracted that do not overlap too strongly in the lower order trait constructs they summarize (Eysenck 1990; Matthews et al. 2003).

The concept of trait hierarchies or trait taxonomies shows that personality factors can explain more diverse behaviors than each of their lower order trait components alone. Thus, factors permit parsimonious and comprehensive descriptions of personality differences. Identifying multiple personality factors increases the possibilities to explain complex observable diversity among individuals. Unique individual configurations of factor scores; that is, individual personality profiles depict unique personalities (Capitanio 2004; Uher 2008a, b, 2011).

3.2.7 Theoretical Concepts for Cross-Species Comparisons

Comparative personality research merges three areas of emphasis: the individuals' uniqueness, their comparability, and their universality. Uniqueness and comparability are studied based on individual- and variable-oriented analyses. Since the individual's relative position on a trait dimension depends on the scores of those individuals to whom it is compared, the differential perspective implies population dependency. If reference populations change, all trait scores very likely change, too. Hence, studies of universality, which address whether particular personality dimension are common to different populations, are based on specifications of the studied reference populations and on methodologies for comparisons of personality variation among populations (Uher 2008a, b).

In humans, reference populations are typically defined by social criteria such as culture, language, or nationality. Quests for human universals thus refer to trait constructs that are applicable to all humans regardless of their cultural, language, or national background. Similarly, populations in nonhuman primates can be defined by their geographical distribution or living environment. For example, the universality of chimpanzee personality trait constructs can be studied by comparing populations living in wildlife sanctuaries with those living in zoological parks (King et al. 2005). When we define reference populations by biological criteria such as breed, subspecies, or species, universality can be studied on more general population levels. Comparing species nested in the biological classification, such as within genera, families, or orders, could show whether some personality constructs can be assumed to describe behavioral variation that is, for example, uniquely macaque, uniquely pan, uniquely hominoid, or uniquely primate (Uher 2008a; King and Weiss 2011).

This suggests that theoretical concepts and methodologies for comparisons of human cultures can be generalized to comparisons of species. Cross-cultural personality research has shown that personality variation as dimensions of stable interindividual behavioral variation can also be conceptualized *across* different populations (Leung and Bond 1989). Generalizations of these concepts yield three basic kinds of personality dimensions (Uher 2008a). Population-specific trait dimensions differentiate individuals of only one particular population, but not those of other populations. Universal trait dimensions, in contrast, differentiate individuals across different populations. This means that individuals of several considered populations differ along these dimensions. Two kinds of universal trait dimensions can be distinguished. Weak universal traits are dimensions on which populations show similar means and variances. Strong universal traits are dimensions on which populations exhibit significant mean level differences. The latter are thus also population-comparative trait dimensions, yet should not be mistaken for behavioral dimensions that differentiate populations without also differentiating individuals. Comparisons of personality differences among populations are ultimately always based on interindividual variability within each population. These basic kinds of traits can be analyzed with population specific, universal, and population-comparative analyses (in particular factor analysis; for details see Uher 2008a, b).

Evolutionary theory provides supportive arguments for the existence of behavioral differences that can be described with these kinds of trait dimensions (see Sih 2011). In evolutionary research, personality differences are understood as behavioral strategy differences based on trade-offs with different costs and benefits that have evolved in patchy and changing environments, and that reduce the pressure of competition among members of a species. Accordingly, population-specific personality traits could reflect behavioral differences that are niche-differentiated adaptations (Tooby and Cosmides 1989). For example, in adaptation to an arboreal life in swampy rainforests where food is difficult to furnish, orangutans could have evolved individual differences in behavior that are not displayed by other primate species, and that can therefore be interpreted as an orangutan-specific personality trait. Phylogenetic hypotheses, in contrast, suggest that some personality traits could reflect interindividual differences in behavior that can be explained as homologies inherited from common ancestors. For example, all living macaque species may show interindividual differences in sociability. That is, individuals within each species differ in their degree of sociability. Observations suggest that at least some macaque species may thereby differ in their average scores, such as bonnet macaques that scored higher on average sociability than pigtailed macaques (Capitanio 2004). This could indicate that sociability is a strong universal trait dimension that differentiates both individuals and species. Such findings may reflect behavioral patterns inherited from a common macaque ancestor that could be illuminative for theories of speciation (Uher 2008a, b).

These three basic kinds of trait dimensions can co-occur in the personality structure of a population. For example, the personality structure of orangutans may comprise some weak and/or strong universal trait dimensions they share with other species as well as some orangutan-specific trait dimensions. Personality variation of populations can thus be compared quantitatively in shared weak and strong universal trait dimensions on which populations may exert different positioning effects. Personality variation can also be compared qualitatively in terms of differences in the populations' hierarchical trait taxonomies, and thus in their personality factors. The results are referred to as the populations' patterning effects (for details see Uher 2008a, b).

Identification of such differences among populations could be informative for theories and models developed to explain the causation, function, adaptation, and phylogeny of individual differences in behavior. Mean level differences among species, for example, could be associated with ecological differences in predation risk or food density thereby indicating possible functionality and adaptivity of individual behavioral differences. Given their uniqueness, species-specific trait or factor variations could be particularly illuminative regarding ecological functions of behavioral variation and processes of speciation (Capitanio 2004; Uher 2008a, b). For example, if the trait construct of conscientiousness could only explain behavioral variation in humans, this could reveal important information about unique antecedents of human evolution. Personality trait dimensions shared by closely related species, in turn, may indicate behavioral strategy differences that could be interpreted as homologs inherited from common ancestors, whereas those shared by distantly related species occupying similar ecological niches could reflect analogs evolved in adaptation to similar environments (Gosling and Graybeal 2007; Uher 2008b).

The concept of hierarchical trait taxonomies emphasizes that species comparisons depend on comprehensive empirical models of the structure of interindividual behavioral variation of species. For example, if indicators of conscientiousness are not studied in a species, empirical results cannot be interpreted as indicating that this trait construct is not applicable to that species (Weiss et al. 2006). This has strong implications for the validity of species comparisons and may bias inferences on possible antecedents of the emergence of behavioral variation explained with that construct. Methodological approaches are necessary that allow to identify comprehensive and ecologically valid models of the species' hypothetical true trait taxonomies (Uher 2008a, b).

3.3 Methodological Approaches to Primate Personality

Establishing representative and comprehensive taxonomic models of interindividual behavioral differences in a population encounters two crucial bottlenecks: comprehensive selection and systematic reduction. First, all potential trait constructs should be selected comprehensively to avoid ignoring important domains of personality variation in the target population. Second, these trait domains should be analyzed empirically for dimensions of test–retest reliable interindividual behavioral variation that must then be reduced systematically to broad personality factors that summarize their shared variance. Bias or arbitrariness in either of these processes reduces the representativeness of empirically identified hierarchical trait taxonomies, which may compromise inferences on patterning and positioning effects of populations. Whereas reduction procedures are largely based on statistical tools, and thus on statistical criteria, selection procedures require stringent rationales to ensure that a comprehensive pool of potential trait constructs and measures is entered into the identification process (Uher 2008a, b).

3.3.1 A Taxonomy of Methodological Approaches

The diversity of behavioral variations within, and especially across, species makes it difficult to decide what to study. How did human personality psychology solve this problem? To ensure comprehensiveness, some of the founders of trait psychology reasoned that "those individual differences that are most salient and socially relevant in people's lives will eventually become encoded into their language; the more important such a difference, the more likely it is to become expressed as a single word" (John et al. 1988, p 174). Hence, natural language is assumed to be a comprehensive pool of human personality descriptors. This approach provides the basis for much of contemporary research on human personality.

Based on this lexical hypothesis, Allport and Odbert (1936) went through about 550,000 words of the 1925 edition of Webster's New International Dictionary, and identified 17,953 terms describing personality differences. From this enormous list, they further extracted 4,500 adjectives that describe observable and lasting traits. This list set the stage for empirical models of the human personality structure based on different reduction methods. The factor analytic reduction to five broad personality factors (Extraversion, Agreeableness, Conscientiousness, Emotional Stability, and Openness), the so-called Big Five, has received substantial empirical support in various languages (Goldberg 1990; John 1990; de Raad and Barelds 2008).

The lexical hypothesis also implies, however, that the human lexica cannot serve as comprehensive pools of animal personality descriptors. There is no reason to assume that humans have codified in their natural language an equally systematic body of trait-related descriptors for interindividual behavioral differences in other species with which they generally interact little or not at all. The English language, for example, evolved primarily in parts of the world that are outside primate habitat regions. How could English-speaking people have developed a systematic vocabulary that describes all salient and socially relevant behavioral characteristics of nonhuman primate individuals when they do not even encounter such individuals regularly in their daily lives? Over the last decades, primate researchers invested considerable efforts to describe the extraordinary variety of primate behavior in comprehensive ethograms. Rarely are single words, especially adjectives, sufficient to describe and differentiate the complex courses of motion and facial expressions of nonhuman species in a way that all other people, including laypeople, can readily understand their meaning without any further explanation. Human trait descriptors, in contrast, can convey precise information, for example, about specific facial expressions of human emotions.

This does not exclude, however, that some lexical trait descriptors can also be useful to describe interindividual differences in primate behavior, as I will show below. Ultimately, any scientific investigation has to rely on human language. But the usage of lexical trait descriptors in subjective ratings as one out of several methods of personality measurement has to be clearly distinguished from the methodological approach that is used in order to decide at first which behaviors shall be studied for interindividual differences in a species. The lexical approach, which turned out to be enormously productive for human personality psychology, therefore fails as a systematic methodological approach to actualize comprehensive selections of trait constructs and trait measures for nonhuman species in order to empirically establish comprehensive trait hierarchies of their interindividual behavioral variation (Uher 2008b; Uher and Asendorpf 2008).

Besides lexical approaches to human personality, various other methodological approaches are used to decide what to study in human and nonhuman populations; they can be taxonomized into five major groups. (1) Nomination approaches rely on human observers who nominate trait constructs or measures based on their perceptions of the individual behavioral characteristics of the target population and on implicit theories they have developed about it. (2) Adaptive approaches derive trait constructs from ecological and evolutionary theories on interactions of populations with their environments to identify domains of interindividual behavioral variation in response to present and/or past adaptive problems. (3) Bottom-up or emic approaches use naturally evolved, complex systems inherent to the species such as language (exclusively in humans), behavioral, or neurobiological systems to derive trait constructs and measures. (4) Top-down or etic approaches import trait constructs and measures from other species to look for differences and similarities in their patterning effects. (5) Eclectic approaches capitalize on findings and methodologies of the other approaches without holding to a single approach. Since these approaches were developed for various aims and purposes, their rationales are not necessarily suited to identify ecologically valid and comprehensive models of trait taxonomies (Uher 2008a, b).

Systematic bottom-up/emic approaches, such as that of the lexical approach to human personality, enable comprehensive selections because they formulate strategy-based rationales for selection that refrain from specifications of, and thus from restrictions to, any particular personality domains (Uher 2008b). For example, the lexical hypothesis proposes the selection of human personality descriptors from the human lexica without confining this selection to any particular domains of behavior and thus of personality differences.

Top-down/etic approaches, in contrast, fail to enable comprehensive selections because they formulate content-based rationales that confine selections to those trait constructs and measures that are imported from other species or populations; thus, they determine a priori the behavioral domains to be studied for personality differences. For this reason, top-down/etic approaches can only reveal evidence for the applicability of trait constructs or measures to other populations within the range of imported personality domains, but not beyond. Yet they may fail to identify population-specific domains of interindividual behavioral variation that those other populations, from which the constructs and measures are imported, do not exhibit. This may result in incomprehensive taxonomic models in which important personality factors are biased or even missing completely (Church 2001; Uher 2008a, b).

For example, top-down/etic approaches based on trait descriptors of the human Five Factor Model yielded different patterning effects both in orangutans (Weiss et al. 2006), and in chimpanzees (King and Figueredo 1997; King et al. 2005; King and Weiss 2011). However, these species differences could be established only within the scope of personality domains described by these trait descriptors, but not beyond. For example, in adaptation to their ecological niches, orangutans and chimpanzees may have developed species-specific domains of interindividual behavioral variation that humans may not show, and that can therefore not be identified with top-down approaches from human personality descriptors. Yet, in chimpanzees, this top-down approach was more comprehensive than a top-down approach based on trait measures originally selected for rhesus macaques (Murray 1998), which could yield only half of the personality domains shown with a top-down/etic approach based on descriptors of human personality factors. This illustrates the substantial impact selection procedures have on the comprehensiveness of empirically identified trait taxonomies.

Bottom-up/emic approaches, in contrast, study personality "as from inside the system" (Pike 1967, p 120). Thus, if they are applied systematically, they enable comprehensive selections of behavioral domains that can be studied for personality differences. Moreover, because they rely on population-specific trait constructs and measures, bottom-up/emic approaches ensure that the behaviors studied for interindividual differences reflect behaviors that actually occur in natural settings; that is, they ensure that the thus derived personality constructs are ecologically valid. Replications of similar personality factors across different populations on the basis of population-specific trait measures provide strong evidence for the universality of the behavioral variation they explain (Church 2001). For example, the lexical bottom-up approach was carried out in human populations speaking English (Allport and Odbert 1936; Goldberg 1990), German (Angleitner et al. 1990), and Dutch (Hofstee et al. 1981), amongst others. Five strongly similar factors emerged in English and German, whereas two additional factors were shown in the Dutch.

Ecological validity cannot be ensured by top-down/etic approaches, however, because they import specific trait constructs and measures from other populations, and thus study personality "as from outside of a particular system" (Pike 1967, p 120). They may sometimes force constructs and measures on a species that may not be applicable to that species (Gosling et al. 2003, p. 283), and that consequently lack ecological validity (Uher 2008a, b; Uher and Asendorpf 2008).

The potentials for comprehensive selections of ecologically valid personality constructs are also limited in nomination approaches (e.g., Stevenson-Hinde and Zunz 1978) that are likewise based on content-based selection strategies. Being nonconspecific outsiders, we have only limited access to nonhuman species; intuitive nominations by a few knowledgeable informants therefore run the risk of overlooking important individual differences that are not salient to human observers or that do not match their implicit personality theories.

Eclectic approaches try to increase comprehensiveness of selections by combining findings and methodologies from different approaches. They mostly rely on top-down/etic approaches from trait constructs developed for different species (e.g., Rouff et al. 2005) that are sometimes also complemented with expert nominations (Freeman et al. 2011). The comprehensiveness of this content-based selection strategy depends not only on the existing knowledge about other species and the trait constructs that have been developed for them, but in particular on the rationales used to select trait constructs across species and studies, and to merge diverse constructs in order to eliminate redundancies. Yet these rationales are rarely described explicitly (Uher 2008a, b).

Adaptive approaches, by contrast, may be suited for comprehensive selections of ecologically valid traits constructs; but to my knowledge, they have not yet been applied to nonhuman primates (Uher 2008a).

3.3.2 The Behavioral Repertoire × Environmental Situations Approach

In human personality research, systematic bottom-up/emic approaches from the lexica proved to be extremely useful to establish ecologically valid and comprehensive trait taxonomies. The behavioral repertoire × environmental situations approach (Uher 2008a, b) can be considered an alternative systematic bottom-up approach that derives trait constructs from inside the behavioral and ecological system of a population. Its rationale is grounded in trait psychology and conceives personality differences as interindividual differences in intraindividually stable patterns of conditional probabilities to display particular categories of behaviors in particular categories of environmental situations. Consequently, the approach proposes compiling all important behavioral categories from the known behavioral repertoires of populations (usually species), and plotting them systematically against all situational categories in which they are typically displayed. The resulting behavior-situation units are used to derive hypothetical personality constructs that are then studied empirically for temporally consistent interindividual variability. These empirical analyses are essential since the trait constructs are construed only theoretically; they need not reflect empirical domains of interindividual variability in the studied population. If individuals show no variability or temporal consistency therein, the particular construct is discarded. Finally, these trait constructs are analyzed for intercorrelations and reduced to a few factors in order to derive a structural personality model that describes the studied population (Uher 2008a, b).

Similarly, Gosling et al. (2003, p 283) postulated that "to ensure comprehensiveness, the range of personality traits studied in a species must fully represent the behavioral repertoire of that species." The behavioral repertoire × environmental situations approach fulfills this requirement and extends beyond it. First, the behavioral repertoire approach considers not only the behavioral repertoire, but also the categories of environmental situations in which certain behaviors are typically displayed. This crucial element is inherent to the rationale of the approach. It is derived from trait psychological findings of consistent interactional patterns between individual and situational features. Explicit incorporation of the individuals' environments also opens up connections to ecological and evolutionary perspectives on personality differences (Uher 2008b).

Second, the behavioral repertoire × environmental situations approach generates theoretical constructs and not trait measures. It is not the behavioral categories compiled in the review that are studied, but theoretical constructs derived from a broad range of behavioral and situational categories. For this reason, the approach can consider behavioral categories of various types and functions that would not fit into the homogeneous and disjunctive categories of one single ethogram. This is, however, necessary to actualize a comprehensive approach. Once trait constructs are generated, measures for empirical investigation are systematically selected, which also helps to keep their number manageable for empirical studies (Uher 2008b).

Third, instead of studying behavior from scratch, the approach generates trait constructs from behavioral and situational categories of known meaning and function. It capitalizes on the expertise behavioral sciences have gained on the behavior of the average individual of the study population, and searches systematically for consistent variation among individuals therein (Uher 2008a, b).

The behavioral repertoire × environmental situations approach has already been applied to the great apes species (Uher 2008a, b; Uher and Asendorpf 2008; Uher et al. 2008). The behavioral and situational categories that were cataloged on a broad and general level for each of these closely related species were strikingly similar.

They were therefore pooled to generate trait constructs that are likely applicable to all great ape species. For initial empirical tests in a small sample of captive individuals, behavioral and situational categories that can only be observed in the wild were excluded. Furthermore, traits involving the same behavioral categories, but more specific situational categories were subsumed within one broader trait construct. For example, arousability in social vs. nonsocial situations was subsumed within one arousability construct. This trait generation procedure yielded 19 qualitatively distinct potential trait constructs (listed in Fig. 3.1). Methodological studies in a sample of 20 zoo-housed great apes, among them the Uher et al. (2008) study already discussed earlier, provide initial empirical evidence for stable interindividual differences that are described by these trait constructs (Uher 2008a, b; Uher and Asendorpf 2008).

The behavioral repertoire × environmental situations approach yielded substantial empirical evidence for temporally reliable interindividual differences in very similar trait domains as those shown by top-down/etic approaches in these species – and could also show some further trait domains beyond. These include food orientation, friendliness to youngsters, or sexual activity that are important for great apes, but that have been excluded during the development of the human Big Five factors (see e.g., Schmitt and Buss 2000). These findings emphasize that top-down/etic approaches may permit first explorations of so far unstudied species, but they ultimately require empirical convergence to bottom-up/emic findings to validate the comprehensiveness and ecological validity of their trait constructs for each particular species; for detailed discussions see Uher (2008a, b).

3.4 Methods of Measurement for Primate Personality

Personality constructs can be measured with various methods. The choice of assessment method is thereby independent of the methodological approach; these are two separate meta-theoretical steps (Uher 2008b). This means that trait constructs of human personality derived with lexical bottom-up approaches can be measured not only with lexical trait descriptors, but also with ethological measures of behavior. And vice versa, constructs derived with the behavioral repertoire x environmental situations approach can also be measured with ratings on lexical trait descriptors as I will show now.

3.4.1 The Diversity of Assessment Methods

In nonhuman research, methods of personality assessment are often classified into two groups with coding or ethological behavior observations labeled as objective methods on the one hand, and ratings labeled as subjective methods on the other hand (Gosling 2001, 2008; Capitanio 2004; Freeman et al. 2011). But in fact, methods of personality assessment span a continuum from records of single behavioral acts to ratings of adjectives as abstract personality descriptors, with methods utilizing elements of both, such as act frequency ratings (Borkenau et al. 2004) or behavior-descriptive verb ratings (Uher and Asendorpf 2008), in the middle.

Ethological methods of behavior measurement (see Altmann 1974; Lehner 1996) are close to the behavioral act pole of this continuum. Since they are based on direct observations of behavior, they seem to suggest greater objectivity than ratings. But no observation of behavior is without abstraction. The observer has to group behavioral events into classes by abstracting properties that recur in more than one event across different levels of behavioral complexity ranging from single muscle movements to more abstract behavior categories (Lehner 1996), and this is inevitably a subjective process. Thus, although the well-defined, homogeneous and independent categories of ethograms should minimize the scope left for subjective decisions, behavioral observations always do have subjective components.

Whereas subjectivity may be lowest in ethological observations, it is highest in abstract ratings that are close to the opposite pole of the objective-subjective continuum of assessment methods. Ratings rely on human ability to differentiate individuals reliably, to perceive individual behavior, to recall observations from multiple occasions in different situations over time, to aggregate this information mentally, and to express overall judgment on predefined sets of personality descriptors (so-called items) in standardized psychometric scales (Funder 1999; Uher and Asendorpf 2008). Hence, all methods of personality measurement are eventually based on observable behavior. They differ only in the *degree* of subjectivity with which they make it possible to capture interindividual behavioral variation.

Not just any measure is per se a useful measure of personality constructs. Personality measures must differentiate well and reliably among individuals, that is, they must have high discriminatory power (Kline 2000; see also Fairbanks and Jorgensen 2011). This must also be shown for rating data; independent raters must agree substantially and provide reliable distinctions between individuals. Interrater reliability can be determined for both the rank order of the individuals on a given personality descriptor (variable-oriented view), and the individual profiles across multiple items (individual-oriented view).

Ratings are often assumed to imply stability in the targets' behavior since they are derived from mental aggregations by the judges. But human observers tend to overestimate stability (Uher and Asendorpf 2008). For instance, a few observations of extreme instances of behavior, such as strong aggression, can bias observers to assume overall high aggressiveness. When later observing mild aggression by the same animal, observers may judge this as an instance of high aggression. These biases can occur even in repeated observations of concrete behaviors, but are more marked in global judgments based only on intuitive aggregation of observed earlier behavior. Such biases may become particularly problematic when observation time is limited. Establishing test–retest reliability for rating measures is thus as important as it is for behavior measures. I now illustrate analyses of interrater and test–retest reliability with rating data obtained with the Great Ape Personality Inventory (GAPI).

3.4.2 *Ratings on Behavior-Descriptive Verbs and Trait-Adjectives: The Great Ape Personality Inventory (GAPI)*

The GAPI is a psychometric instrument to assess in captive Great Apes (bonobos, chimpanzees, gorillas, and orangutans) personality traits that were derived with the behavioral repertoire × environmental situations approach (see above, Uher 2008a, b). It is available in two complementary formats that are useful for validation. The behavior-descriptive verb form (*GAPI-B*) describes observable, trait-indicating behaviors in circumscribed situations using verbs only. Food orientation, for example, is described with "When there is food, *Name* is (often) quickly on the spot." Thirty-four items were constructed, of which ten are reversed in their meaning to reduce the effects of response sets. The trait adjective form (*GAPI-A*) describes the trait constructs with single trait adjectives in everyday language such as "*Name* is (very) gluttonous." None of the 17 items is reversed in meaning. English translations of the original German items are provided in Tables 3.1 and 3.2.

Table 3.1 Great ape personality inventory – behavior-descriptive verb items (GAPI-B)[a]

Personality trait construct	Items *GAPI* – behavior-descriptive verb items	Code	Interrater reliability ICCt$_1$	Interrater reliability ICCt$_2$	Temporal reliability *r*
Aggressiveness to humans	*Name* (often) jumps at the grate or window when persons stay in front of it	AG2	0.86	0.78	0.94***
	Name (often) spits or throws objects from the enclosure	AG3	0.84	0.88	0.94***
	Name (often) tries to scratch persons through the grate	AG4	0.92	0.69	0.82***
Anxiousness	When *name* is alone in a room he/she (often) moves about continuously, and sometimes has diarrhea	AX2	0.79	0.90	0.95***
	When one comes close to the grate near to *name*, he/she (often) shies away quickly	AX3	0.06	0.25	0.86***
Arousability	Prior to the feeding, *name* (often) moves about a lot	AR2	−0.21	0.62	0.59**
	When being fed, *name* (often) makes many sounds	AR3	−0.17	0.82	0.92***
Curiosity	*Name* (often) touches new objects, such as enrichment items, at great length	CU2	0.76	0.60	0.91***
	Confronted with novel food, *name* (mostly) ignores it	CU3	0.88	0.78	0.61**
Distractibility	When *name* is busy with something, he/she (often) disrupts his/her activity as soon as something else is going on	DI2	0.78	0.73	0.72***

(continued)

Table 3.1 (continued)

Personality trait construct	Items *GAPI* – behavior-descriptive verb items	Code	Interrater reliability ICCt_1	Interrater reliability ICCt_2	Temporal reliability r
Dominance	In the group, *name* is (most often) the first to get to the food	DO2	0.96	0.85	0.98***
	Name is (most often) the last to get to the food	DO3	0.98	0.82	0.96***
Food orientation	When there is food, *name* is (often) quickly on the spot	FM2	0.79	–0.50	0.69***
	Between feeding times, one (hardly ever) sees *name* eating	FM3	0.69	0.70	0.64**
Friendliness to humans	When called, *name* (often) comes to the grate closely	FR2	0.89	0.27	0.20
	(At times), *name* even allows close contact with humans	FR3	0.84	0.43	0.88***
Friendliness to conspecifics	*Name* (often) grooms other group members	FR5	0.69	0.82	0.90***
	Name has (hardly ever) body contact with other group members	FR6	0.71	0.86	0.89***
Friendliness to youngsters	*Name* spends (a lot of) time with youngsters	CH2	0.75	0.96	0.91***
	Name (often) plays with youngsters	CH3	0.57	0.49	0.89***
Gregariousness	*Name* (often) withdraws from his/her conspecifics in the indoor or outdoor enclosure	GR2	0.94	0.90	0.89***
	Name sits together with his/her conspecifics (a lot)	GR3	0.93	0.89	0.88***
Impulsiveness	When he/she does not get his /her food immediately, *name* (often) quickly knocks at the grate or window	IM2	0.04	0.42	0.68***
	Name (often) waits calmly until it is his/her turn to get his/her food	IM3	–0.28	0.68	0.89***
Persistency	With dealing with enrichment materials, *name* (often) gives up easily	PE2	0.11	0.64	0.79***
	Name can keep him-/herself busy with something (for a long time)	PE3	0.41	0.56	0.94***
Physical activity	In the indoor or outdoor enclosure, *name* keeps walking or brachiating (most of the time)	AC2	0.93	0.93	0.98***
	(Most of the time), *name* is sitting or lying	AC3	0.96	0.90	0.90***

(continued)

Table 3.1 (continued)

Personality trait construct	Items *GAPI* – behavior-descriptive verb items	Code	Interrater reliability ICCt_1	ICCt_2	Temporal reliability *r*
Playfulness	*Name* (often) plays on his/her own with objects such as enrichment items	PL2	0.76	0.42	0.88***
	Name (rarely) plays with adolescent or adult members of the group	PL3	0.82	0.77	0.79***
Sexual activity	*Name* (often) establishes sexual contact with his/her conspecifics	SX2	0.77	0.96	0.96***
	Name (often) stimulates him-/herself sexually	SX3	0.79	0.72	0.90***
Vigilance	*Name* (often) notices small changes in the cages or enclosures quickly	VI2	−0.69	0.17	0.85***
	Name (often) watches everything around him/her very closely	VI3	−0.12	0.49	0.83***
Mean			0.72	0.74	0.88

Note: these are translations of the original German items with which the presented data were collected. The German items can be obtained from the author. Some items can be reversed in meaning depending on whether they are used as *agreement* scales from (1) *strongly disagree* to (5) *strongly agree*, for which the statements of frequency given in parenthesis should be included in the item text, or as *frequency* scales from (1) *hardly ever* to (5) *very often* on items presented without the frequency quantifying expressions provided in parentheses. Variable-oriented interrater reliability rated with the *great ape personality inventory (GAPI) – behavior-descriptive verb items* (B) in test periods t_1 and t_2 was computed with *ICC* (3,*k*). It depicts reliability of the mean ratings on the basis of $k=4$–5 independent raters per ape (Shrout and Fleiss 1979). For analyses of test–retest reliability, the scores were aggregated over all raters within each rating period. Variable-oriented test–retest reliability of these aggregated scores over the 5 weeks between rating periods t_1 and t_2 was computed with Pearson correlation *r*. $^{***}p<0.001$, $^{**}p<0.01$. Mean reliability scores across the 34 items were computed with *r*-to-*Z* transformation

[a]For captive samples

Ten keepers rated the same 20 individuals studied behaviorally by Uher et al. (2008) on a computer-based interface. On each format, they specified their level of agreement with the statements given in the items on five-point Likert agreement scales from (1) strongly disagree to (5) strongly agree. For this reason, the items contained statements of frequency (such as “often” or “hardly” in *GAPI-B*) or of degree of intensity (such as “very” in *GAPI-A*; given in parentheses in Tables 3.1 and 3.2). Alternatively, ratings could be indicated on frequency scales from (1) hardly ever to (5) very often on items presented without the frequency quantifying expressions provided in parentheses. This could facilitate understanding of the items, but would hinder inversions of item meanings, thus increasing probabilities of response sets.

For comparisons among methods, ratings were scheduled to parallel the behavioral data collection of the Uher et al. (2008) study. All individuals were rated twice by four to five raters, with an interval of 5 weeks (for details see Uher and Asendorpf 2008). This design allowed analyses of interrater reliability for each data collection

Table 3.2 Great ape personality inventory – trait adjective items (GAPI-A)[a]

Personality trait construct	Items *GAPI* – trait adjective items	Code	Interrater reliability		Temporal reliability
			$ICCt_1$	$ICCt_2$	r
Aggressiveness to humans	To humans, *name* is (very) aggressive	AG1	0.90	0.61	0.80***
Anxiousness	*Name* is (very) anxious	AX1	0.73	0.73	0.81***
Arousability	*Name* is (quickly) excited	AR1	0.82	0.80	0.89***
Curiosity	*Name* is (very) curious	CU1	0.47	0.74	0.87***
Distractibility	*Name* is (very) distractible	DI1	0.75	0.65	0.80***
Dominance	*Name* is (very) dominant	DO1	0.97	0.94	0.98***
Food orientation	*Name* is (very) gluttonous	FM1	0.86	0.83	0.83***
Friendliness to humans	To humans, *name* is (very) friendly	FR1	0.54	0.82	0.89***
Friendliness to conspecifics	To her conspecifics, *name* is (very) friendly	FR4	0.78	0.61	0.93***
Friendliness to youngsters	To youngsters, *name* is (very) friendly	CH1	0.03	0.35	0.67***
Gregariousness	*Name* is (very) gregarious	GR1	0.89	0.88	0.91***
Impulsiveness	*Name* is (very) impulsive	IM1	0.82	0.50	0.81***
Persistency	*Name* is (very) persistent (such as with enrichment materials)	PE1	0.39	0.73	0.90***
Physical activity	*Name* is physically (very) active	AC1	0.98	0.92	0.92***
Playfulness	*Name* is (very) playful	PL1	0.95	0.92	0.91***
Sexual activity	*Name* is sexually (very) active	SX1	0.78	0.95	0.96***
Vigilance	*Name* is (very) vigilant	VI1	−0.28	0.53	0.73***
Mean			0.79	0.79	0.88

Note: these are translations of the original German items with which the presented data were collected. The German items can be obtained from the author. Variable-oriented interrater reliability rated with the *great ape personality inventory* (*GAPI*) – *trait adjective items* (A) in test periods t_1 and t_2 was computed with *ICC* (3,*k*). It depicts reliability of the mean ratings on the basis of $k=4$–5 independent raters per ape (Shrout and Fleiss 1979). For analyses of test–retest reliability, the scores were aggregated over all raters within each rating period. Variable-oriented test–retest reliability of these aggregated scores over the 5 weeks between rating periods t_1 and t_2 was computed with Pearson correlation r. $^{***}p<0.001$, $^{**}p<0.01$. Mean reliability scores across the 17 items were computed with r-to-Z transformation

[a]For captive samples

period, and analyses of test–retest reliability between periods. Interrater reliability was substantial in both variable-oriented and individual-oriented analyses. In the first rating period, the mean variable-oriented reliability among the $k=4$–5 independent raters per ape as indicated by ICC(3,*k*) (Shrout and Fleiss 1979) was 0.72 for behavior-descriptive verbs and 0.79 for trait adjectives. Mean individual-oriented interrater agreement was ICC(3,*k*)=0.84 for behavior-descriptive verbs, and 0.85 for trait adjectives. Results on the item level are given in Tables 3.1 and 3.2 those on the individual level are given in Table 3.3, separately for the two periods of data collection.

Table 3.3 The subjects and individual-oriented analyses of interrater reliability, test–retest reliability, and validity of personality profiles rated with the GAPI

Subjects				Interrater reliability *GAPI*				Temporal reliability *GAPI*		Validation		
				Behavior-descriptive verb items *B*		Trait-adjective items *A*		Behavior-descriptive verb items *B*	Trait-adjective items *A*	Cross-method coherence between		
Species	Name	Age	Sex							B–A	E–B	E–A
				$ICCt_1$	$ICCt_2$	$ICCt_1$	$ICCt_2$	*r*	*r*	*r*	*r*	*r*
Bonobo	Joey	22	M	0.76	0.86	0.85	0.87	0.87***	0.85***	0.53*	0.60**	0.22
	Kuno	8	M	0.92	0.91	0.92	0.94	0.97***	0.97***	0.34	0.94**	0.31
	Limbuko	9	M	0.76	0.85	0.85	0.84	0.93***	0.83***	0.55*	0.40	–0.01
	Ulindi	11	F	0.87	0.80	0.86	0.80	0.93***	0.92***	0.66#	0.49*	0.39
	Yasa	7	F	0.68	0.77	0.88	0.89	0.83***	0.97***	0.51*	0.70**	0.28
Chimpanzee	Dorien	24	F	0.65	0.80	0.19	0.70	0.83***	0.71**	0.47	0.69**	0.35
	Fraukje	28	F	0.76	0.82	0.86	0.80	0.91***	0.90***	0.87***	0.51*	0.34
	Frodo	11	M	0.76	0.85	0.80	0.76	0.79***	0.90***	0.28	0.61**	–0.05
	Robert	29	M	0.85	0.82	0.76	0.84	0.89***	0.90***	0.44#	0.62**	0.06
	Sandra	11	F	0.70	0.82	0.82	0.84	0.88***	0.96***	0.70**	0.79***	0.54*

Gorilla	Bebe	25	F	0.80	0.76	0.88	0.90	0.89***	0.92***	0.62**	0.51*	0.27
	Gorgo	23	M	0.77	0.81	0.50	0.66	0.96***	0.93***	0.58#	0.74**	0.22
	N'diki	27	F	0.75	0.78	0.83	0.75	0.87***	0.93***	0.62**	0.28	0.46#
	Ruby	7	F	0.86	0.78	0.89	0.93	0.90***	0.94***	0.31	0.73**	–0.06
	Viringika	9	F	0.77	0.78	0.82	0.87	0.84***	0.91***	0.73***	0.68**	0.40
Orangutan	Bimbo	24	M	0.84	0.84	0.87	0.85	0.95***	0.91***	0.63#	0.54*	0.05
	Dokana	16	F	0.91	0.93	0.93	0.91	0.96***	0.96***	0.72**	0.80**	0.77**
	Dunja	31	F	0.85	0.87	0.87	0.89	0.95***	0.96***	0.34	0.58*	0.13
	Padana	7	F	0.86	0.87	0.84	0.88	0.93***	0.89***	0.47#	0.46#	0.48#
	Pini	16	F	0.90	0.89	0.83	0.85	0.95***	0.92***	0.55*	0.79***	0.47#
Mean				0.81	0.84	0.83	0.85	0.91	0.92	0.57	0.67	0.30

Note: details on the subjects' species, their age in years, and sex (F=female, M=male). Reliability of the individuals' personality profiles rated with the *great ape personality inventory* (*GAPI*) *behavior-descriptive verb items* (B) and *trait adjective items* (A) was computed with *ICC* (3,*k*) separately for the two non-overlapping test periods t_1 and t_2. It depicts reliability of the mean ratings on the basis of $k=4–5$ independent raters per ape (Shrout and Fleiss 1979). For analyses of test–retest reliability, the scores were aggregated over all raters within each rating period. Individual-oriented test–retest reliability of these aggregated scores over the 5 weeks between rating periods t_1 and t_2 was computed with Pearson correlation r. $^{***}p<0.001$, $^{**}p<0.01$. Coherence between the individuals' personality profiles across 17 traits rated with the *GAPI – behavior descriptive verb items* (B), with the *GAPI – trait adjective items* (A), and with ethological behavior measures (E) obtained from observations in 14 laboratory tests and group situations was computed with Pearson correlations r. $^{***}p<0.001$, $^{**}p<0.01$, $^{*}p<0.05$, $^{\#}p<0.10$. To further increase the reliability of the personality profiles obtained with each method, they are based on data that were each aggregated on the trait level across the two studied time periods spanning about 6 weeks. Corresponding data from variable-oriented analyses are reported in Uher et al. (2008), and Uher and Asendorpf (2008)

Since ratings showed high interrater reliability, mean rating scores were calculated across keepers within each rating period. Test–retest reliabilities between these averaged ratings were substantial. Over 5 weeks, variable-oriented correlations were $r=0.88$ for both behavior-descriptive verb and trait adjective items; individual-oriented correlations were $r=0.91$ for behavior-descriptive verb and 0.92 for trait adjective items. Results on the item level are given in Tables 3.1 and 3.2, and those on the individual level are given in Table 3.3. Comparisons of test–retest reliability scores between different personality measures showed that those obtained with ratings were significantly higher than those obtained with ethological methods; the effect sizes were large ranging from $d=0.73$ to 0.91 (Uher and Asendorpf 2008). These results should be kept in mind when interpreting temporal reliability or temporal stability of personality differences based on rating methods.

3.4.3 Validation in Personality Research

Personality measures must not only be reliable, they must also be valid. That is, it must be shown that they measure what they are supposed to measure. Establishing empirical validity is crucial for research on theoretical constructs such as personality traits. The central concern is thus to link a theoretical concept with empirical findings. This is the purpose of validation through nomological networks. A nomological network includes a theoretical framework that represents the basic features of the trait construct in question, an empirical framework how this shall be measured, and specification of the interrelationships among and between these two frameworks (Cronbach and Meehl 1955). For example, if one is interested in curiosity, a weak approach is to study it with only one method, whether by rating or by ethological measure. A stronger approach is to do both and to show coherence between the different measures of the construct of curiosity. Converging evidence from different methods establishes a strong case of construct validation for the studied personality construct (Cronbach 1988).

I illustrate the use of nomological networks with data from great apes. I analyzed the construct validity of personality traits derived with the behavioral repertoire × environmental situations approach in these species (Uher 2008a, b) with three different assessment methods. That is, for each trait construct, I specified a priori several ethological behavior measures (Uher et al. 2008), two behavior-descriptive verb items, and one trait adjective item that theoretically should reflect that construct well. These measures span a nomological network around each trait construct. For most traits, the theoretical relations among these measures could be substantiated empirically. The mean variable-oriented correlation across 17 trait constructs between behavior-descriptive verb ratings and trait adjective ratings was $r=0.71$; between behavior-descriptive verb ratings and composite ethological behavior measures it was $r=0.56$; and between trait adjective ratings and composite ethological behavior measures it was $r=0.35$ (Uher and Asendorpf 2008). Mean individual-oriented correlations across 20 individual personality profiles were

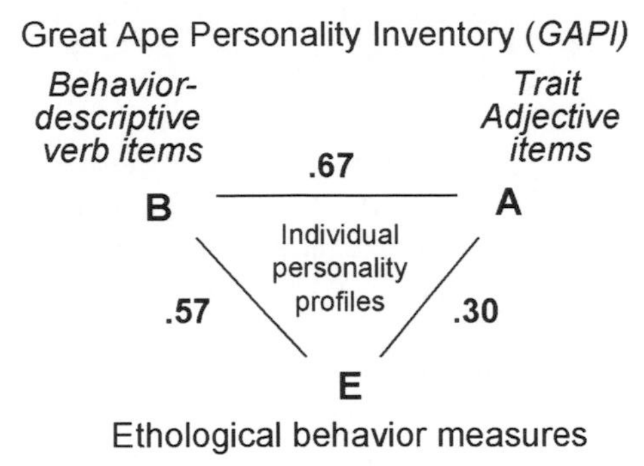

Fig. 3.3 Individual-oriented coherence across 20 individuals (mean Pearson correlations *r* computed with *r*-to-*Z* transformation) among individual personality profiles rated with the great ape personality inventory (*GAPI*) – *behavior-descriptive verbs* (B) and the *GAPI* – *trait adjectives* (A), and measured with ethological methods of behavior measurement (E) in a series of 14 laboratory tests and group observations. To further increase the reliability of the personality profiles obtained with each method, they are based on data that were each aggregated on the trait level across the two studied time periods spanning about 6 weeks. The ethologically measured behavior profiles of four individuals were incomplete since the subjects could not be tested in the laboratory

virtually identical (Fig. 3.3). This established substantial evidence for the construct validity of personality trait constructs derived with the behavioral repertoire × environmental situations approach.

These studies are also useful to explain the processes of validating psychometric instruments for personality ratings in nonhuman species. Standard inventories of human personality are based on (1) a theoretical foundation. They are developed using iterative procedures of empirical testing and statistical item selections. For the resulting instruments, empirical evidence for sufficient (2) interrater agreement and (3) test–retest reliability for each single item, (4) validity for each single personality construct as well as their (5) empirical intercorrelations and factor structure are routinely shown in large samples, but these characteristics are generally taken for granted in later applications (Kline 2000). These standard criteria are documented in application manuals together with (6) norm distributions for specific reference populations.

Surprisingly, these essential and well-established methodological foundations of instrument development have received only very little attention in primate personality research. The *GAPI* is one of the first published primate personality inventories for which the first four of these six essential standard steps of instrument development have been accomplished. Except for top-down/etic approaches from rating items of the human Five-Factor Model (King and Figueredo 1997; Weiss et al. 2006), which are grounded in phylogenetic theory, to my knowledge, no other rating list published to date for taxonomic personality research is based on a theoretical foundation. Interrater reliability is almost always analyzed, but test–retest reliability is rarely studied (for exceptions see McGuire et al. 1994; Stevenson-Hinde et al. 1980; Uher and Asendorpf 2008).

First steps towards validation have already been made for some rating lists by showing empirical relations to single behavior measures (McGuire et al. 1994;

Capitanio 1999; Pederson et al. 2005; Kuhar et al. 2006). However, the behavior measures were often selected without a priori specification of their theoretical relationships to the studied trait constructs. Many of these behaviors were selected from ethograms that are used for research questions other than personality differences. As already noted earlier, however, every behavior is not per se a useful measure of personality. Since most of these studies failed to analyze the test–restest reliability of their behavioral measures, it remains unclear whether they are sufficiently aggregated to represent in fact reliable personality measures. Unless test–restest reliability is shown for behavioral measures, coherence with rating measures, and thus validation, may be compromised. This could explain why many studies show only low to moderate correlations between ratings and ethological behavior measures of nonhuman primate personality.

The use of nomological networks in the Uher and Asendorpf (2008) study, and empirical test–restest reliability of all obtained measures, that is of ethological personality measures and two kinds of personality rating measures, allow systematic analyses of the validity of the GAPI. Item analyses are important since it is the items that activate the raters' pertinent knowledge, that initiate their mental assessment processes, and that provide the frameworks in which the raters can indicate their resulting judgments (Funder 1999; Uher and Asendorpf 2008). Item analyses are particularly relevant for inherently anthropocentric trait adjective items. Their use for personality ratings in humans is theoretically (Goldberg 1990) and empirically well founded (Kenrick and Funder 1988), but evidence for their validity in nonhuman species is rarely provided, despite their popularity. Yet without systematic validation, the behaviors they actually refer to in particular species remain unclear as well as what they are actually measuring (Uher 2008a, b; Uher and Asendorpf 2008).

Trait adjectives can have implicit connotations for raters that are not obvious from their general meaning. For example, in great apes, "friendly to his/her conspecifics" was surprisingly uncorrelated with both behavior-descriptive verb ratings and ethological behavior measures of grooming and body contact. This finding could indicate that keepers base their judgments of individuals as "friendly" not on prosocial behaviors, such as grooming, but instead on low aggression. This would have significant implications for predictions of behavior in particular situations, such as in group introductions, because low aggressiveness may not necessarily imply high prosociality. Differences in interpretation like these, which are neither obvious nor intended, are obscured by items that complement trait adjectives with "clarifying behavioral definitions" as frequently used for primate ratings, such as defining "gentle" with "responds to others in an easy, kind manner" (Stevenson-Hinde and Zunz 1978; McGuire et al. 1994; King and Figueredo 1997; Weiss et al. 2006). Separate analyses of trait adjectives, and their supposed behavioral definitions, are thus important for validation (Uher and Asendorpf 2008).

Trait adjectives require large inferences from observable behavior, and may therefore be prone to anthropomorphic interpretations of behavior. Behavior-descriptive verb items, in contrast, are less inferential and less susceptible to biases and subjectivity than trait adjective items since they require the raters to focus on specific, perceivable behaviors. Validation analyses of the GAPI show that coherence with ethological behavioral measures of personality is substantially higher for behavior-descriptive verbs than for trait adjectives (see Fig. 3.3; Tables 3.1–3.3). This may be because behavior-descriptive verbs are both behaviorally based, whereas trait adjectives as abstract personality descriptors may have broader predictive ranges of behaviors and situations. However, a study on human personality could not clearly support the hypothesis that trait adjectives generally refer to more exemplars than verbs (Borkenau and Müller 1991). Rather, the relation between the grammatical form of personality-descriptive categories and the number of their exemplars was mediated by category breadth. If category breadth was held constant, grammatical form correlated significantly with the rated trait prototypicality of their exemplars. That is, verbs describe more accurately how individuals are actually behaving than adjectives (Borkenau and Müller 1991).

The empirical results presented in this chapter and in the Uher and Asendorpf (2008) study square nicely with these findings. They underscore the particular utility of non-trait-adjective rating methods, such as behavior-descriptive verb or act-frequency ratings, which combine the greater accuracy of behavior prediction with the economy of rating methods.

It is obvious that ratings constitute economic methods of personality assessment (Vazire et al. 2007; Gosling 2008; Freeman et al. 2011), but they do so only if their validity is evidenced empirically. As I have shown, psychometric validation requires substantial empirical and statistical work that in nonhuman species ultimately includes coherence with observable behavior. Thus initially, ratings are much more labor-consuming methods of personality assessment than ethological methods. But as validated psychometric instruments, they allow economic measurements of personality. For the GAPI, four of six essential steps of standard instrument development have already been accomplished. Further steps require empirical studies in larger samples to analyze the species' factor structures and their norm distributions. They could also include psychometric analyses of larger item pools for iterative processes of statistical item selections.

In conclusion, there is no single method of personality assessment that is generally inferior or superior to others. The question of method selection should therefore not be polarized by premature recommendations (as in Vazire et al. 2007; Gosling 2008; Freeman et al. 2011) that obscure the diversity of assessment methods and the important functions this very diversity serves for construct validation. Instead, the advantages and disadvantages of the different methods of measurement should be weighed selectively for their relevance to the particular research questions at hand (Uher 2008a, b).

3.4.4 Establishing Comparability of Trait Constructs Across Populations

Populations, such as species, can also show population-specific behaviors that are not shown in other populations. This must be considered when personality variation is compared among populations. Cross-population comparisons presuppose comparability of trait constructs even if they are measured with different behaviors. A first step is analysis of functional equivalence of behaviors used to measure a trait construct (Mehta and Gosling 2008). Comparability analyses of meaning and functions of behaviors have been established in ethology and rely on fine-grained contextual prepost analyses of behavioral sequences (Preuschoft 1992; Preuschoft and van Hooff 1995).

Yet functional equivalence of behavioral measures alone is insufficient to conclude that personality constructs are comparable across populations. Personality variation can be compared only on the construct level, not on the level of single measures (Uher 2008b); comparability of trait constructs therefore has to be established empirically as structural equivalence. Methodologies for statistical comparison of factorial structures of functionally equivalent, yet nonidentical trait measures across different population levels have been established in cross-cultural research (Vijver and Poortinga 2002). They can be generalized to other population comparisons such as among species (for details see Uher 2008b).

Since all ratings necessarily rely on human language, researchers using trait adjective ratings are tempted to assume that identical items also imply comparability across the different species to which they are applied (Weiss and Adams 2008). But because trait adjectives can have fairly different implicit connotations in other species, their "functional" equivalence has to be established first through empirical convergence with behavioral measures. That factorial structures of personality constructs obtained with identical items can differ among species has already been shown descriptively (King and Figueredo 1997; Weiss et al. 2006). But so far, structural equivalence of such factors has been analyzed statistically only between two different populations of captive chimpanzees (Weiss et al. 2007); statistical analyses of their structural equivalence or nonequivalence across species are still pending.

3.5 Conclusions

Human personality psychology provides a rich and solid foundation of theoretical concepts, methodological approaches, and methods of assessment with unquestionable suitability for nonhuman primate personality research. Many concepts and methodologies for within-population research are directly applicable to nonhuman primates. Those established for cross-cultural comparisons of human population can be generalized systematically to comparisons of nonhuman populations including species. There is much for us to learn from human personality psychology; its knowledge and experiences in solving many puzzling research issues can give nonhuman personality research a competitive edge to head for new advances in the near future.

Acknowledgments I am grateful to the editors and to Wendy Johnson, Jochen Fahrenberg, and Jens Asendorpf for valuable comments on the manuscript. I thank the zookeepers at the Wolfgang Köhler Primate Research Center in Leipzig, Germany, for their kind cooperation and for rating the apes; Josep Call from the Max Planck Institute for Evolutionary Anthropology, Leipzig, Germany, for supporting this study; and Josefine Kalbitz for recording behavior for the reliability analyses.

References

Allport GW (1937) Personality: A psychological interpretation. Macmillan, New York
Allport GW, Odbert HS (1936) Trait names: A psycholexical study. Psychol Monogr 47:1
Altmann J (1974) Observational study of behavior: Sampling methods. Behaviour, 49:227–267
Angleitner A, Ostendorf F, John OP (1990) Towards a taxonomy of personality descriptors in German: A psycho-lexical study. Eur J Pers 4:89–118
Asendorpf JB (1988) Individual response profiles in the behavioral assessment of personality. Eur J Pers 2:155–167
Asendorpf JB, van Aken MAG (1999) Resilient, overcontrolled, and undercontrolled personality prototypes in childhood: Replicability, predictive power, and the trait-type issue. J Pers Soc Psychol 77:815–832
Block J (1971) Lives through time. Bancroft, Berkeley
Borkenau P, Müller B (1991) Breadth, bandwidth, and fidelity of personality-descriptive categories. Eur J Pers 5:309–322
Borkenau P, Mauer N, Riemann R et al. (2004) Thin slices of behavior as cues of personality and intelligence. J Pers Soc Psychol 86:599–614
Buss DM (1991) Evolutionary personality psychology. Annu Rev Psychol 42:459–491
Cairns RB, Bergman LR, Kagan J (1998) Methods and models for studying the individual. Sage Publications, Thousand Oaks
Capitanio JP (1999) Personality dimensions in adult male rhesus macaques: Prediction of behaviors across time and situation. Am J Primatol 47:299–320
Capitanio JP (2004) Personality factors between and within species. In: Tierry B, Singh M, Kaumanns W (eds) Macaque societies. Cambridge University Press, Cambridge
Caspi A, Roberts BW (1999) Personality continuity and change across the life course. In: Pervin LA, John OP (eds) Handbook of personality: Theory and research (2nd ed). Guilford Press, New York, NY
Cervone D, Pervin LA (2008) Personality: Theory and research (10th edn). Wiley, Hoboken
Church AT (2001) Personality measurement in cross-cultural perspective. J Pers 69:979–1006
Crawford MP (1938) A behavior rating scale for young chimpanzees. J Comp Psychol 26:79–91
Cronbach LJ (1988) Five perspectives on the validity argument. In Wainer H, Brown HI (eds) Test validity. Erlbaum, Hillsdale
Cronbach LJ, Meehl P (1955) Construct validity in psychological tests, Psychol Bull 52:281–302
De Raad B, Barelds DPH (2008) A new taxonomy of Dutch personality traits based on a comprehensive and unrestricted list of descriptors. J Pers Soc Psychol 94:347–364.
Eysenck HJ (1990) Biological dimensions of personality. In: Pervin L (ed). Handbook of personality theory and research. Guilford, New York
Fairbanks LA, Jorgensen MJ (2011) Objective behavioral tests of temperament in nonhuman primates. In: Weiss A, King JE, Murray L (eds) Personality and temperament in nonhuman primates. Springer, New York
Fleeson W (2004) Moving personality beyond the person-situation debate: The challenge and the opportunity of within-person variability. Curr Dir Psychol Sci 13:83–87
Freeman H, Gosling SD, Shapiro SJ (2011) Comparison of methods for assessing personality in nonhuman primates. In: Weiss A, King JE, Murray L (eds) Personality and temperament in nonhuman primates. Springer, New York
Funder DC (1999) Personality judgment: A realistic approach to person perception. Academic, San Diego
Funder DC (2007) The personality puzzle (4th edn). W.W. Norton & Co, New York
Funder DC (2006) Towards a resolution of the personality triad: Persons, situations and behaviors. J Res Pers 40:21–34
Funder DC, Colvin CR (1991) Explorations in behavioral consistency: Properties of persons, situations, and behaviors. J Pers Soc Psychol 60:773–794

Galton F (1869) Hereditary genius: An inquiry into its laws and consequences. Macmillan, London
Goldberg LR (1990) An alternative "description of personality": The Big-Five factor structure. J Pers Soc Psychol 59:1216–1229
Goodall J (1986) The Chimpanzees of Gombe: Patterns of behavior. Harvard University Press, Cambridge
Gosling SD (2001) From mice to men: What can we learn about personality from animal research? Psychol Bull 127:45–86
Gosling SD (2008) Personality in nonhuman animals. Soc Pers Psych Compass 2:985–1001
Gosling SD, Graybeal A (2007) Tree thinking: A new paradigm for integrating comparative data in psychology. J Gen Psychol 134:259–277
Gosling SD, Lilienfeld SO, Marino L (2003) Personality. In: Maestripieri D (ed) Primate psychology, Harvard University Press, Cambridge, Massachusetts
Hartshorne H, May MA (1928) Studies in the nature of character. Vol 1. Studies in deceit. Macmillan, New York
Hebb DO (1949) Temperament in chimpanzees: I. Method of analysis. J Comp Physiol Psychol 42:192–206
Hofstee WKB, Brokken FB, Land H (1981) Constructie van een Standaard-Persoonlijkheids-Eigenschappenlijst (S.P.E.L.). Nederlands Tijdschrift voor de Psychologie 36:443–452
John OP (1990) The "Big Five" factor taxonomy: Dimensions of personality in the natural language and in questionnaires. In Pervin LA (ed) Handbook of personality: Theory and research. Guilford, New York
John OP, Angleitner A, Ostendorf F (1988) The lexical approach to personality: A historical. review of trait taxonomic research. Eur J Pers 2:171–203
Johnson W (2007) Genetic and environmental influences on behavior: Capturing all the interplay. Psychol Rev 114:423–440
Kenrick DT, Funder DC (1988) Profiting from controversy: Lessons from the person-situation debate. Am Psychol 43:23–34
King JE, Figueredo AJ (1997) The Five-Factor Model plus Dominance in chimpanzee personality. J Res Pers 31:257–271
King JE, Weiss A (2011) Personality from the perspective of a primatologist. In: Weiss A, King JE, Murray L (eds) Personality and temperament in nonhuman primates. Springer, New York
King JE, Weiss A, Farmer KH (2005) A chimpanzee (*Pan troglodytes*) analogue of cross-national generalization of personality structure: Zoological parks and an African sanctuary. J Pers 73:389–410
Kline P (2000) The handbook of psychological testing (2nd ed). Routledge, London
Kuhar CW, Stoinski TW, Lukas KE, Maple TL (2006) Gorilla Behavior Index revisited: Age, housing and behavior. Appl Anim Behav Sci 96:315–326
Lehner PN (1996) Handbook of ethological methods (2nd ed). Cambridge University Press, Cambridge
Leung K, Bond MH (1989) On the empirical identification of dimensions for cross-cultural comparisons. J Cross Cult Psychol 20:133–151
Magnusson D (1988) Individual development from an interactional perspective: A longitudinal study. Erlbaum, Hillsdale
Martin JE (2005) The influence of rearing on personality ratings of captive chimpanzees (*Pan troglodytes*). Appl Anim Behav Sci 90:167–181
Marwitz M, Stemmler G (1998) On the status of individual response specificity. Psychophysiology 35:1–15
Matthews G, Deary IJ, Whiteman MC (2003) Personality traits (2nd Ed). Cambridge University Press, Cambridge
McGuire M, Raleigh M, Pollack D (1994) Personality features in vervet monkeys: The effects of sex, age, social status, and group composition. Am J Primatol 33:1–13
Mehta PH, Gosling SD (2008) Bridging human and animal research: A comparative approach to studies of personality and health. Brain Behav Immun 22:651–661

Mervielde I, Asendorpf JB (2000) Variable-centred and person-centred approaches to childhood personality. In: Hampson SE (ed) Advances in personality psychology, Vol. 1. Psychology Press, Hove

Mischel W (1968) Personality and assessment. Wiley, New York

Mischel W (1977) The interaction of person and situation. In: Magnusson D, Endler NS (eds) Personality at the crossroads: Current issues in interactional psychology. Erlbaum, Hillsdale

Mischel W, Peake PK (1982) Beyond déjà vu in the search for cross-situational consistency. Psychol Rev 89:730–755

Mischel W, Shoda Y (1995) A cognitive-affective system theory of personality: Reconceptualizing situations, dispositions, dynamics, and invariance in personality structure. Psychol Rev 102:246–268

Mischel W, Shoda Y, Mendoza-Denton R (2002) Situation-behavior profiles as a locus of consistency in personality. Curr Dir Psychol Sci 11:50–54

Mischel W, Shoda J, Smith RE (2003) Introduction to personality: Toward an integration (7th ed). Wiley, New York

Murray LE (1998) The effects of group structure and rearing strategy on personality in chimpanzees (*Pan troglodytes*) at Chester, London and Twycross Zoos. Int Zoo Yearb 36:97–108

Murray LE (2002) Individual differences in chimpanzee (*Pan troglodytes*) personality and their implications for the evolution of mind. In: Harcourt D, Sherwood B (eds) New perspectives in primate evolution and behaviour. Otley, Westbury Publishing

Pederson AK, King JE, Landau VI (2005) Chimpanzee (*Pan troglodytes*) personality predicts behavior. J Res Pers 39:534–549

Pike KL (1967) Language as behavior and emic and etic standpoints for the description of behavior. In: Borgatta EF (ed) Social psychology: Readings and perspective. Rand McNally, Chicago

Preuschoft S (1992) 'Laughter' and 'smile' in Barbary macaques (*Macaca sylvanus*). Ethology 91:200–236

Preuschoft S, van Hooff JARAM (1995) Homologizing primate facial displays: A critical review of methods. Folia Primatol 65:121–137

Rouff HJ, Sussman RW, Strube MJ (2005) Personality traits in captive lion-tailed macaques (*Macaca silenus*). Am J Primatol 67:177–198

Rushton JP, Brainerd CJ, Pressley M (1983) Behavioral development and construct validity: The principle of aggregation. Psychol Bull 94:18–38

Schmitt DP, Buss DM (2000) Sexual dimensions of person description: Beyond or subsumed by the big five? J Res Pers 34:141–177

Shrout PE, Fleiss JL (1979) Intraclass correlations: Uses in assessing rater reliability. Psychol Bull 2:420–428

Sih A (2011) Behavioral Syndromes: A behavioral ecologist's view on the evolutionary and ecological implications of animal personality. In: Weiss A, King JE, Murray L (eds) Personality and temperament in nonhuman primates. Springer, New York

Stern W (1911) Die differentielle Psychologie in ihren methodischen Grundlagen (2. Auflage). [Differential Psychology in its methodological foundations (2nd ed)]. Barth, Leipzig

Stevenson-Hinde J, Zunz M (1978) Subjective assessment of individual rhesus monkeys. Primates 19:473–482

Stevenson-Hinde J, Stillwell-Barnes R, Zunz M (1980) Individual differences in young rhesus monkeys: consistency and change. Primates 21:498–509

Suomi SJ, Novak MA, Well A (1996) Aging in rhesus monkeys: Different windows on behavioral continuity and change. Dev Psychol 32:1116–1128

Tett RP, Guterman HA (2000) Situation trait relevance, trait expression, and cross-situational consistency: Testing a principle of trait activation. J Res Pers 34:397–423

Tooby J, Cosmides L (1989) Adaptation versus phylogeny: The role of animal psychology in the study of human behavior. Int J Comp Psychol 2:175–188

Uher J (2008a) Comparative personality research: Methodological approaches (Target article). Eur J Pers 22:427–455

Uher J (2008b) Three methodological core issues of comparative personality research. European Eur J Pers 22:475–496
Uher J (2011) Individual behavioral phenotypes: An integrative meta-theoretical framework. Why 'behavioral syndromes' are not analogues of 'personality'. Dev Psychobiol, published online Mar 22, 2011, doi:10.1002/dev.20544
Uher J, Asendorpf JB (2008) Personality assessment in the Great Apes: Comparing ecologically valid behavior measures, behavior ratings, and adjective ratings. J Res Pers 42:821–838
Uher J, Asendorpf JB, Call J (2008) Personality in the behaviour of great apes: Temporal stability, cross-situational consistency, and coherence in response. Anim Behav 75:99–112
van de Vijver FJR, Poortinga YH (2002) Structural equivalence in multilevel research. J Cross Cult Psychol 141:141–156
Vazire S, Gosling SD, Dickey AS et al. (2007) Measuring personality in nonhuman animals. In: Robins RW, Fraley RC, Krueger RF (eds) Handbook of research methods in personality psychology. Guilford Press, New York
Weiss A, Adams MJ (2008) Species of nonhuman personality assessment. Eur J Pers 22:472–474
Weiss A, King JE, Hopkins WD (2007) A cross-setting study of chimpanzee (*Pan troglodytes*) personality structure and development: Zoological parks and Yerkes National Primate Research Center. Am J Primatol 69:1264–1277
Weiss A, King JE, Perkins L (2006) Personality and subjective well-being in orangutans (*Pongo pygmaeus* and *Pongo abelii*). J Pers Soc Psychol 90:501–511
Yerkes RM (1939) The life history and personality of the chimpanzee. Am Nat 73:97–112

Chapter 4
Personality from the Perspective of a Primatologist

James E. King and Alexander Weiss

Abstract Although concerns about anthropomorphism and subjectivity have limited the widespread use of subjective personality ratings in primate research, recent developments show that subjective personality ratings of primates have acceptable interrater reliabilities and display good evidence of construct validity. The role of human personality dimensions or factors has typically been one of defining the taxonomy of human personality and examining relationships between those factors and multiple human traits and outcomes. We describe three additional areas in which subjective personality ratings may be useful within the context of primatology. First, when human personality factors are extended to nonhuman primates, interspecies differences in factor structure afford opportunities to infer changes that have occurred in the underlying behavioral correlations during evolutionary development. In other words, personality factors can be interpreted as evolutionary characters. Our studies show a close correspondence between the personality factor structure of chimpanzees and orangutans with one notable exception. We suspect that a factor in chimpanzees reflecting an association between aggression, emotionality, and unpredictability is homologous to the Conscientiousness factor in humans. Such a Conscientiousness factor is completely absent in orangutans. This chimpanzee factor may be a consequence of intense intragroup competition in male chimpanzees. Second, the highly aggressive disposition of wild male chimpanzees may also reflect a spectrum of personality differences between male and female chimpanzees. Male chimpanzees score higher than females on traits related to aggression, unpredictability, and emotionality, a pattern unlike that displayed by humans. Third, subjective personality ratings are potentially useful in addressing the disparity between current measures of

J.E. King (✉)
Department of Psychology, University of Arizona, Tucson, AZ 85721, USA
e-mail: kingj@u.arizona.edu

A. Weiss et al. (eds.), *Personality and Temperament in Nonhuman Primates*,
Developments in Primatology: Progress and Prospects, DOI 10.1007/978-1-4614-0176-6_4,

psychological well-being in nonhuman primates that focus on low levels of well-being including pathological behaviors, and measures of happiness or subjective well-being in humans that focus on positive subjective states.

4.1 Introduction: Anthropomorphism and Scientific Psychology

Animal characters with the gift of language and human-like personalities, ranging from endearing to evil, are often featured in folktales, children's stories, and cartoons. This genre goes back at least to the time of Aesop's fables that illustrated moral truths through the antics of a colorful variety of animals that were, e.g., shrewd, conniving, or impulsive. The appeal of human-like or anthropomorphic traits is also evident in the earliest writings in scientific primatology during the early twentieth century. Wolfgang Köhler in his classic book *The Mentality of Apes* (Köhler 1925) used the word *personality* to describe individual differences among his nine chimpanzee subjects, although the exclusive focus of his studies was the problem solving and intelligence of the chimpanzees. Similarly, Robert Yerkes (1925, 1929, 1943) often commented on the distinctive personalities of his chimpanzees. For example, Yerkes described Chim (now known to have been a bonobo) as "sanguine, venturesome, trustful, friendly, and energized," while describing Chim's companion, Panzee, as "distrustful, retiring, and lethargic" (Yerkes 1929, p. 281). More generally, he noted that different chimpanzees are "good and ill-natured, stable and unstable, calm and excitable, industrious and lazy" (Yerkes 1939, p. 112).

4.1.1 In Defense of Anthropomorphism

In spite of the highly notable individual differences in chimpanzee personality and temperament, neither Köhler, Yerkes, nor any other primatologist in the early part of the twentieth century ventured beyond the informal application of personality descriptors to individuals. The first attempt at quantitative measurement of nonhuman primate personality was made by Crawford (1938) who measured individual differences in chimpanzee personality with rating scales closely tied to specific behaviors. In 1942, Karl Lashley, a usually vigorous proponent of a rigorously objective approach to psychology, assumed the directorship of the Yerkes Laboratory of Primate Biology in Orange Park, Florida (Dewsbury 2006). Interestingly, his outline of proposed research for the laboratory included "Individual Differences in Temperament." Unfortunately, this promise of systematic inquiry into chimpanzee temperament was not fulfilled. The problem was that primatologists as well as psychologists have traditionally had an ambivalent attitude toward attributing personalities to animals. As an alternative to the use of anthropomorphic personality descriptors, euphemisms including "salients" (Billingslea 1941), "psychological individual" (Carr and Kingsbury 1939a), "directional disposition" (Carr and Kingsbury 1939b), and, most recently, "individuality" (Jones 2005) have been proposed as replacements.

Anthropomorphism can be defined as the description of an animal's behavior or personality in terms ordinarily used to describe human behavior or personality (Schilhab 2002). The term would apply particularly to descriptions based on subjective judgments of animal behavior. Anthropomorphic descriptions can be applied to short-term psychological states such as "angry" or to intentional states implied by statements that the animal "knows that" or is "afraid of." Relatively permanent traits including personality descriptors may also have an anthropomorphic component. Critics argue that anthropomorphic description implies a false assumption that whatever trait is being described in the animal is fundamentally similar to the corresponding trait in humans (Kennedy 1992). This criticism has sometimes reached levels that motivated one observer to note that it was treating "anthropomorphism as a kind of disease, which scientists are especially qualified to cure" (Mynott 2009, p. 24). The most direct answer to blanket criticisms of anthropomorphism is that use of personality descriptive terms with animals is subject to the same psychometric tests for validity that have been used in human personality research for decades (Vazire et al. 2007).

Although application of human personality descriptors to animals was described as anthropomorphic, paradoxically, attribution of those same descriptors to humans became an important component of research human individual differences (Winter and Barenbaum 1999). In fact, as Andrews (2009) pointed out, subjective personality ratings of humans not capable of self-reports are generally accepted as is the case for the *Child Behavior Checklist* for children of ages from 1½ to 5 years. Furthermore Kwan et al. (2008) showed that human projection of their own personality to other humans was significantly stronger than their projection of their own personality to dogs.

A related issue is whether human personality-descriptive terminology should be used to describe personality in nonhuman species. Might it be more scientifically rigorous to apply a different set of descriptors to each species thereby avoiding the implication that the same descriptor in two species refers to the same trait? The problem with this strategy is that the universe of personality descriptors would soon become an absurdly cumbersome Tower or Babel. Therefore, we believe that current use of trait names based originally on human personality is a mostly feasible strategy in naming traits as well as personality factors. However, this amplifies the need for construct validation data to clarify the trait or factor meaning.

4.1.2 *Validation of Subjective Personality Ratings*

The first step before validation of personality ratings is the demonstration of acceptable between-rater agreement or interrater reliability. This is usually interpreted as agreement among raters on an individual's trait values. A review of interrater reliabilities from subjective personality ratings from across multiple studies of mammalian and nonmammalian species by Gosling (2001) showed reasonably high ratings in most cases. Reliabilities for animal ratings are generally within the range of comparable reliabilities for human personality ratings (e.g., King and Figueredo 1997). Furthermore, ratings of human personality are usually based on self-ratings, rater-reports by at least one person familiar with the target, or a combination of self- and rater-reports. Ratings by more than one or two people familiar with a human subject are not usually practicable. In contrast, when animals are rated, multiple raters are often available for each subject thereby leading to increased reliabilities of mean scores and compensation for relatively low sample sizes.

The second basic requirement for personality ratings is validity, particularly construct validity (Cronbach and Meehl 1955; Campbell and Fiske 1959; Messick 1989). Construct validity is applicable when a measure such as a personality trait based on subjective ratings is based on an expected network of correlations with other measures based on either a theory or the inherent meaning (sometimes called the face validity) of the trait. The potential set of predicted correlations should include those expected to be reasonably large and statistically significant (convergent validity) as well as those expected to be nonsignificant (discriminant validity). The cluster of predicted correlations between personality ratings and other independent measures thus become a direct measure of the personality trait's construct validity.

The predicted correlation would be simple and direct when the personality trait and the correlated behavior are semantically similar. The positive correlation between ratings of *playful* and frequency of play behavior in rhesus monkeys reported by Stevenson-Hinde et al. (1980) is one example. More recent examples are described by Uher (2011) and by Uher and Asendorpf (2008). A more complex and indirect verification occurs if the personality measure has a less specific, more open-ended definition exemplified by *sensitive* or *protective* (King and Figueredo 1997) or if correlated traits are combined into factors such as Extraversion (King and Figueredo 1997) or Sociability (Capitanio 1999).

For example, Kuhar et al. (2006) showed that zoo-housed gorillas with higher scores on an Extraversion factor displayed higher affiliative and higher contact aggression behaviors. Similarly, Pederson et al. (2005) showed that an Extraversion factor measured in zoo-housed chimpanzees was positively correlated with the specific behaviors social approach and gymnastic activities, while an Agreeableness factor was negatively correlated with agonistic behaviors. In contrast, an Openness factor had a nonsignificant correlation with agonistic behavior. Vazire et al. (2007), in a study of captive chimpanzees, found a mean absolute correlation of 0.22 between 60 pairs of trait ratings and behavior frequencies that were predicted to be strong.

A large body of experimental literature (Kenny 1994; Funder 1995; Letzrins et al. 2006; Ross et al. 2007) has shown that personality ratings of humans are quite accurate especially in contexts outside of laboratory settings. Therefore, when viewed in a wider, evolutionary perspective, the accuracy of human judgments about personality traits of nonhuman primates should not be surprising.

The increased application of personality measures to research on nonhuman primates has mainly occurred within the context of nonhuman primate models of important human phenomena including development (King et al. 2008), immune function (Capitanio 2011), and abnormal behavior (Lilienfeld et al. 1999). However, the focus of this chapter will be to show how subjective personality ratings can be used to address questions arising from within primatology itself.

4.2 Personality Factors

4.2.1 Personality Factors as Latent Variables

The historical foundations of factor analytic descriptions of personality ratings of nonhuman primates lie in the development of the taxonomic approach to human personality description. In 1936 Allport and Odbert identified 17,953 English words that described long lasting human personality traits (Allport and Odbert 1936). Even after exclusion of rare and vague words, several thousand remained. Following approximately 40 years of research and debate about the number of dimensions underlying this vast number of descriptors, a widespread but not universal agreement was reached that human personality could be best described at the most general level by five dimensions, the Five-Factor Model (Digman 1996). These five factors, Extraversion, Agreeableness, Conscientiousness, Neuroticism, and Openness, have been the basis for a large body of human personality research, mainly based on one or more of the dimensions serving as predictors of a large number of observed human characteristics. Moreover, the Five-Factor Model has often been used as a framework to synthesize multiple personality studies in humans (Roberts et al. 2007) and nonhuman animals (Gosling and John 1999).

Extension of the taxonomy of personality-descriptive characteristics beyond humans to one or more nonhuman species opens a new vista of questions and issues. These new issues begin with the conceptualization of personality dimensions or factors as latent variables (Gorsuch 1983; Borsboom et al. 2003). A latent variable is present when a set of quantities or items such as personality ratings are intercorrelated. If a significant portion of the intercorrelations among items is attributable to the correlation of each item with a theoretical construct based on all items in the set, then the construct is a latent variable. Ideally, the items would be uncorrelated with each other except for their common correlations with the latent variable. There are obvious practical benefits that accrue from using the latent variable as a parsimonious substitute for the entire list of intercorrelated items (see Fig. 4.1).

A list of personality descriptors defining a factor usually reveals multiple items with semantic similarity to other items. Examples include *nervous* and *anxious* within the Neuroticism descriptors or *talkative* and *garrulous* within the Extraversion descriptors. Other correlated descriptors, while not fully synonymous, nevertheless reflect traits that are commonly understood to covary, e.g., *assertive* and *adventurous* within the Extraversion descriptors. Is it possible that personality factors therefore are nothing more than an expression of semantic similarity of some items in

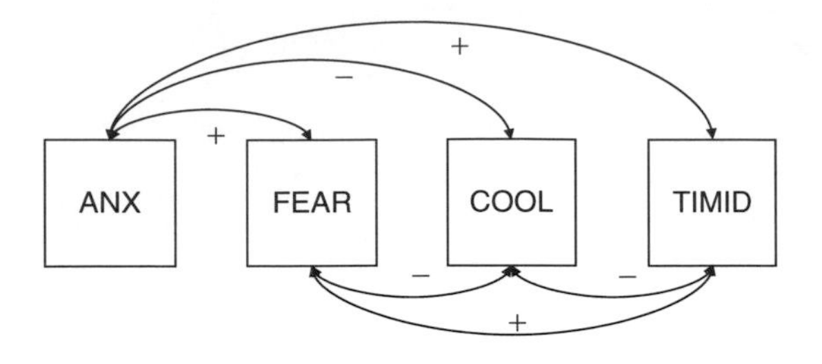

Bivariate correlations:

6 parameters are needed to describe relationships among the traits Anxious, Fearful, Cool, and Timid.

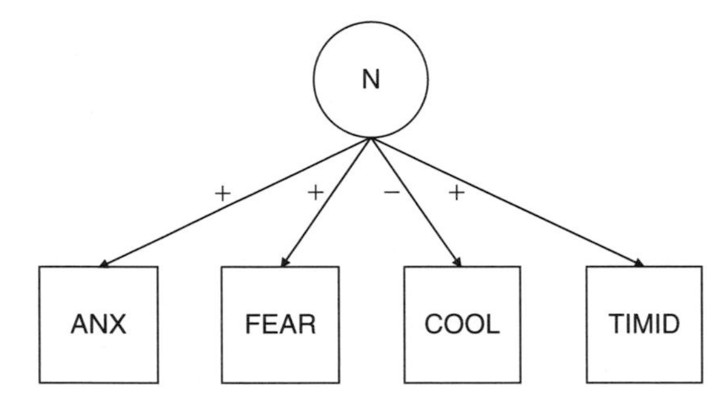

Factor analysis:

4 parameters are needed to describe relationships among the same traits.

Fig. 4.1 Use of a latent variable or factor simplifies relationships among correlated personality descriptors

combination with knowledge about linked human characteristics or an implicit theory of personality? After considerable debate about this issue, it is now clear that human personality factors capture reliable individual differences and are not just expressions of semantic similarities or an implicit theory of personality (Weiss and Mendelson 1986; Borkenau 1992; Funder 1995; Letzrins et al. 2006).

For purposes of interspecies comparison of factor structure, the most interesting aspect of personality factors is the presence of intercorrelated items within factors that have little similarity based on either semantic meaning or common beliefs about personality. For example, although *impulsive* and *depressed* are both indicators of human Neuroticism, there is little semantic overlap between the two words. An important point is that the convention of labeling latent variables or personality factors with a single word such as "Neuroticism" or "Agreeableness" can be misleading. Lists of items defining personality factors contain a heterogenous assortment of specific descriptors. In fact, a commonly used measure of the human Five-Factor Model, the NEO-PI-R (Costa and McCrae 1992) divides each of the five factors into six facets. Facets within a single factor are differentiated by their correlations with external measures (Costa and McCrae 1998). In addition, the facets are independently heritable (Jang et al. 1998). The facets are therefore not simply alternative indices of a single homogenous latent variable but reflect instead partially independent entities with biological underpinnings. Specifically, facets can be viewed as a pattern of interrelated traits that constitute a pattern species-specific to humans.

A further necessary condition for treating a collection of correlated traits as a unitary latent variable is that the rank order of individual differences on that latent variable remains constant across varying environmental and social conditions as well as temporal intervals (van Oers et al. 2005; Penke et al. 2007). However, the assumption of rank order invariance across environments is rarely addressed in personality research.

4.2.2 Personality Factors as Evolutionary Characters

If the five human personality factors have a biological basis, evidence of their presence should be seen in nonhuman species, especially nonhuman primates. Gosling and John (1999) in a review of 19 studies with 12 mammalian and nonmammalian species presented the first evidence that these factors are well represented among nonhuman species. Factors similar to Extraversion, Agreeableness, and Neuroticism were reported in most species with Openness somewhat less prevalent than the other three factors. A factor similar to Conscientiousness was only reported in chimpanzees (King and Figueredo 1997). However, any conclusions drawn from the failure to find evidence of Conscientiousness or any other factor should be tentative; the apparent absence of a factor may simply be a result of insufficient items that are sensitive to that factor in a given study or even that, while the set of items is sufficient for assessing the factor in one species, it is not sufficient for assessing the factor in a different species (see Uher 2011). Support for the absence of a factor emerges if the factor occurs in one species but not the other after both species have been rated on identical item sets.

In the King and Figueredo study, zoo workers rated 100 chimpanzees on 43 personality-descriptive adjectives that had been selected to represent the five human personality factors. A Dominance factor emerged in addition to five factors that resembled the five human factors, including an analogue of Conscientiousness. Items with high loadings on chimpanzee Conscientiousness included those that were later divided into two facets (King et al. 2008). The first facet, named *tameness*, contained five descriptors including not *defiant*, not *irritable*, and not *aggressive*. The second, named *predictability*, contained five descriptors including *predictable*, not *reckless*, and not *erratic*. The chimpanzee Conscientiousness factor might appear to be unlike the human Conscientiousness factor that includes facets labeled *competence*, *order*, *dutifulness* (Costa and McCrae 1998), none of which are applicable to chimpanzees. However, two other facets of human Conscientiousness, *self-discipline* and *deliberation*, are semantically similar to the combined meaning of the two chimpanzee facets. Furthermore, other researchers have reported human factors similar to Conscientiousness, namely Constraint (Tellegen 1985), Dependability (Almagor et al. 1995), and Self-control (Conn and Rieke 1994). Therefore, we believe that the human Conscientiousness factor can be interpreted as an evolutionarily modified version of the human Conscientiousness factor. A further speculative possibility is that both the human and chimpanzee Conscientiousness factors are evolutionarily modified versions of a factor present in a common ancestor.

If the personality structure of a species is to be a useful evolutionary character for understanding behavioral evolution it should remain relatively invariant across different environments and rearing conditions. Between-species comparison of personality structure would be meaningless if that structure were so labile that it varied significantly across different habitats. The similarity of human personality structure across a wide range of Eastern and Western cultures (Church 2001; McCrae et al. 2005) gives reason for optimism that personality structure is largely constant within nonhuman species as well.

King et al. (2005) examined generalization of the six chimpanzee factors described by King and Figueredo (1997) in a comparison between zoo-housed chimpanzees and chimpanzees living within a naturalistic environment at a rehabilitation and reintroduction sanctuary in the Republic of the Congo (Congo, Brazzaville). In addition, most of the rating forms used to assess the sanctuary chimpanzees were written in French instead of English. A between group comparison of internal consistency alpha values across factors showed that only Extraversion differed significantly between the zoo-housed and Congo chimpanzees. In addition, the data from the chimpanzees in the sanctuary and a new set of zoo chimpanzees not used in the original study were pooled. The factor structure from these data was directly compared to the original factor structure described in King and Figueredo (1997) via targeted orthogonal Procrustes rotation (McCrae et al. 1996). This analysis revealed that the factor structure generalized across samples and that the Dominance, Extraversion, Conscientiousness, and Agreeableness factors clearly replicated across samples. The failure of the Neuroticism and Openness factors to generalize was probably the result of the small number of adjectives defining these factors: three Neuroticism and two Openness items. Similarly, in cross-cultural studies of human personality structure, the Neuroticism and Openness factors do not display as robust a pattern of generalization as the other three human factors (Saucier and Goldberg 2001). A later study (Weiss et al. 2007) based on larger sample sizes and incorporating between group comparisons with Procrustes rotation also showed generalization of Dominance, Extraversion, Agreeableness, and Conscientiousness from zoo settings to a laboratory setting and from one sample of zoo-housed chimpanzees to another independent sample of zoo-housed chimpanzees. Most recently, the factor structure of chimpanzees from zoos in the United States and Australia was compared with that from chimpanzees in Japan rated on Japanese language questionnaires (Weiss et al. 2009). Again, despite the language differences in the rating forms for the two samples, the Dominance, Extraversion, Agreeableness, and Conscientiousness factors generalized across the two cultures.

A second condition for optimal comparison of personality structure across species is the use of a common set of items for rating the different species. Although any factor will incorporate a potentially large number of descriptive items, the associations between items and factors as measured by factor loadings are far from constant. The use of a consistent set of items will therefore avoid a confounding of species and items.

Weiss et al. (2006) obtained ratings of 152 zoo-housed orangutans on 48 adjectives including the 43 adjectives used in the previous assessment of chimpanzee personality (King and Figueredo 1997). The orangutans displayed five factors: Dominance, Extraversion, Agreeableness, Neuroticism, and Intellect (see Table 3 in Weiss et al. 2006). The composition of Extraversion, Agreeableness, and Neuroticism were mostly similar in chimpanzees and orangutans. However, Dominance was more narrowly defined in orangutans, being composed of only low-Agreeableness markers, e.g., *stingy*. Also, in striking contrast to chimpanzee personality structure, items related to the chimpanzee Conscientiousness factor did not define a distinct factor in orangutans. Instead, some of these items combined with items in the chimpanzee Openness factor to define the orangutan Intellect factor.

The between-species differences among great apes and humans can be conceptualized by assuming that there are six distinct families of similar personality factors corresponding to the human five factors in addition to Dominance. The defining items for any factor will differ somewhat among species but will nevertheless be distinct from other factors. Therefore, we will refer to these related factor families as domains that have similarities to "fuzzy sets" (Smithson and Verkuilen 2006). In other words, the items defining a factor (or set) will vary somewhat across species. In addition, item membership in a particular factor (or set) set is expressed as a continuous probability (or factor loading) instead of a simple yes or no. The two principal between-species differences are the absence of the Dominance factor in humans and the absence of the Conscientiousness factor in orangutans. These findings do not mean that humans have low levels of traits associated with Dominance or that orangutans have low levels of traits associated with Conscientiousness. The presence of a factor in one species but not another simply indicates that items defining the factor are intercorrelated in the former but not the latter species independently of the mean levels of the items.

In addition, a clear difference between human personality domains or factors and those of nonhuman species is that the human domains encompass a far more extensive set of traits. For example, the trait "gullible," defining high Agreeableness in humans is meaningless in nonhumans. Species differences in the breadth of personality domains are a potential future source of comparative analysis.

Both chimpanzees and orangutans displayed a Dominance factor in addition to an Extraversion factor. Likewise, Gosling and John's (1999) review of factor analytic studies of personality in 12 species showed evidence of a Dominance factor separate from Extraversion in 7 of the 19 studies. Descriptors relating to Dominance in chimpanzees and orangutans are subsumed into other domains in humans, particularly Extraversion. For example, *assertiveness* is one of the six facets of human Extraversion (Costa and McCrae 1992).

These between-species differences in factor structure suggest that personality factors or domains can be usefully conceptualized as evolutionary characters, viz. characteristics that change to varying degrees within an evolutionary lineage (Harvey and Pagel 1991; Gosling and Graybeal 2007). However, the evolutionary characters represented by personality factors are a correlated pattern of traits or a latent variable instead of being a single trait. The factor loadings or the correlations between the items and the overall factor define the character and the change in the pattern of those loadings across species is an index of evolutionary change.

If the between-species change in factor loadings is sufficiently large, a factor may either appear or disappear during the evolution of a lineage. This is illustrated by the factor loadings for Dominance, Extraversion, Agreeableness, and Conscientiousness in orangutans (Weiss et al. 2006) and chimpanzees (King and Figueredo 1997). The loadings of the first three factors display relatively modest between-species variation whereas the Conscientiousness factor has seemingly emerges in chimpanzees from relatively uncorrelated adjectives in orangutans. An interesting but as yet unanswered question is what aspects of chimpanzee evolutionary adaptation led to the correlation of the aggressive, emotional, and unpredictable components of the Conscientiousness factor. One possible explanation is that the Conscientiousness factor has an evolutionarily recent origin occurring over the most recent six million years of human evolution. The presence of Conscientiousness in humans and chimpanzees but not orangutans is consistent with this possibility because humans shared a common ancestor with chimpanzees about six million years ago while the common ancestor of humans and orangutans occurred approximately 14 million years ago at about the time the lineage leading to orangutans diverged from lineages leading to the African apes (Raaum et al. 2005). The most evolutionarily parsimonious explanation would be that a proto-Conscientiousness factor was present in the common ancestor of humans and chimpanzees and then persisted in the human and chimpanzee lineages until the present time. In that case, the similarities between the Conscientiousness factor in humans and chimpanzees would be attributable to homology or "identity by descent" (Harvey and Pagel 1991). Another possible, although less parsimonious, explanation is that the Conscientiousness factor evolved independently in the human and chimpanzee lineages through parallel evolution. The large differences in habitat and social structure of chimpanzees and recent humans (Boehm 1999) lessens the likelihood of parallel evolution. In either case, the Conscientiousness factor would be an evolutionarily recent, derived characteristic.

Personality data on the two other great ape species, gorillas and bonobos, would provide clarification of this tentative evolutionary scenario based on homology. Since the bonobo and chimpanzee lineages diverged only about 2.5 million years ago, it seems likely that bonobos should also display the Conscientiousness factor. The divergence of the gorilla lineage from that of humans and chimpanzees occurred more recently than the comparable divergence of the orangutan lineage. Therefore, the presence of a Conscientiousness factor in gorillas would suggest that this factor predates the common ancestor of humans and chimpanzees. Absence of a gorilla Conscientiousness factor would imply a more recent origin of the Conscientiousness domain, one closer to the common ancestor of humans and chimpanzees.

A more general hypothesis is that some aspect of social organization throughout the primate order resulted in selection for a Conscientiousness factor on multiple occasions through parallel evolution. In this case the Conscientiousness factor might be related to aspects of social structure including fission–fusion grouping, male bonding, dominance structure, and group size. Personality data on Old World monkeys, especially macaques, which are highly speciated with variable social structure, would yield a strong test of this hypothesis.

This discussion about the possible evolutionary development of personality domains in the great apes is tentative and based on a relatively small amount of supporting data. Comparative ratings data on additional markers of the six personality domains discussed here will be essential for more conclusive verification of our hypothesis about the evolution of personality domains. In addition, collection of data from primates living in their natural habitat is fundamentally important. Although as noted earlier, the evidence for generalization of personality domains across different habitats is encouraging, data on personality organization in the wild as well as the relationship of personality domains to life-history variables including reproduction and mortality is essential to incorporate personality data into a coherent account of the evolution of personality.

4.3 Personality and the Behavior of Wild Chimpanzees

The study of personality in the context of behavior in a naturalistic setting has notable potential for illuminating the effects of personality variables on evolutionary fitness and life-history strategies. The evolutionary significance of variation in human personality has been extensively discussed (Nettle 2006; Penke et al. 2007), particularly in the context of attractiveness and mating strategies (Buss and Duntley 2006). However, two fundamental and related problems complicate efforts by evolutionary psychologists to understand the role of personality differences in shaping the evolutionary emergence of humans over the past few million years. First, current human environmental and social conditions are vastly different from those existing during the evolutionary development of modern humans and during most of the approximately 150,000 years since modern humans first appeared. Second, as a result of the radically changed environments, inferences about the effects of human personality variations of evolutionary fitness require an assumed generalization from modern conditions to the environment of evolutionary adaptation or EEA (Symons 1990), viz. the environment in which humans evolved into their present state. Although recent data indicate that the human genome has rapidly changed over the past 40,000 years (Hawks et al. 2007) any current human environment remains markedly disconnected from the evolved adaptations of the species. In contrast, nonhuman primates in their natural habitats are usually living in a close approximation of their EEAs and therefore afford a more direct opportunity for assessing the effects of personality variables on life-history characteristics that may relate to evolutionary adaptation.

4.3.1 Dominance and Evolutionary Adaptation

Dominance, a central trait in interpreting animal behavior, has been defined in multiple ways, the most common of which include priority to food, success in agonistic encounters, and access to mates (Richards 1974). Dominance also has a complex association with endocrine function (see Anestis 2011). Although these classic definitions are still useful, it is now clear that dominance can be expressed by behaviors more subtle than overt aggression and displacement, including vocalizations, direct stares, arm gestures, and assertive postures. Dominance hierarchies, whether linear or not, are usually defined by the results of interactions between pairs of animals. However, as noted earlier, dominance is also closely related to a more general personality trait or latent variable defined by several intercorrelated traits including some that are not obviously related to inter-animal competition. For example, the 12-item Dominance factor identified in zoo-housed chimpanzees (King and Figueredo 1997) included, in addition to *dominant* and not *submissive*, the descriptors *persistent*, *independent*, and not *fearful*. The chimpanzee Dominance factor with the items *dominant* and *submissive* removed had a correlation of 0.52 with a traditional ordinal ranking of dominance (Schneider and King 2008). The Dominance factor is also heritable (Weiss et al. 2000).

In the wild, dominance is expressed differently in female than in male chimpanzees. Wild chimpanzees live in distinct communities ranging from 20 to 110 individuals (Goodall 1986). Females, but not males, usually migrate out of their natal groups around the age of sexual maturity. Females engage in semi-solitary foraging sometimes accompanied by one or more offspring or other mothers and their offspring. Although interactions between adult females are infrequent, observations over a sufficiently long period show a dominance hierarchy. There is also evidence that high ranking females sometimes kill infants of lower ranking females (Goodall 1986; Pusey et al. 1997). Pusey et al. (1997) used submissive pant grunts observed over a 22-year period to estimate dominance ranks among female chimpanzees at the Gombe National Park, Tanzania. Observations of life-history data over a 35-year period showed that females with higher ranks lived longer, had a higher proportion of infants surviving to at least 5 years, had shorter inter-birth intervals, and had faster maturing daughters than the less dominant females.

A later study of the Gombe chimpanzees based on 33 years of weight measurements showed that body mass varied positively with food availability and total area within the community range (Pusey et al. 2005). Body mass was also positively related to dominance among females but not males, although the more dominant members of both sexes displayed more stable body weight. It was interesting that body mass was not related to male dominance despite males displaying considerably more linear and intensely competitive hierarchies than females.

A subsequent related study (Murray et al. 2006) showed that the higher and less variable weights of more dominant females could be attributed to higher quality and more consistent diets although these females also invested less time in foraging. Effects of female dominance on feeding-related behaviors have also been observed in West African populations. Lehmann and Boesch (2005) observed that dominant female chimpanzees range over a greater portion of the community range and forage in the peripheral areas more than less dominant chimpanzees.

The pervasive relationships between female dominance in chimpanzees and several variables directly related to evolutionary fitness is somewhat surprising in light of the infrequent dominance-related interactions among females and their usually benign expression in terms of vocalization or mild aggression. Furthermore, females are less social than males, spending about 50% of their time either alone or with dependent offspring (Pusey 2001). Yet, despite the limited contact among females, a mildly expressed dominance relationship seems to be highly predictive of success in reproduction and feeding. The reason for this correspondence between dominance and strong indicators of evolutionary fitness remains to be explained. We suspect that some further aspect of personality variation lies behind a proximate explanation of this puzzle.

4.3.2 Aggression and Personality in Male Chimpanzees

Male chimpanzees in the wild have a linear dominance hierarchy with the alpha male often playing a despotic role over other chimpanzees in the population. The behavioral repertoire of dominant males includes aggressive displays that include a belligerent blend of charging, piloerection, arm waving, swaying, object throwing, and vocalizations (Goodall 1986). Male, but not female, dominance rank is also often maintained by alliances with other males. Not surprisingly, recent studies using DNA testing to confirm paternity have verified that dominant male chimpanzees sire more offspring in captive (Klinkova et al. 2005) as well as in wild East African populations (Constable et al. 2001). Similarly, Wroblewski et al. (2009) found that dominant male chimpanzees sired most but not all of the 34 offspring over a 22-year period at Gombe. Chagnon's (1988) classic study of violence based on revenge and kinship relationships in the Yanomamö may be a dramatic illustration of a related phenomenon in human tribal societies. Men who had engaged in a revenge killing had more wives and more offspring than men who had not participated in a killing.

Aggression by male chimpanzees in the wild goes well beyond the domination of other males within a male dominance hierarchy. Male chimpanzees at five sites in both West and East Africa engage in lethal intergroup aggression (Goodall 1986; Wilson and Wrangham 2003). Small groups of adult males, referred to as patrols, quietly approach their community's boundary sometimes stopping to sniff the ground or destroy nests made by chimpanzees in the adjoining community. If the patrol encounters a single male from the other community, it will typically attack the male causing either death or severe injuries. In addition, females without sexual swellings are sometimes killed and if that female has an infant, the infant may be killed and eaten (Goodall 1986; Watts et al. 2002).

Intragroup nonlethal aggression by male chimpanzees is common and often directed toward other males as well as females. One of the most dramatic expressions of this aggression is male aggression toward adult females. Using a long-term database from Kibale National park, Muller et al. (2007) addressed the question of whether the aggression was related to the promiscuous mating pattern of female chimpanzees. Males directed their attacks mainly toward those females who were most likely to conceive, viz. those who displayed sexual swelling and had previous infants. The attacks increased the likelihood that the males would subsequently copulate with the females. Unlike other male aggression, the attacks on fecund females were unrelated to the male dominance. This type of male aggression therefore appears to be part of a male reproductive strategy.

Anthropologists and primatologists have sometimes speculated about levels of aggression in human and chimpanzee societies. Boehm (1999) proposed a theory stating that during recent human evolution, selection favored reduction of nonlethal aggression and despotic dominance hierarchies within groups along with emergence of an egalitarian social structure. However, the intensity of between-group lethal aggression was assumed to have remained relatively undiminished. Because of the close phylogenetic relationship between humans and chimpanzees, Boehm's theory implies that chimpanzees, although sharing a recent common ancestor with humans, should not have undergone the evolutionary pacification of humans, and therefore should have retained the despotic dominance hierarchies, along with high levels of within-group nonlethal aggression and high levels of between-group lethal aggression.

Boehm's prediction was tested by Wrangham et al. (2006) who assembled published data on levels of inter- and intragroup aggression in human hunter-gatherer societies and in wild chimpanzee populations. Estimates of death rates from lethal aggression between communities were within the same order of magnitude for chimpanzees and humans. However, the median rate of nonlethal attacks within communities was 384 times higher for male chimpanzees than for male humans. For females, the median attack rate was only 182 times higher in chimpanzees than in humans.

The pervasive expression of aggressive behaviors by male chimpanzees toward conspecifics is dramatic. Male chimpanzee aggression comprised a major portion of the book *Demonic Males: Apes and the Origins of Human Violence* (Wrangham and Peterson 1996) that described the propensity for male violence in male humans, chimpanzees, gorillas, and orangutans. Presumed increases in evolutionary fitness accompanying that aggression are the most common basis for explanations of male chimpanzee aggression (e.g., Wilson and Wrangham 2003; Muller et al. 2007). However, we believe that understanding of male chimpanzee aggression can also be advanced by viewing it in the proximate context of traits underlying personality domains.

We recently identified some possible clues to the particularly aggressive male behavior based on the pattern of male–female differences in personality traits among zoo-housed chimpanzees (King et al. 2008). There was a striking disparity between males and females on five of the six personality dimensions rated. Males were rated as higher than females on the Dominance and Neuroticism factors and on an activity-related facet of Extraversion. But the males were rated as lower on each of the two facets of Conscientiousness: predictability defined by items including (not) *impulsive*, (not) *erratic*, and *predictable* and tameness defined by items including (not) *aggressive*, (not) *defiant*, and (not) *irritable*. The males were also lower on the Agreeableness factor. In other words, male chimpanzees, relative to females, displayed a broad syndrome of enhanced predispositions toward high activity, aggression, unpredictability, and emotionality. Since unpredictability and emotional instability would be expected to broaden the contexts in which aggression occurs, as well as its intensity, this syndrome may indeed be part of the proximate structure supporting male chimpanzee violence in their natural habitats noted above. Human males have lower scores on the Agreeableness factor but, unlike male chimpanzees, also have lower scores on the Neuroticism factor including the impulsiveness facet (Costa et al. 2001). Thus, the human personality profile does not support unrestrained aggression as strongly as does the chimpanzee personality profile.

Clearly, we base this conjecture upon an assumed generalization from zoo to wild habitats. Nevertheless, we believe that just as conventionally defined dominance has been shown to predict important life-history traits, other personality dimensions, including those that are related to or support dominance, will eventually be shown to have similarly important consequences for evolutionary adaptation.

4.4 Happiness as a Personality Dimension in Primates

4.4.1 Psychological Well-Being and Quality of Life in Primates

A 1985 amendment to the Animal Welfare Act (Public Law 99–198) required research institutions that maintain nonhuman primates to make provision for their "psychological well-being." In the aftermath of the amendment, there was a surge in research involving enrichment of the environmental and social conditions of captive primates (Wolfle 1999; Honess and Marin 2006; Baker et al. 2007). To justify the enrichment, a corresponding body of research emerged showing its beneficial effects on various objective indices of well-being including reduction of cortisol, aggression, and stereotyped behaviors (Brent et al. 1989; Boinski et al. 1999; Jones and Pillay 2004).

The term "quality of life" has been applied to the overall effect of environmental and social enrichment (McMillan 2005) on an animal's well-being. Quality of life is affected by an animal's health and well-studied variables involving environmental and social enrichment. A varied, interesting, and comfortable environment is necessary for an animal to experience a satisfactory quality of life. However, as McMillan notes, quality of life is most fundamentally determined by the animal's reaction to that environment, its feelings, affect, and overall subjective experience. Importantly, different animals may respond differently to objectively identical environments. Therefore, an indicator of an animal's reaction to its environment that potentially includes the animal's subjective experience is required to make a full assessment of its quality of life, particularly if the assessment is going to be applicable not only to pathologically low quality of life but also to the higher levels as well.

Low well-being and low quality of life may be manifested by many possible objective indicators of distress including, but not limited to, stereotyped behavior, hair plucking, and elevated cortisol levels. A larger challenge is to identify indices of high well-being. Mere absence of pathological behaviors will not be sufficient. One possibility is that the presence of behaviors common in a species' natural habitat (i.e., species-specific behaviors) is indicative of psychological well-being, an idea that is often expressed in both the scientific (Maki and Bloomsmith 1989) and popular (Hearn 1994) literature. In other words, an animal doing what is "natural" should be happy. Although, the linkage between naturally occurring behaviors and happiness has considerable intuitive appeal, it rests on the untested assumption that behavior shaped by natural selection also incorporates inherent happiness. One limitation of this approach is that the absence of species-typical behaviors may not necessarily be an indicator of distress and the presence of those behaviors may not necessarily indicate high levels of psychological well-being. This would seem to be particularly true if those species-specific behaviors were relatively stereotyped and adaptive since evolutionary adaptation would probably have made those behaviors resistant to disruption by all but the most stressful circumstances. In addition, captive settings may limit the range of species-specific behaviors that are possible. We suggest an alternative approach based on the human ability to assess emotional and affective states in other humans, i.e., empathy (Ickes 2001, 2003).

The understanding that human and animal subjective experiences have similarities is an old idea in psychology noted early in the twentieth century by Margaret Floy Washburn: "We are obliged to acknowledge that all psychic interpretation of animal behavior must be on the analogy of human experience" (Washburn 1909, quoted in Bekoff 2005). In a similar spirit, Burghardt (1997) proposed that "private experience" including empathy be added as a fifth member of Tinbergen's (1996) four basic questions about animal behavior. These original questions related to a behavior's proximate cause, its evolutionary history, its contribution to evolutionary adaptation, and its ontological development. Questions about the private experience of nonhuman primates, including subjective happiness or psychological well-being, will inevitably be met with some skepticism. However, one avenue that can be exploited to gauge an animal's subjective happiness is the fact that humans are accurate at judging the happiness of other humans (Diener et al. 1999). We need only apply this empathic ability with acceptable interrater reliability and construct validity accuracy to nonhuman species. In the case of nonhuman primates, particularly the great apes, this is a reasonable assumption.

Wemelsfelder et al. (2001) described an interesting approach based on human interspecies empathy referred to as "free choice profiling" to describe the well-being of domestic pigs. During observation of individual animals, observers independently generated their own descriptive words that seemed applicable to that animal. Examples were "relaxed," "playful," and "steady." She used generalized Procrustes analysis to quantify the extent to which different observers viewing the same animal at the same time generated a common set of descriptors. There was a high degree of between-observer agreement as well as within-observer agreement when videotaped sequences were observed a second time by observers. After the behavioral descriptors were arranged along axes, the positive pole of one axis was interpreted as reflecting "confidence." This was a measure of the animals' welfare or psychological well-being.

4.4.2 Direct Subjective Assessment of Happiness in Nonhuman Primates

A purely empathic approach to measuring psychological well-being was devised by King and Landau (2003) and later used by Weiss et al. (2006). They asked zoo workers to complete a four-item questionnaire relating to the psychological well-being of chimpanzees and orangutans, respectively. The first item asked the rater to judge the relative amount of time the subject was in a happy, positive mood as opposed to an unhappy, negative mood. The second item asked about the extent to which social interactions were a positive experience for the subject. The third item asked about the extent to which the subject was successful in achieving its goals or wishes. These three items were based on components of human measures of happiness or subjective well-being that assess an overall positive affect (Diener et al. 1999) and success in meeting life's goals (Cantor and Sanderson 1999). The fourth item asked the raters to imagine that they were the target subject for a week in which they would experience and feel the world the same way as the subject. The raters then indicated how pleasant or enjoyable that experience would be. Rating apes on this item was based on the assumption of a more direct empathic ability of the raters than any of the other three items. The summed total for all four items based on a seven-item Likert scale was then designated as a subjective well-being score.

For both the chimpanzees (King and Landau 2003) and the orangutans (Weiss et al. 2006) each of the four items and the total subjective well-being scores had high interrater reliabilities that were comparable to or higher than comparable reliabilities reported in human subjective well-being studies. Since the four items formed a single factor for both species, the summed total was designated as subjective well-being.

One avenue for validating the subjective well-being factor was based on the well-established relationships between subjective well-being and personality in humans. Similarities between humans and apes in the relationships between personality factors and subjective well-being constitute evidence for the comparability of subjective well-being measures across species. Human subjective well-being is negatively related to Neuroticism and positively related to Extraversion, Agreeableness, and Conscientiousness. However, Extraversion and Neuroticism have the strongest relationship to human subjective well-being (Diener et al. 1999; Steel et al. 2008). These findings are largely congruent with the chimpanzee and orangutan studies. Higher chimpanzee subjective well-being was related to higher Dominance, Extraversion, and Conscientiousness (King and Landau 2003) while higher orangutan subjective well-being was related to lower Neuroticism and higher Extraversion and Agreeableness (Weiss et al. 2006).

Chimpanzee subjective well-being scores were also negatively correlated with submissive behaviors in zoo settings (Pederson et al. 2005). This result is consistent with high positive correlation between Dominance and subjective well-being, especially since the two adjectives *dominant* and *submissive* had positive and negative loadings on the Dominance factor.

We believe that these data on the relationships between personality and subjective well-being show that the higher levels of happiness or subjective well-being in nonhuman primates can be successfully measured through subjective assessment. The studies described above also demonstrated that personality differences account for about half of the variance in subjective well-being within zoo environments, a result consistent with human findings (Diener et al. 1999). However, the high correlations between personality factors and subjective well-being should not be interpreted as diminishing the important role of environmental and social conditions on subjective well-being. Unquestionably, primates housed in sparse, unstimulating, or stressful conditions experience lowered subjective well-being. Analogously, human happiness is significantly lower in highly impoverished (Diener et al. 1995) and less free (Inglehart et al. 2008) societies.

4.5 Conclusions

This chapter describes three domains within current primatological research that, we believe, may be profitably addressed with subjective personality measures. First, the evolution of mind, behavior, and personality has been a persistent and beguiling issue from the beginning of animal behavior research in the late nineteenth century (Romanes 1884; Hobhouse 1901). However, the obvious lack of fossil or any other hard evidence for most past psychological phenomena imposes a formidable obstacle to progress, especially in taxa such as primates that lack a wide variety of closely related and extant species. We believe that the comparison of interspecies differences in the patterns of correlated personality dimensions (i.e., factor structures) described in this chapter offers a new approach to this problem. Behavioral ecologists have shown recent interest in the evolutionary significance of adaptations expressed as patterns of correlated behaviors in the wild, referred to as suites, constellations, or behavioral syndromes (Sih 2011; Sih et al. 2004). The use of comparable personality factors across species takes this strategy one step further by incorporating the potential to document changes within the same factor as shown in the differences between chimpanzees and orangutans in the Conscientiousness factor.

The second potential use of personality factors is in identifying proximate personality variables that affect behavior in the natural habitat. A large literature in human personality psychology describes relationships between personality and diverse human behaviors ranging from aggression (Burt and Donella 2008) to altruism (Ashton et al. 1998). No comparable data on personality and behavior of wild primates exist. However, the data on male–female differences in personality suggest that the highly aggressive nature of male chimpanzees in the wild may be a consequence of a particularly broad spectrum of higher male scores on personality traits related to aggression, viz. aggressiveness, unpredictability, and emotionality. We hope that in the future individual personality assessment will become a standard and accepted procedure for primate field studies.

The third context in which subjective personality ratings are useful is in addressing the striking disparity between traditional measures of psychological well-being in nonhuman primates and measures of happiness or subjective well-being in humans. Whereas absence of pathological behaviors, presence of normal species-specific behaviors, and normal physiological parameters traditionally defined psychological well-being in nonhuman primates, long-term balance of positive and negative affect and global life satisfaction defined human subjective well-being (Diener et al. 1999). No doubt, concern about possible unscientific aspects of anthropomorphism impeded earlier attempts to measure traits involving an animal's subjective experience and moods. However, it is only a modest and reasonable assumption that humans and other primates, especially the great apes, share similar feelings related to happiness. Consequently, the ability of the former to assess those feeling in the latter should not be surprising or denied.

References

Allport GW, Odbert HS (1936) Trait names: A psycho-lexical study. Psychol Monogr 47 (No. 212)

Almagor M, Tellegen A, Walker NG (1995) The Big Seven model: A cross-cultural replication and further exploration of the basic dimensions of natural language trait descriptors. J Pers Soc Psychol 69:300–307

Andrews K (2009) Politics or metaphysics? On attributing psychological properties to animals. Biol Philos 24:51–63

Anestis SF (2011) Primate personality and behavioral endocrinology. In: Weiss A, King JE, Murray L (eds) Personality and temperament in nonhuman primates. Springer, New York

Ashton MC, Paunonen SV, Helmas E et al. (1998) Kin altruism, reciprocal altruism, and the Big-Five personality factors. Evol Hum Behav 19:243–255

Baker KC, Weed JL, Crockett CM (2007) Survey of environmental enhancement programs for laboratory primates. Am J Primatol 69:377–394

Bekoff M (2005) The question of animal emotions: An ethological perspective. In: McMillan FD (ed) Mental health and well-being in animals. Blackwell, Ames

Billingslea FY (1941) The relationship between emotionality and various other salients of behavior in the rat. J Comp Psychol 31:69–77

Boehm C (1999) Hierarchy in the forest: The evolution of egalitarian behavior. Harvard University Press, Cambridge

Boinski S, Swing SP, Gross TS, Davis JK (1999) Environmental enrichment of brown capuchins (*Cebus apella*): Behavioral and plasma and fecal cortisol measures of effectiveness. Am J Primatol 48:49–68

Borkenau P (1992). Implicit personality and the five factor model. J Pers 60:295–327

Borsboom D, Mellenbergh GJ, van Heerden J (2003) The theoretical status of latent variables. Psychol Rev 110:203–219

Brent L, Lee DR, Eichberg JW (1989) Evaluation of two environment enrichment devices for singly caged chimpanzees (*Pan troglodytes*). Am J Primatol Suppl 1:65–71

Burghardt GM (1997) Amending Tinbergen: A fifth aim for ethology. In: Mitchell RW, Thompson NS, Miles HL (eds) Anthropomorphism, anecdote, and animals: The emperor's new clothes? SUNY Press, Albany NY

Burt SA, Donella MB (2008) Personality correlates of aggressive and non-aggressive antisocial behavior. Pers Ind Diff 44:53–63

Buss DM, Duntley JD (2006) The evolution of aggression. In: Schaller M, Simpson JA, Kenrick DT, Madison CT (eds) Evolution and social psychology. Psychology Press, New York

Campbell DT, Fiske DW (1959) Concurrent and discriminant validation by the multitrait-multimethod matrix. Psychol Bull 56:81–105
Cantor N, Sanderson CA (1999) Life task participation and well-being: The importance of taking part in daily life. In: Kahneman D, Diener E, Schwartz N (eds) Well-being: The foundations of hedonic psychology. Russell Sage Foundation, New York
Capitanio JP (1999) Personality dimensions in adult male rhesus macaques: Prediction of behaviors across time and situation. Am J Primatol 47:299–320
Capitanio JP (2011) Nonhuman primate personality and immunity: Mechanisms of health and disease. In: Weiss A, King JE, Murray L (eds) Personality and temperament in nonhuman primates. Springer, New York
Carr HA, Kingsbury FA (1939a) The concept of the individual. Psychol Rev 46:199–225
Carr HA, Kingsbury FA (1939b) The concept of directional disposition. Psychol Rev 46:359–382
Chagnon NA (1988) Life histories, blood revenge, and warfare in a tribal population. Science 239:985–992
Church AT (2001) Personality measurement in cross-cultural perspective. J Pers 69:979–1006
Conn SR, Rieke ML (1994) 16PF fifth edition: Technical manual. Institute for Personality and Ability Testing, Champaign
Constable JL, Ashley MV, Goodall J et al. (2001) Noninvasive paternity assignment in Gombe chimpanzees. Mol Ecol 10:1279–1300
Costa PT Jr., McCrae RR (1992) Revised NEO Personality Inventory (NEO-PI-R) and NEO Five-Factor Inventory (NEO-FFI) professional manual. Psychological Assessment Resources, Odessa
Costa PT Jr., McCrae RR (1998) Six approaches to the explication of facet level traits: Examples from conscientiousness. Eur J Pers 12:117–134
Costa PT Jr., Terracciano A, McCrae RR (2001) Gender differences in personality traits across cultures: Robust and surprising findings. J Pers Soc Psychol 81:322–331
Crawford MP (1938) A behavior rating scale for young chimpanzees. J Comp Psychol 26:79–92
Cronbach LJ, Meehl PE (1955) Construct validity in psychological tests. Psychol Bull 52: 281–302
Dewsbury DA (2006) Monkey farm: A history of the Yerkes Laboratories of Primate Biology, Orange Park, Florida, 1939-1965. Bucknell University Press, Lewisburg
Diener E, Suh EM, Lucas RE et al. (1999) Subjective well-being: Three decades of progress. Psychol Bull 125:276–302
Diener E, Diener MC, Diener C (1995) Factors predicting the subjective well-being of nations. J Pers Soc Psychol 69:851–864
Digman JM (1996) The curious history of the five-factor model. In: Wiggins JS (ed) The five-factor model of personality: Theoretical perspectives. Guilford, New York
Funder DC (1995) On the accuracy of personality judgment: A realistic approach. Psychol Rev 102:652–670
Goodall J (1986) The chimpanzees of Gombe: Patterns of behavior. Harvard University Press, Cambridge
Gosling SD (2001) From mice to men: What can we learn about personality from animal research? Psychol Bull 127:45–86
Gosling SD, Graybeal A (2007) Tree thinking: A new paradigm for integrating comparative data in psychology. J Gen Psychol 134:259–277
Gosling SD, John OP (1999) Personality dimensions in nonhuman animals: A cross-species review. Curr Dir Psychol Sci 8:69–75
Gorsuch RL (1983) Factor analysis. Saunders, Hillsdale
Harvey PH, Pagel MD (1991) The comparative method in evolutionary biology. Oxford University Press, New York
Hawks J, Wang ET, Cochran GM et al. (2007) Recent acceleration of human adaptive evolution. Proc Natl Acad Sci USA 104:20753–20758
Hearn V (1994) Animal happiness. HarperCollins, New York
Hobhouse LT (1901) Mind in evolution. Macmillan, London

Honess PE, Marin CM (2006) Enrichment and aggression in primates. Neurosci Biobehav Rev 30:413–436

Ickes W (2001) Measuring empathic accuracy. In: Hall JA, Bernier FJ (eds) Interpersonal sensitivity: Theory and measurement. Erlbaum, Mahwah

Ickes W (2003) Everyday mindreading: Understanding what other people think and feel. Prometheus Books, Amherst

Inglehart R, Foa R, Peterson C et al. (2008) Development, freedom, and rising happiness. Persp Psychol Sci 3:264–285

Jang KL, McCrae RR, Angleitner A et al. (1998) Heritability of facet-level traits in a cross-cultural twin sample: Support for a hierarchical model of personality. J Pers Soc Psychol 74:1556–1565

Jones CB (2005) Behavioral flexibility in primates: Causes and consequences. Springer, New York

Jones M, Pillay N (2004) Foraging in captive hamadryas baboons: Implications for enrichment. Appl Anim Behav Sci 88:101–110

Kennedy JS (1992) The new anthropomorphism. Columbia University Press, New York

Kenny DA (1994) Interpersonal perception. Guilford Press, New York

King JE, Figueredo AJ (1997) The Five-Factor Model plus Dominance in chimpanzee personality. J Res Pers 31:257–271

King JE, Landau VI (2003) Can chimpanzee (*Pan troglodytes*) happiness be estimated by human observers? J Res Pers 37:1–15

King JE, Weiss A, Farmer KH (2005) A chimpanzee (*Pan troglodytes*) analogue of cross-national generalization of personality structure: Zoological parks and an African sanctuary. J Pers 73:389–410

King JE, Weiss A, Sisco MM (2008) Aping humans: Age and sex effects in chimpanzee (*Pan troglodytes*) and human (*Homo sapiens*) personality. J Comp Psychol 122:418–427

Klinkova E, Hodges JK, Fuhrmann K (2005) Male dominance rank, female mate choice, and male mating and reproductive success in captive chimpanzees. Int J Primatol 26:357–384

Köhler W (1925) The mentality of apes. (Winter E, Trans.) Harcourt Brace, New York (Original work published in 1924)

Kuhar CW, Stoinski TS, Lucas KE et al. (2006) Gorilla behavior index revisited: Age, housing, and behavior. Appl Anim Behav Sci 96:315–326

Kwan VS, Gosling SD, John OP (2008) Anthropomorphism as a special case of social perception: A cross-species social relations model analysis of humans and dogs. Soc Cog 26:129–142

Lehmann J, Boesch C (2005) Bisexually bonded ranging in chimpanzees (*Pan troglodytes verus*). Behav Ecol Sociobiol 57:525–535

Letzrins TD, Wells SM, Funder DC (2006) Information quantity and quality affect the realistic accuracy of personality judgments. J Pers Soc Psychol 91:111–123

Lilienfeld SO, Gershon J, Duke M et al. (1999) A preliminary investigation of the construct of psychopathic personality (psychopathy) in chimpanzees (*Pan troglodytes*). J Comp Psychol 113:365–375

Maki S, Bloomsmith MA (1989) Uprooted trees facilitate the psychological well-being of captive chimpanzees. Zoo Biol 8:79–87

McCrae RR, Terracciano A, 78 Members of the Personality Profiles of Cultures Project (2005) Universal features of personality traits from the observer's perspective: Data from 50 cultures. J Pers Soc Psychol 88:547–561

McCrae RR, Zonderman AB, Costa PT Jr. et al. (1996) Evaluation replicability of factors in the Revised NEO Personality Inventory: Confirmatory factor analysis versus Procrustes rotation. J Pers Soc Psychol 70:552–566

McMillan FD (2005) The concept of quality of life in animals. In: McMillan FD (ed) Mental health and well-being in animals. Blackwell, Ames

Messick S (1989) Validity. In: Linn RL (ed), Educational measurement. American Council on Education and National Council on Measurement, Washington

Muller MN, Kahlenberg SM, Thompson ME et al. (2007) Male coercion and the costs of promiscuous mating for female chimpanzees. Proc Biol Sci 274:1009–1014

Murray CM, Eberly LE, Pusey AE (2006) Foraging strategies as a function of season and rank among wild female chimpanzees. Behav Ecol 17:1020–1028

Mynott J (2009) Birdscapes: Birds in our imagination and experience. Princeton University Press, Princeton

Nettle D (2006) The evolution of personality variation in humans and other animals. Am Psychol 61:622–631

Pederson AK, King JE, Landau VI (2005) Chimpanzee (*Pan troglodytes*) personality predicts behavior. J Res Pers 39:534–549

Penke L, Denissen, JJA, Miller G. (2007) The evolutionary genetics of personality. Eur J Pers 21:549–587

Pusey AE (2001) Of genes and apes: Chimpanzee social organization and reproduction. In: de Waal FBM (ed) Tree of origin: What primate behavior can tell us about human social evolution. Harvard University Press, Cambridge

Pusey AE, Oehlert GW, Williams JM et al. (2005) Influence of ecological and social factors on body mass of wild chimpanzees. Int J Primatol 26:3–31

Pusey AE, Williams J, Goodall J. (1997) The influence of dominance rank on the reproductive success of female chimpanzees. Science 277:828–831

Raaum RL, Sterner KN, Noviello CM et al. (2005) Catarrhine primate divergence dates estimated from complete mitochondrial genomes: Concordance with fossil and nuclear DNA evidence. J Hum Evol 48:237–257

Richards SM (1974) The concept of dominance and methods of assessment. Anim Behav 22: 914–930

Roberts BW, Kuncel NR, Shiner R, Caspi et al. (2007) The Power of personality: The comparative validity of personality traits, socioeconomic status, and cognitive ability for predicting important life outcomes. Perspect Psychol Sci 2:313–345

Romanes GJ (1884) Mental evolution in animals. Appleton, New York

Ross L, Nisbett RE, Funder DC (2007) Issue 3: Are our social perceptions often inaccurate? In: Nier JA (ed) Taking sides: Clashing views in social psychology (2nd edn). McGraw-Hill, New York

Schneider SM, King JE (2008). Dominance in chimpanzees (*Pan troglodytes*): Personality, position and beyond. Paper presented at XXII Congress of the International Primatological Society, Edinburgh, Scotland

Saucier G, Goldberg LR (2001) Lexical studies of indigenous personality factors: Promises, products, and prospects. J Pers 69:847–879

Schilhab TSS (2002) Anthropomorphism and mental state attribution. Anim Behav 63: 1021–1026

Sih A (2011) Behavioral syndromes: A behavioral ecologist's view on the evolutionary and ecological implications of animal personality. In: Weiss A, King JE, Murray L (eds) Personality and temperament in nonhuman primates. Springer, New York

Sih A, Bell AM, Johnson JC et al. (2004) Behavioral syndromes: An integrative overview. Q Rev Biol 79:241–277

Smithson MJ, Verkuilen J (2006) Fuzzy set theory: Application in the social sciences. Sage Publications: Newbury Park

Steel P, Schmidt J, Shultz J (2008) Refining the relationship between personality and subjective well-being. Psychol Bull 134:138–161

Stevenson-Hinde J, Stillwell-Barnes R, Zunz M (1980) Subjective assessment of rhesus monkeys over four successive years. Primates 21:66–82

Symons D (1990) Adaptiveness and adaptation. Ethol Sociobiol 11:427–444

Tellegen A (1985) Structures of mood and personality and their relevance to assessing anxiety, with an emphasis on self-report. In: Tuma AH, Maser JD (eds) Anxiety and anxiety disorders. Erlbaum, Hillsdale

Tinbergen N (1996) On aims and methods of ethology. In: Houk LD, Drickamer LC (eds) Foundations of animal behavior: Classic papers with commentaries. University of Chicago Press, Chicago

Uher J (2011) Personality in nonhuman primates: What can we learn from personality psychology. In: Weiss A, King JE, Murray L (eds) Personality and temperament in nonhuman primates. Springer, New York
Uher J, Asendorpf JB (2008) Personality assessment in great apes: Comparing ecologically valid behavior measures, behavior ratings, and adjective ratings. J Res Pers 42:821–838
van Oers K, de Jong G, van Noordwijk AJ et al. (2005) Contribution of genetics to the study of animal personalities: A review of case studies. Behaviour 142:1185–1206
Vazire S, Gosling SD, Dickey et al. (2007) Measuring personality in nonhuman animals. In Robins RW, Fraley RC, Krueger RF (eds) Handbook of research methods in personality psychology. Guilford Press, New York
Washburn, MF (1909) The animal mind: A textbook of comparative psychology. Macmillan, London
Watts DP, Mitani JC, Sherrow HM (2002) New cases of inter-community infanticide by male chimpanzees at Ngogo, Kibale National Park, Uganda. Primates 42:263–270
Weiss DS, Mendelson GA (1986) An empirical demonstration of the implausibility of the semantic similarity explanation of how trait ratings are made and what they mean. J Pers Soc Psychol 50:596–601
Weiss A, King JE, Figueredo AJ (2000) The heritability of personality factors in chimpanzees (*Pan troglodytes*). Behav Genet 30:213–221
Weiss A, King JE, Hopkins WD (2007) A cross-setting study of chimpanzee (*Pan troglodytes*) personality structure and development: Zoological parks and Yerkes Regional primate Research Center. Am J Primatol 69:1264–1277
Weiss A, King JE, Perkins L (2006) Personality and subjective-well being in orangutans (*Pongo pygmaeus* and *Pongo abelii*). J Pers Soc Psychol 90:501–511
Weiss A, Inoue-Murayama M, Hong, K-W et al. (2009) Assessing chimpanzee personality and subjective well-being in Japan. Am J Primatol 71:283–292
Wemelsfelder F, Hunter TEA, Mendl MT et al. (2001) Assessing the "whole animal": A free choice profiling approach. Anim Behav 62:209–220
Wilson ML, Wrangham RW (2003) Intergroup relations in chimpanzees. Annu Rev Anthropol 32:363–392
Winter DG, Barenbaum NB (1999) History of modern personality theory and research. In Pervin LA, John OP (eds) Handbook of personality and research (2nd ed). Guilford Press, New York
Wolfle, TL (1999) Psychological well-being on nonhuman primates: A brief history. J Appl Anim Welf Sci 2:297–302
Wrangham RW, Peterson D (1996) Demonic males: Apes and the origins of human violence. Houghton Mifflin, Boston
Wrangham RW, Wilson MI, Muller MN (2006) Comparative rates of violence in chimpanzees and humans. Primates 47:14–26
Wroblewski EE, Murray CM, Keele BF et al. (2009) Male dominance rank and reproductive success in chimpanzees, (*Pan troglodytes schweinfurthii*). Anim Behav 77:873–885
Yerkes RM (1925) Almost human. The Century Company, New York
Yerkes RM (1929) The great apes: A study of anthropoid life. Yale University Press, New Haven
Yerkes RM (1939) The life history and personality of the chimpanzee. Am Nat 73: 97–112
Yerkes RM (1943) Chimpanzees: A laboratory colony. Yale University Press, New Haven

Part III
Empirical Studies

Chapter 5
Objective Behavioral Tests of Temperament in Nonhuman Primates

Lynn A. Fairbanks and Matthew J. Jorgensen

Abstract Objective behavioral tests can be useful tools in studies of the causes and consequences of individual differences in the temperament of nonhuman primates. This chapter begins by presenting guidelines for designing effective behavioral tests to produce valid and reliable measures of trait-like individual differences. It follows with a review of several well-characterized procedures for measuring individual differences in nonhuman primates on dimensions of neonatal temperament, anxiety, novelty seeking and impulsivity. Key points that emerge from this review include the value of establishing the test–retest reliability of traits, the importance of context for interpreting behavioral results, and the need for greater attention to specifying and validating the connection between the behavioral measures and the temperament constructs of interest. Attention to these features will enhance our ability to identify genetic, physiological and environmental causes of individual differences in temperament, and the influence of temperament on outcomes of interest in biology, psychology and psychiatry.

5.1 Introduction

Temperament is an important area of study for the psychologist interested in the social and developmental consequences of individual differences in response tendencies, and for the evolutionary biologist interested in ecological influences and fitness outcomes of behavior. There is evidence that temperament and its physiological underpinnings are related to immune functioning and disease susceptibility, and

L.A. Fairbanks (✉)
Department of Psychiatry and Biobehavioral Sciences, Semel Institute,
University of California, 760 Westwood Plaza, Los Angeles, CA 90095, USA
e-mail: lfairbanks@mednet.ucla.edu

A. Weiss et al. (eds.), *Personality and Temperament in Nonhuman Primates*, Developments in Primatology: Progress and Prospects, DOI 10.1007/978-1-4614-0176-6_5,

that an individual's characteristic response style influences the probability of early mortality, the likelihood of successful reproduction and the ability to form cooperative social relationships (Capitanio et al. 1999; Sih et al. 2004). Temperament is also important to the geneticist, the physiologist and the neurobiologist interested in elucidating the mechanisms that lead to behavioral differences, and to the biomedical scientist for the development of effective animal models of human psychiatric and biobehavioral disorders. To make scientific progress in all of these areas, it is essential to have effective means for measuring individual differences in temperament dimensions. Several chapters in this volume describe the use of subjective rating systems for measuring temperament and personality traits. Others use physiological indicators as markers for individual differences in temperament. This chapter will review objective methods that use quantitative coding of behavior in specified contexts to measure temperament in nonhuman primates.

Many of the original studies using objective tests of individual differences in nonhuman primates derived from developmental psychology, and much of the research in this area has focused on the early appearance of traits relevant to vulnerability for psychopathology (e.g., Hinde and Spencer-Booth 1971; Kalin 2003). Because of the early emphasis on infancy, research in this area has typically used the term temperament to indicate a relatively consistent set of response dispositions that underlie an individual's behavioral style. Temperament is generally considered to appear early and to be more biologically based than personality, although the two terms are often used interchangeably. The concept of temperament also includes implied emotional and cognitive components and characteristic sets of physiological responses (Clarke and Boinski 1995).

In much of the early animal behavior research, temperament was treated as a single dimension, labeled shy-bold (Wilson et al. 1993). In nonhuman primates, experience has shown that it is more accurate to perceive temperament as varying in multiple dimensions. Subjective rating systems that are designed to reveal the structure of personality across a broad array of traits typically identify multiple independent dimensions, including traits related to fearfulness, confidence, excitability, sociability and aggressiveness in nonhuman primate populations (e.g., Stevenson-Hinde et al. 1980a; Capitanio 1999; Weiss et al. 2006). Among studies using objective behavioral coding methods, constructs that have received the most attention have been anxiety and emotional reactivity, novelty seeking and impulsivity. These dimensions of temperament are relevant to vulnerability for psychological and psychiatric disorders in humans and are also relevant to animal life history and evolution. Individual differences in temperament in these domains affect the balance between approach and avoidance in situations involving food getting, predator avoidance and competition for mates. As such they have important implications for safety, survival and reproductive success (Sih et al. 2004).

This review begins with an outline of methods for developing and scoring objective behavioral tests of temperament. It then follows with examples of several well-characterized methods for assessing dimensions of temperament in nonhuman primates, including tests for novelty seeking and impulsivity that we have developed at the Vervet Research Colony. Attention is paid to the particular temperament constructs that are being measured, to the consistency, reliability and validity of outcome measures according to the criteria outlined below, and to their usefulness as research tools.

5.2 Guidelines for Objective Behavioral Tests for Temperament

A variety of test situations and behavior coding systems have been used to measure dimensions of temperament in nonhuman primates with different attributes, outcome measures and results. Subjective tests for measuring personality in nonhuman primates have benefited from the high standards established for evaluating human personality inventories, including statistical methods for measuring scale cohesion, reliability, and validity (Gosling 2001), but these high standards have not been as consistently applied to objective measurement procedures. This section outlines steps to take in the development of effective objective behavioral tests to measure valid and reliable dimensions of temperament.

5.2.1 Step 1: Designing the Test Situation

Objective behavioral tests of temperament are generally designed with a specific temperament construct in mind. Most primate researchers adopt an ethological approach, using information about species socioecology and behavior to design a test situation that will elicit behavior that is relevant to the trait in question. For example, placing a monkey in an unfamiliar and threatening situation elicits species-typical defense reactions that are adaptive responses to danger (Kalin and Shelton 1998; Bardi et al. 2005). The degree of danger is an important ingredient in designing a test for anxiety-related traits. A test that is too threatening will produce fear and avoidance in all subjects, while a test that is not alarming enough may produce unreliable results due to lack of attention to the test stimuli. The ideal test will elicit a broad range of responses, and expose individual differences in the relevant temperament dimension. Different components of the test may provide opportunities for different kinds of behavioral responses and thus allow measurement of independent dimensions within a general temperament construct, such as reactive vs. avoidant anxiety.

5.2.2 Step 2: Designing the Scoring System

The scoring method used to measure the animals' response to the test situation influences the nature of the outcome variables. Scoring systems used in primate behavioral tests for temperament typically begin with knowledge of the species' typical behavioral responses to the test conditions. Some of the studies have selected a small number of informative behaviors to code, while others use a detailed and exhaustive list of behavior categories (e.g., Kalin and Shelton 1998; Bethea et al. 2004). Computer-based recording systems produce rates of behavior that can vary from zero to an undefined upper limit, with a high likelihood of skewed distributions and outliers. Coding systems that measure a behavior as occurring or not occurring within a set of timed intervals have a circumscribed upper bound, but score distributions may still be skewed (Fairbanks 2001). Systems for measuring behavioral traits that rate items on an ordinal intensity scale produce outcome scores that reflect predetermined levels of severity and significance (Schneider et al. 2001). Each of these methods has advantages and disadvantages, and implications for statistical analysis and interpretation.

5.2.3 Step 3: Constructing the Outcome Measures

Outcome measures of temperament often begin with a number of coded behaviors that are scored multiple times under different test conditions. These individual scores are likely to be correlated with one another and do not represent independent outcome measures. Many measures in behavioral tests of temperament are structurally dependent (e.g., frequency and duration of a behavior, or latency to approach and time near) and others may be functionally correlated. Because separate testing of large numbers of items is likely to produce false positive and unreliable results, several methods have been used to combine or reduce behavioral measures. If initial screening shows that two measures are highly intercorrelated, one of them may be selected for further analysis (Spencer-Booth and Hinde 1969; Mason et al. 2006). Behaviors that are correlated highly across test conditions can be summed or averaged to produce a single measure. Data reduction procedures such as factor analysis or multidimensional scaling can also be used to identify correlated groups of behaviors and to define independent behavioral dimensions within a dataset (Watson and Ward 1996; Fairbanks 2001; Williamson et al. 2003; Bardi et al. 2005). All of these methods require attention to the number and distribution of the items entered into the test and to the number of subjects evaluated. For example, normally distributed variables and a sample of at least 100 subjects are generally recommended to produce a stable and reliable factor structure (e.g., Guadagnoli and Velicer 1988), but factor analysis is still useful in smaller samples to identify an independent set of measures for further analysis. After dimensions of behavior are identified, a summary score may be designed to standardize the outcome measure for further use. This may be a simple sum, which tends to give more weight to high frequency behaviors, or it may be an adjusted sum to weigh items equally or differentially based on importance (Santillan-Doherty et al. 2006). The mean rates and distributions of the resulting outcome measures will influence the ability of the measures to differentiate reliably among individuals.

5.2.4 Step 4. Scale Reliability

Assessment of the internal consistency, or scale reliability, is an essential step in developing outcome scales for subjective personality rating systems. The same methods can be applied to composite outcome measures for objective behavioral tests of temperament. The assessment of scale reliability is based on the correlations among individual measures that make up the scale relative to the variances of the measures, and it reflects the degree to which the items in the scale hold together. The most commonly used index of scale reliability is Cronbach's alpha (Cronbach 1951). Software for assessing scale reliability also includes calculation of reliability with and without individual items, providing information that is useful in selecting the most appropriate items to go into each summary scale (SPSS 2005).

5.2.5 Step 5: Test–Retest Reliability

The concept of temperament implies some degree of consistency over time and across situations. All of the steps outlined here can be followed and a score can be derived with high internal consistency and construct validity that reflects a temporary *state*, and not a temperamental *trait*. Test–retest reliability is an essential step to demonstrate that an objective behavioral test is measuring a dimension of temperament. The Pearson correlation coefficient is frequently used to measure the correlation between scores at two time points, but this method is insensitive to a mean difference between scores. A measure that reflects the absolute and not just the relative consistency of scores is the intra-class correlation coefficient (ICC). The ICC measures the contribution of variance between subjects to the total variance and, as such, is a better measure of the true test–retest reliability of a scale (Shrout and Fleiss 1979). The ICC can also be used to assess reliability across more than two test sessions (SPSS 2005).

5.2.6 Step 6: Construct Validity

Construct validity refers to how well the individual or composite measure captures the meaning of the theoretical construct of interest. In the case of tests designed for an animal model of a human disorder, the elements involved in diagnosis of the disorder must be considered. If temperament constructs are derived from an ethological perspective, then care must be used in labeling outcome measures. A construct like novelty seeking is relatively easy to operationalize in behavioral terms, but anxiety is more difficult to define. The dictionary definition of the term "anxiety" is apprehension of an impending or anticipated ill, while "fear" is a response to current danger. The common definition of the term "shy" implies hesitancy to approach and interact in a social situation, while the term "timid" is a more general word for hesitancy to act in social and nonsocial contexts. Scoring for "behavioral inhibition" in children includes active behaviors such as "crying, sobbing and fretting" and "facial expression and vocalizations of distress" in addition to long latencies to approach and inhibition of play (Garcia Coll et al. 1984). "Behavioral reactivity" is a term that has been used to refer to behavioral overreaction to stressful circumstances, including high levels of activity, distress vocalization and aggressive behavior, and also to inhibition of behavior. Because of the diversity of these behavioral responses, contradictory reactions have been interpreted as indicators of anxious temperament. In developing an objective behavioral index of temperament, it is important to consider the meaning of the temperament constructs, and to pay careful attention to labeling of scores derived from the tests.

5.2.7 Step 7: Convergent Validity

Convergent validity of a behavioral index is determined by the degree to which the index varies with external variables that are expected to be related to the construct. Convergent validity is frequently established through relationships with physiological and neurobiological traits that are related to the construct (e.g., Kalin et al. 1998; Kalin and Shelton 2003). Developmental and life history factors, such as age, sex, dominance rank and age of emigration, can also be used as indicators of convergent validity. For example, a measure of impulsivity would be expected to be higher in adolescents than in adults (Fairbanks et al. 2004a). Comparison of subjective rating scales with objective behavioral measures can help to support the validity of both methods in measuring temperament (Capitanio 1999; Uher and Asendorpf 2008).

5.2.8 Step 8: Discriminative Validity

Discriminative validity refers to the degree to which a behavioral index differentiates between opposing constructs. This is particularly important to demonstrate for traits derived from behavioral tests that are superficially similar, such as response to novel objects in the home cage and in an unfamiliar environment. It is also crucial to demonstrate discriminative validity for traits that use the same behavior to represent different constructs. For example, rates of aggression have been used to measure anxiety and impulsivity, and latency to approach is included in measures of anxiety, impulsivity and novelty seeking (e.g., Kalin and Shelton 1998; Fairbanks 2001, Williamson et al. 2003). Behavioral indices with good discriminative validity should not be highly correlated with measures that are purported to reflect another dimension. More attention needs to be paid to discriminative validity in the development and use of behavioral measures for temperament constructs.

5.2.9 *Step 9: Predictive Validity*

Finally, a well-developed index for a temperamental trait will be useful in predicting future successes and vulnerabilities of individuals that differ in the trait. Predictive utility has been demonstrated for subjective personality scales and for physiological indicators of temperament (Capitanio et al. 1999; Higley et al. 1996a), but there are only a few studies that have used behavioral indices of temperament to predict important outcomes later in life (e.g., Schneider et al. 2001; Fairbanks et al. 2004b).

5.3 Behavioral Tests for Measuring Temperament in Nonhuman Primates

5.3.1 *Neonatal Temperament*

The Primate Neonatal Neurobehavioral Assessment (PNNA) was originally developed by Schneider et al. (1991) to study environmental influences on early infant development in rhesus monkeys. The assessment is a 20 min battery of developmental tests adapted for rhesus monkey infants from items on the Brazelton Neonatal Assessment and the Bayley scales of infant development (Bayley 1969; Brazelton 1973).

The PNNA is a standardized set of tests with procedures and a scoring system that are clearly defined and consistently implemented. During the assessment, the infant monkey is removed from the mother or from the nursery and held by the examiner. The tests include visual orienting, visual following and duration of attention to a plastic toy or a sound. Motor ability is tested by noting the infant's reflexes, righting response, activity and coordination when the infant is held in different positions. Assessments are made of the infant's emotional state throughout the battery. Each of the 41 items is scored as 0, 1 or 2, with descriptions of the responses that qualify for each rating. This type of scoring system insures equal weighting of each scored item. There is some subjective judgment involved in the scoring, but the definitions are clearly linked to specific acts by the infant. For example, fearfulness is defined as the occurrence of fear grimaces or trembling. This is not substantially different from the observer judgments that are required for any "objective" behavior coding system. The test is typically administered multiple times per infant over the first month of life to provide a robust measure of individual differences and developmental trends.

Factor analysis was initially used to reduce the large number of individual items into four factors labeled Orientation, State Control, Motor Maturity and Activity (Schneider et al. 1991). These factors parallel item groupings that have emerged from assessments of human infants, supporting the construct validity and cross-species generalizability of this assessment method. Individual differences in emotionality are reflected in the State Control cluster, which includes irritability, consolability, struggle during testing and predominant state from alert to agitated. While the scores on the four clusters reflect the rapid maturational changes that occur over the first few weeks of life, there is evidence that individual differences are stable over time, particularly after 14 days of age (Schneider and Suomi 1992).

Scores from these four clusters of the PNNA have been used as outcome measures for prenatal and postnatal experimental interventions (Schneider et al. 1992, 1997; Champoux et al. 2002a; He et al. 2004). Deficits in orientation and motor maturity in early infancy predicted difficulties with a learning task years later (Schneider et al. 2001) and identified individuals with later neurological problems (Champoux et al. 2000). PNNA scores have also been able to differentiate subpopulations of rhesus monkeys (Indian vs. Chinese-Indian hybrids), and individuals differing in a polymorphism of the serotonin transporter gene (Champoux et al. 1994, 2002b). There is evidence of genetic heritability for all four clusters (Champoux et al. 1999), indicating that this assessment method measures trait-like dimensions of individual differences in early temperament, cognitive and motor functioning.

This test illustrates many properties of a well-designed objective behavioral measurement system for assessing temperament in nonhuman primates. The PNNA has been administered and scored uniformly within and across laboratories. Initial attention focused on the construct validity of the four outcome clusters, labeled Orientation, State Control, Motor Maturity and Activity. The method of constructing the outcome scales has been consistently applied across studies. The scores translate readily to constructs used in clinical research with human infants, and have been shown to be informative as predictors and as outcome measures for developmental research. One disadvantage is that the scores are affected by the infant's acute response to separation from the mother for testing, making it more difficult to administer to mother-reared infants. This results in mother-reared infants scoring "worse" than nursery-reared infants, a finding that is contrary to other indicators of social competence and well-being for infants reared under these two conditions. This is a serious problem for interpretation of studies that combine mother-reared and nursery-reared subjects. Otherwise, the PNNA is an excellent example of an objective behavior scoring system that has the features of a well-developed test, including attention to rater reliability, test–retest reliability, construct validity, predictive validity and utility as a research tool.

5.3.2 Anxiety

The majority of objective behavioral tests of temperament in nonhuman primates have been designed to measure anxiety-related traits. Because anxiety is a subjective, emotional response that involves anticipation of negative consequences, the challenge has been to find overt behavioral manifestations that represent this internal state. Two different approaches have been developed: the first uses behavioral clues of anxiety under normal social circumstances and the second uses fear-inducing standardized challenge situations to provoke anxious behavior.

5.3.2.1 Noninvasive Measures of Anxiety

A noninvasive method for measuring anxiety using an ethological approach is based on the hypothesis that scratching is a displacement activity reflecting levels of emotionality and motivational conflict (Diezinger and Anderson 1986; Maestripieri et al. 1992a). This method has been validated by measuring changes in scratching and other self-directed "displacement" behaviors across contexts that are known or expected to elicit different levels of anxiety. Thus, Diezinger and Anderson (1986) found that scratching was more common in social than in nonsocial contexts, and particularly in situations that involved anticipation of potential conflict before a feeding competition test in a stable group of socially living rhesus monkeys. Research by Maestripieri and his colleagues demonstrated that scratching by mothers of young infants increased when they were separated from their infants and when the infant was at greater risk of harm by other group members (Maestripieri et al. 1992b; Maestripieri 1993a, b). Scratching increases in other anxiety-producing social contexts, such as proximity to a higher ranking monkey or following aggression in captivity and in the field (Aureli and van Schaik 1991; Pavani et al. 1991; Castles et al. 1999). Drugs that increase anxiety cause an increase in rates of scratching, while a drug that decreases anxiety reduces the rate of scratching in macaques, thus providing experimental evidence of the link between scratching and anxious mood (Schino et al. 1996)

These studies demonstrate that rate of scratching is a valid and reliable measure of anxious states that can be used to measure short-term effects of different contexts or experimental manipulations. There is less evidence that scratching and other self-directed behaviors reflect stable and trait-like individual differences in anxious temperament. One study by Maestripieri (2000) on five adult female rhesus monkeys showed a high rate of consistency in individual differences from one birth season to the next, with less individual consistency between the birth season and the mating season. In another study, Troisi et al. (1991) found that rates of postpartum scratching were positively correlated with individual differences in maternal possessiveness for seven macaque mothers. The sample sizes for these studies were small but they provide preliminary support for the use of this measure as an indicator of anxious temperament.

An advantage of a noninvasive behavior coding approach is that the animals do not have to be disturbed, and the measure reflects individual differences in response to circumstances within the normal range of the animal's experience. This method can be used to study individual differences in the field and in different social settings where manipulation of the animals is not feasible or desirable, and it is applicable to animals of all ages, although most of the validation has been with adults. Use of scratching as a behavioral index of anxiety is a simple method with demonstrated convergent validity, ecological relevance and some indication of trait reliability. Disadvantages of measuring temperament through rates of behavior in an uncontrolled environment are that it takes more observation time to produce a reliable measure of individual rates of behavior, particularly for relatively rare behaviors.

5.3.2.2 Standardized Behavioral Tests for Anxiety

Many studies have used response to novel objects and people in unfamiliar settings to indicate emotionality and anxious temperament in nonhuman primates. Most of this research is modeled after studies of behavioral inhibition in children (Kagan et al. 1988). Behavioral inhibition is assessed by observing young children's responses to unfamiliar people and toys in a novel environment. Classification of a child as inhibited is based on long latencies to approach the unfamiliar person or object, suppressed play and active displays of negative affect including crying, fretting and facial expressions or vocalizations of distress (Garcia Coll et al. 1984). There is evidence that children classified as inhibited using this test paradigm have an increased likelihood of developing anxiety disorders later in life.

The most thoroughly characterized test of this type for nonhuman primates is the Human Intruder Test, developed by Kalin and Shelton (1998) to measure anxious temperament and behavioral inhibition in infant rhesus monkeys. In this test, the infant is removed from the mother and placed alone in a small test cage. The experimenter leaves the room and the infant's behavior is recorded onto a videotape. For the first 10 min, the infant is simply alone in the room. For the No Eye Contact portion of the test, a person enters the room, stops at a specified distance in front of the test cage and presents his/her profile to the monkey. This is followed by a second Alone period, and then by a 10 min period when the experimenter reenters the room and stares directly at the test subject. Behavior is recorded using a relatively simple coding system that measures the frequency of coo vocalizations, freezing, movement and aggressive behavior (hostility and barking). These behaviors were selected because they were commonly observed and because they represented different adaptive responses to the test situation. During the alone phase, cooing was the most common response. Freezing was the predominant behavior in the No Eye Contact condition, and defensive aggressive behavior was observed during the Stare portion of the test. Kalin and Shelton (1998) interpret the coo as representing a call for help, freezing as a detection avoidance strategy and defensive aggression as an appropriate response to the direct stare. The magnitude of each response is considered to be a reflection of the degree of anxiety experienced by the infant. The subjects were exposed to repeated testing over a 4 day period, with the final behavior score representing an average across days. This produces a more robust and reliable estimate of individual differences than can be obtained from a single test. Inter-rater reliability for each item was also determined.

When infants were tested again at 4, 8 and 12 months, Kalin and Shelton (1998) found that individual differences in rates of freezing, but not cooing, were consistent over that age range. This led to the conclusion that coo and freeze reflect different motivational and neurobiological systems. Longitudinal assessments at 1 and 2 years of age demonstrated that freezing in the No Eye Contact condition and defensive aggression in the Stare condition were trait-like characteristics that were consistent over time (Kalin et al. 2001). A recent report that freezing in the No Eye Contact is a heritable trait also suggests that this measure reflects an underlying dimension of temperament (Rogers et al. 2008). Research by Kalin and his colleagues has demonstrated that these three defensive responses (cooing, freezing and defensive aggression) are controlled by different neuroanatomical and physiological systems (Kalin et al. 1998, 2005, 2007).

Kalin's Human Intruder Test is a good example of a standardized method for assessing temperament that is relatively easy to administer and score. It also produces reliable trait-like individual differences in response to an anxiety-provoking situation. The scored behaviors have face validity as measures of temperament reflecting different ecologically relevant responses to threat (call for help, avoid detection, defensive aggression). Test–retest data demonstrate that individual differences in freezing and defensive aggression are trait-like and consistent over time while research into the neurobiological mechanisms of these traits suggests that they represent independent dimensions of anxious temperament. By focusing on a small number of valid and consistent measures, Kalin and his colleagues have been able to build a coherent body of research focused on understanding the neurobiology of anxiety (Kalin and Shelton 2003). Modified versions of the Human Intruder Test have been successfully used in other laboratories to measure genetic and environmental influences on behavioral traits (Williamson et al. 2003; Bethea et al. 2004; Capitanio et al. 2006).

Another variation of the novelty test for anxious temperament involves removing an individual from its home environment and placing it in a test cage with a variety of novel objects that vary from neutral to threatening. Studies using this general method differ in the number and timing of test trials, in the behaviors scored and in data reduction methods. Objects used include simple stationary toys, objects with flashing or moving parts, talking or remote-controlled toys, objects with predator features, mirrors, unfamiliar food items and preferred food items next to threatening objects. Several studies have found positive correlations in the latency to approach and interact with different kinds of novel objects (Spencer-Booth and Hinde 1969; Watson and Ward 1996; Mason et al. 2006) indicating that these tests are measuring an underlying response disposition. When additional behaviors are recorded, such as locomotor activity, expressions of fear or attempts to escape, they are typically found to be independent of the object approach behaviors. For example, Watson and Ward (1996) found that behavior in a series of novelty tests produced four independent factors labeled boldness, activity, curiosity and escape in juvenile and adult small-eared bushbabies. Movement, distress vocalization and interacting with a novel fruit fell on separate factors in a battery of novelty tests for infant rhesus monkeys (Williamson et al. 2003). In an individual cage test, Bardi et al. (2005) found that cortisol levels at the end of the test period were positively correlated with locomotion and vocalization, but not with displacement activities or vigilance as indicators of anxiety in juvenile baboons. These results suggest that behavioral responses to novelty test conditions reflect more than one dimension of temperament. The behaviors coded and methods used differ across studies, but there appears to be a tendency for overt measures of activity and agitation to represent a different aspect of temperament than latency to approach and interact with novel objects. How these different dimensions relate to the construct of anxious temperament has not been consistently defined. Both the presence and absence of specific behaviors, such as movement and vocalization, have been interpreted as indicators of anxiety. Greater attention to construct validity and to the consistency and reliability of outcome measures from novel object tests for anxious temperament is needed.

5.3.3 *Novelty Seeking*

The simple shy-bold dimension from early animal behavior research cast active, novelty seeking behavior at the opposite pole of the fearful, anxious dimension of temperament. Human personality inventories, in contrast, place anxiety and novelty seeking on independent dimensions of temperament (Cloninger 1987). Research with nonhuman primates, described here, indicates that the emotional response to novelty is separable from the tendency to actively seek new information and experiences.

The primary difference between standardized tests that measure emotional reactivity and those that measure novelty seeking in nonhuman primates is the test location. For the Human Intruder Test and novel object tests described above, the subject was removed from its home cage and placed in a novel environment (inescapable novelty test). This alone causes acute distress with accompanying physiological and behavioral stress responses (Stevenson-Hinde et al. 1980b; Laudenslager et al. 1999; Bardi et al. 2005). The tests described in this section measure free-choice novelty seeking by presenting the animals with the option of approaching and exploring a novel object or place within or adjacent to the familiar home environment.

Spencer-Booth and Hinde (1969) tested a variety of home group novelty test stimuli for consistency and reliability in assessing individual differences among rhesus monkey infants housed with their mothers in small social groups. The home group challenge tests included placement of a novel object at the edge of the home cage (a moving propeller), being looked at by an unfamiliar person at the edge of the cage, and direct stare from a familiar and unfamiliar person when the infants were 6 months old, designed to assess the infants' responses to mildly stressful situations. The behavioral measures in each test were the latency of the infant to leave the mother, total time off, and the percentage of time the infant was more than two feet from the mother during the test period. These measures were correlated and time off the mother was selected as the primary dependent variable. The results of the 6 month tests demonstrated that responses to all three tests were positively correlated, and were also positively correlated with the time the infant spent near its mother during undisturbed observations in the home group, suggesting that these tests reflect a cohesive dimension of individual differences.

Similar novelty challenge tests conducted at 12 months of age involved attachment of a filter cage, with an opening that the infant but not the mother could enter, to the outside of the home cage. The tests included response to a mirror, to grass placed in front of the mirror, and to a preferred food item (bananas) placed at the far end of the filter cage. The 12 month test results indicated positive correlations across the four dependent measures and a high degree of consistency across days for the same test, particularly for the amount of time spent in the filter cage. Additional stimuli were tested for animals at older ages with similar procedures to assess the cohesiveness, consistency and reliability of the measures. Each test was assessed for test–retest reliability, and for the independence and reliability of each of the outcome measures. Consistency of individual differences was measured across tests to determine whether the tests were measuring similar or different constructs, and each was related to measures of undisturbed behavior in the home group.

These procedures identified several tests that produced consistent individual differences in successive tests at the same age. Scores on these tests were significantly related to individual differences in behavior in the home group and they differed for infants raised under different social conditions, either alone with their mothers or in larger social groups. Infants who spent more time off the mother in the undisturbed group setting had higher scores on the home group novelty tests. Spencer-Booth and Hinde (1969) refrained from naming temperament dimensions except as general indicators of emotional or motivational characteristics, but in a later paper from this group, the term "enterprising" was used to refer to infants with higher rates of entering the filter cage to retrieve preferred food items (Simpson et al. 1989).

Discriminative validity of the free-choice novelty tests as reflecting a different dimension of temperament from those measured in the inescapable novelty tests was demonstrated by comparison of results from a similar set of tests in an unfamiliar indoor test cage at 12 and 30 months (Spencer-Booth and Hinde 1969). In the first few days of testing, individual differences in test response were not consistent between the home group and indoor test environments, but the correlations improved with repeated testing as the subjects habituated to the indoor cage over a 6-day period (Spencer-Booth and Hinde 1969). Interestingly, results from the indoor testing environment, with its higher stress level, were better than the home group environment in detecting the long-term effects of brief separations from the mother at 6 months of age (Hinde and Spencer-Booth 1971; Spencer-Booth and Hinde 1971). This indicates that involuntary novelty tests measure anxiety and emotional reactivity, while free-choice home group novelty tests tap into a dimension that reflects active curiosity, novelty seeking and enterprise.

At the Vervet Research Colony, we have developed a simple standardized test to measure free-choice novelty seeking in the home cage for animals of all ages. The vervets live in large outdoor cages in stable, matrilineal social groups. During the test, a novel and potentially threatening object is placed inside or on the edge of the home cage, within reach of the animals but away from any of the preferred sitting or resting places. Novel objects are salient enough to arouse interest, curiosity, and some potential for fear. An object that is too threatening will cause the average subject to stay at a distance, while an object that is too boring will not arouse enough interest to stimulate approach. Objects are prescreened on nonsubject animals to determine if they fall into the middle range. An object is ideal if the animals attend to the object and almost all subjects eventually approach and examine it. The latency to approach and the number of 1-min intervals that each animal is within 1 m of the object or touches the object are scored for a 30-min test session. The area within 1 m of the object only occupies a small portion of the home enclosure, and it takes a positive act by the monkeys to approach the object. A team of observers familiar with identifying individual monkeys makes a consensus determination of who is within 1 m for each interval.

Early research at the Vervet Research Colony demonstrated that latency to approach a novel object or enter a novel space under these circumstances was affected by early experience with the mother. Mothers who were highly protective and who restricted their infant's early explorations produced juvenile offspring who had significantly longer latencies to enter a tunnel into a new area and longer latencies to approach a potentially threatening novel object in the home group (Fairbanks and McGuire 1988, 1993). Statistical genetics analysis using the multigenerational pedigree at the Vervet Research Colony indicates that this trait is highly heritable. After controlling for age, approximately 50% of the total trait variance in latency to approach can be explained by genetic relationships among individuals (Bailey et al. 2007). A portion of this genetic variance is due to a polymorphism in the dopamine D4 receptor gene, a gene that has also been implicated in novelty seeking and attention deficit disorder in humans (Ebstein et al. 1996; Benjamin et al. 1996).

The discriminative validity of the Home Group Novelty Test as a measure of novelty seeking is demonstrated by a comparison with scores obtained in an involuntary novelty test. The Vervet Research Colony animals described above were also tested in an individual cage novelty test paradigm, similar to those described in Sect. 5.3.2, involving two baseline assessment periods and presentation of three novel objects. Responses were highly correlated across conditions, so summary scores were calculated for engage object (touch or inspect object), displacement behaviors (scratch, self-directed behavior), aggression (threaten object or observer) and agitation (locomotion, vocalization, stereotyped movements and abnormal postures). Figure 5.1 shows the multidimensional scaling results for the four measures from the individual cage novelty test and the three measures described above from the Home Group Novelty Test for a sample of 145 two-year-old vervet monkeys tested in both paradigms a few months apart. We used an alternating least-squares algorithm to calculate a two-dimensional Euclidean distance solution (SPSS 2005, Alscal). The resulting two-dimensional map shows that the three Home Group Novelty Test measures formed a cohesive cluster that was associated with the individual cage score for engagement with the novel objects. This cluster was clearly differentiated from agitation, anxious and aggressive behaviors observed in the individual cage test. Cortisol levels measured 2 h after the individual cage tests were significantly correlated with agitation scores ($r=0.41$, $p<0.001$). These results provide evidence that the involuntary individual cage novelty test measures emotional reactivity, while behavior in the home group test reflects individual differences in curiosity and in motivation to explore.

Combining latency to approach and time near the object from the Home Group Novelty Test produces a reliable index that has face validity as a simple and straightforward measure of novelty seeking (Fairbanks et al. 2011). The Novelty Seeking Index [(30 – latency)+intervals within 1 m] for well-chosen stimulus objects, is normally distributed. The demographic properties of this index are consistent with

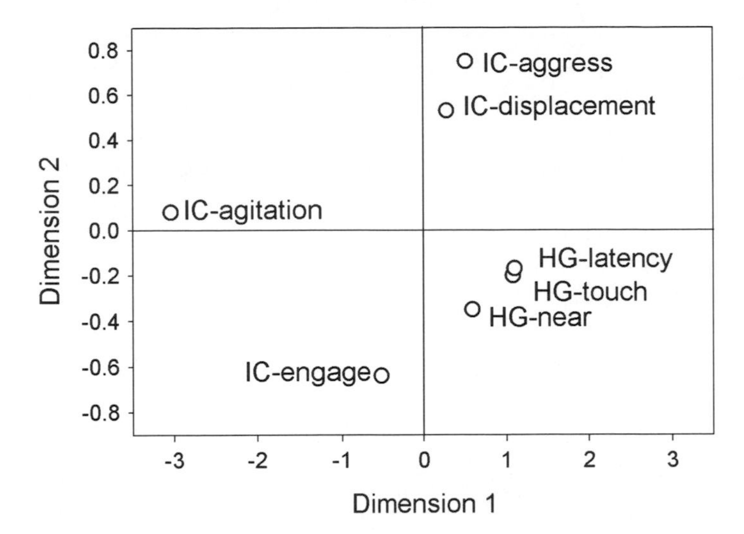

Fig. 5.1 Multidimensional scaling model of behavior in Individual Cage (IC) and Home Group (HG) Novelty tests, using SPSS Alscal Euclidean distance model. Subjects were 145 juvenile vervets tested under both conditions

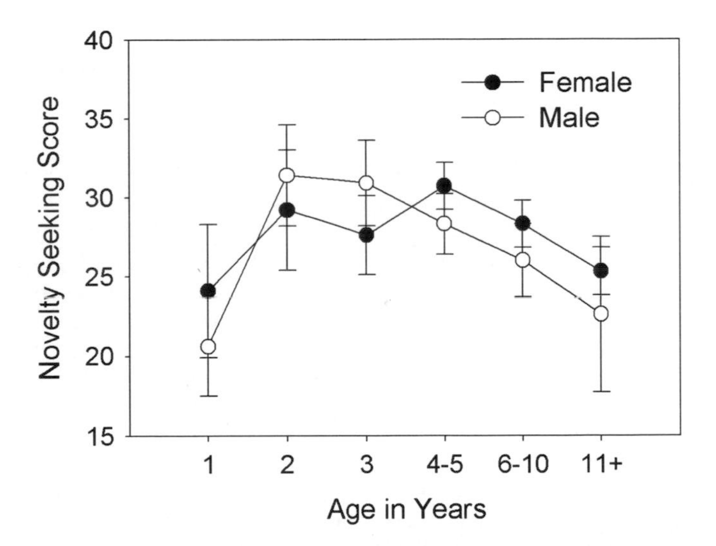

Fig. 5.2 Mean (±standard error) Novelty Seeking Index by age and sex (age: $F=2.56$, $p<0.03$; sex: $F=0.30$, n.s.; age × sex: $F=0.57$, n.s.) for 2003 Home Group Novelty Test (Bailey et al. 2007)

the construct of novelty seeking: NS scores are highest for juveniles and adolescents, then decline with age in adulthood (Fig. 5.2). Males do not differ from females in mean scores or in developmental trends for this index. Test–retest reliability from 1 year to the next is high for each of the components and for the index as a whole.

The ICC for the Novelty Seeking Index for Home Group Novelty tests conducted in five successive years was 0.73. This year-to-year consistency indicates that the Novelty Seeking Index measures trait-like individual differences in this dimension of temperament.

Santillan-Doherty et al. (2006) further differentiated behavioral tests for novelty seeking into those involving risk-taking and those reflecting curiosity. They compared both with a subjective rating scale designed to measure novelty seeking in socially housed adult stumptailed macaques. Nine presentations of the risk stimulus (a kennel used for animal transport) and nine different curiosity stimuli (including toys and food-related enrichment devices) provided robust measures of individual responses. The ethogram included 17 behaviors for the risk items and 11 for the curiosity items. Instead of considering the behaviors separately, they rated each behavior in terms of degree of risk or curiosity on a four-point scale. For the risk scale, level one was for behaviors that occurred at a distance from the object, level two for close proximity, level three for touching the exterior and level four for partial or complete entering of the kennel. An index was constructed that weighted behaviors by risk or curiosity level, then averaged scores across the nine replications. This method accounts for low frequencies of higher level behaviors, and produces a final score that reflects the dimension from low to high for risk-taking and curiosity more precisely than a simple sum. The constructed indices had high internal reliability and the intra-class correlation coefficient for the two indices together was 0.88.

Next Santillon-Doherty et al. (2006) developed a subjective rating scale by adapting items from Cloninger's Novelty Seeking scale for relevance to stumptail macaque behavior. Comparison of the risk and curiosity indices with the temperament rating demonstrated that all three were significantly correlated with one another. Together the three scales produced a combined Cronbach's alpha of 0.78, indicating a high degree of scale cohesion. This study illustrates methods for producing and testing reliable and valid behavioral indices for psychological traits. It also demonstrates that curiosity and risk-taking both contribute to the trait dimension of novelty seeking.

5.3.4 *Impulsivity*

Impulsivity is a personality construct that has particular relevance for animal models of human psychiatric and sociobehavioral disorders. Impulsivity is defined as a tendency to act quickly, on impulse, and without regard to consequences.

5.3.4.1 Behavior in Semi-Natural Settings

Higley, Mehlmann and colleagues have used measures of behavior in undisturbed groups to assess impulsivity and its relationship to serotonin neurochemistry in rhesus monkeys (Mehlman et al. 1994; Higley et al. 1996a, b). For much of this research, the behavioral measures were derived from field observations of the animals in semi-natural settings. Measures used to define impulsivity and impulsive aggression include the ratio of contact aggression to total aggression, and the ratio of long leaps between branches to total leaps. These ratios were selected because they measured the tendency for the animals to go overboard and engage in unchecked versions of risky behaviors. They have face validity as indicators of impulsivity, and have been shown to have ecological relevance for the animals (Higley et al. 1996b). For males, the age of emigration from the natal group has also been used as an indicator of impulsivity in semi-free-ranging rhesus monkeys (Kaplan et al. 1995; Mehlman et al. 1995).

These observational measures of behavioral impulsivity have the advantage of measuring ecologically relevant behaviors that are expressed within the range of expected environmental circumstances. They can be used to predict significant life history events, such as reproductive success and early mortality. A limitation of ethological methods to measure temperament is the length of time it takes to collect enough data to obtain reliable measures of individual differences in undisturbed settings, particularly for relatively rare behaviors like contact aggression. Measures like long leaps and age of emigration are relevant in the field, but they are difficult to translate to other research settings.

5.3.4.2 Intruder Challenge Test

The Intruder Challenge Test was developed at the Vervet Research Colony to provide an efficient method of measuring impulsivity and impulsive aggression that could be used to assess large numbers of animals in a relatively short time (Fairbanks 2001). The Intruder Challenge Test uses the resident–intruder paradigm to assess the behavioral response of an individual to an unfamiliar conspecific on the periphery of the subjects' home enclosure. It differs from novelty tests used to measure anxiety because the subject is on its home territory and baseline levels of anxiety are relatively low. It differs from home group tests for novelty seeking because the stimulus is social and a member of the subject's own species. The Intruder Challenge Test creates an ecologically relevant situation for animals that would typically defend their home range from outsiders in the wild.

At the Vervet Research Colony, we conduct the Intruder Challenge Test by isolating small groups of 2–5 subjects in the outdoor portion of the home cage, with all other group members confined to the indoor section. An unfamiliar adult vervet in an individual cage is placed at the edge of the subjects' home cage for a 30-min test period. The subject animals are free to respond or not, but an unfamiliar conspecific is a potent stimulus and all animals that have been tested have responded in some way. Some attend and look toward the intruder, but never interact directly with the stimulus animal. Others threaten and display to the intruder, but keep their distance. Almost all subjects eventually approach and engage the intruder to some degree. The most active monkeys approach immediately, extend their hands through the fence to touch the intruder or its cage, and press their faces through the chain link barrier, within the intruder's reach. They also threaten and display to the intruder.

Each subject is observed throughout the 30-min test session, with 17 coded behaviors scored as present or absent within successive 1-min intervals. These include the behaviors directed toward the intruder and other behaviors that reflect general activity, agitation, anxiety or aggressiveness. This scoring system is readily learned and inter-rater reliability for two observers watching the same subject have been consistently high (>0.94) for each of the coded behaviors (Fairbanks 2001). Factor analysis of the scored behaviors for an initial sample of 128 adolescent and adult male vervets indicated that the behaviors directed toward the intruder fell into a single strong factor (eigenvalue=4.29) with high loadings (>0.50) for latency to approach, time near, sit near, touch, muzzle, threaten and display to the intruder. After reversing the latency score (30-latency), the seven items were summed to produce a Social Impulsivity Index score that combines rapidity of approach, risk-taking behavior, and aggressiveness toward a stranger. Cronbach's alpha for the Social Impulsivity Index was 0.84, indicating that this scale is internally cohesive (Fairbanks 2001).

Test–retest reliability for the Social Impulsivity Index was initially assessed for 70 males and 56 females, tested 1 month to 1 year later with a different intruder and different test companions (Fairbanks et al. 2004a). The ICC for males was 0.83 and for females it was 0.89, indicating that the scale measures a stable dimension of individual temperament. Statistical genetics analysis using the multigenerational pedigree at the Vervet Research Colony indicated significant heritability for the Social Impulsivity Index and its subscales (approach and aggressiveness), providing further evidence that this index is measuring a trait-like dimension of temperament (Fairbanks et al. 2004a).

The content validity of the Social Impulsivity Index is supported by its ability to capture key features of the trait. Moeller et al. (2001) defined impulsivity as a predisposition toward rapid, unplanned reactions to internal or external stimuli without regard to the negative consequences of those reactions to the impulsive individual or to others. Socially, impulsivity is often associated with unplanned escalation to aggression in situations of potential conflict. The Social Impulsivity Index has content validity to the extent that it includes the essential elements of the construct of social impulsivity, including rapidity of response and apparent lack of attention to risk or negative consequences of the behavior. It also has ecological validity in that it measures the natural response of the animal to a species-relevant situation.

Convergent validity of the Social Impulsivity Index is established by its relationship to other factors that are known or believed to be related to impulsive temperament. Scores on Social Impulsivity vary with age and sex the way we would expect if this index is reflecting the concept of impulsivity. Social Impulsivity scores are higher for adolescents than for adults, and males score higher than females, particularly during adolescence (Fairbanks 2001; Fairbanks et al. 2004a). Figure 5.3 shows the Social Impulsivity scores for males and females tested as juveniles, at 2 years of age, and again as adolescents. As the figure indicates, males are significantly higher than females and the adolescent increase is greater for males than for females. Cognitive features that are typically associated with impulsivity (inattention and the inability to tolerate delay) are also related to the scores on the Social Impulsivity

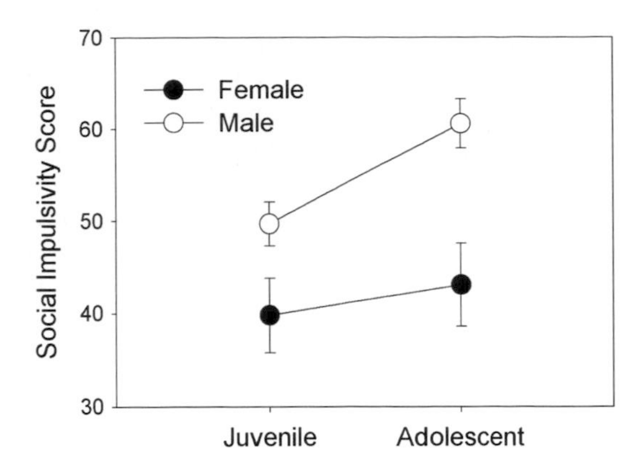

Fig. 5.3 Mean (±standard error) Social Impulsivity Index by age category and sex (age: $F = 8.96, p < 0.01$; sex: $F = 9.92, p < 0.01$; age × sex: $F = 2.50, p < 0.05$)

Index. Recent results from the Vervet Research Colony indicate that males scoring high in Social Impulsivity are just as capable of learning an association task as males scoring low in Social Impulsivity, but they make significantly more mistakes when the learning task involves a delay between the stimulus and the response (James et al. 2007). This difference in cognitive style is consistent with interpretation of the index as a measure of trait impulsivity.

Further evidence for the convergent validity of the Social Impulsivity Index as a measure of impulsive temperament is provided by its relationship to serotonin activity. Serotonin is a neurotransmitter involved in behavior regulation and response inhibition. Reduced serotonin activity is associated with impulsivity and impulsive aggression in humans and nonhuman primates (Linnoila et al. 1993; Higley et al. 1996b; Manuck et al. 1998). Studies using the Intruder Challenge Test have found that scores on the Social Impulsivity Index are negatively related to serotonin activity for male vervets and for female cynomolgus monkeys (Fairbanks et al. 2001; Manuck et al. 2003).

Discriminative validity for the Social Impulsivity Index as a measure of impulsivity, and not anxiety, was provided by experimental manipulation of the serotonin system with fluoxetine, a serotonin reuptake inhibitor that is used to treat anxiety and depression. Following a 9-week course of fluoxetine, the males on fluoxetine appeared calmer than the control animals. They took longer to approach and were less likely to engage and challenge the intruder, resulting in significantly lower Social Impulsivity scores than the controls (Fairbanks et al. 2001). This is the opposite of the response that would be expected if avoidance of the intruder was a measure of fearful-anxious behavior, but is consistent with the interpretation of the index as a measure of impulsivity. Interestingly, a study using an intruder paradigm with marmosets that were removed from their home environment and tested in an unfamiliar cage produced different results (Kinnally et al. 2006). Fluoxetine treatment reduced extreme avoidance of an unfamiliar conspecific under these circumstances. The difference between these two studies is similar to the difference between free-choice and inescapable novelty tests with novel objects. The unfamiliar cage environment produced a higher level of anxiety that was alleviated to some degree by the fluoxetine treatment, thus allowing the subject to interact more calmly with its environment. The results of these two studies underscore the importance of the context in interpreting an animal's response to an unfamiliar conspecific.

Discriminative validity of the Social Impulsivity Index as a measure of impulsivity and not simply novelty seeking is derived from comparison of results from the Intruder Challenge Test and the Home Group Novelty Test. Social Impulsivity and Novelty Seeking scores at the Vervet Research Colony are positively correlated, with Pearson correlations typically about 0.20, suggesting that there is some overlap between the two dimensions, but correlations between the two measures are always lower than correlations within the same measure over time. The social stimulus in the Intruder Challenge Test elicits aggressive and challenging behavior that is not observed in the Home Group Novelty Test. The Social Impulsivity Index also has a different demographic profile from the Novelty Seeking score (Figs. 5.2 and 5.3). Impulsivity scores are higher for males than for females and higher for adolescents than for juveniles. Novelty Seeking scores do not differ between males and females and are consistently high through the juvenile and adolescent period, indicating that this index reflects curiosity more than impulsivity.

The utility of the Social Impulsivity Index in predicting future life history events was demonstrated in a longitudinal study of males at the Vervet Research Colony (Fairbanks et al. 2004b). All males are tested in the Intruder Challenge Test as adolescents, before they are removed from the natal group. Comparison of adolescent impulsivity scores with adult dominance attainment demonstrated that males scoring high in social impulsivity as adolescents were significantly more likely to achieve alpha male status in their adult breeding groups. Adolescence is a time in the life history when emigrating males have been selected to take risks for a high potential gain. One year after successful integration into the new group, the highest-ranking animals scored in the moderate range, indicating that the Social Impulsivity Index is not a simple reflection of current dominance status (Fairbanks et al. 2004b).

The Intruder Challenge Test has proven to be a useful objective behavioral test to measure stable and consistent dimensions of temperament. The Social Impulsivity Index derived from this test has demonstrated its stability, reliability, construct validity, and discriminative and predictive utility as a research tool for ethological, neurocognitive and biomedical research.

5.4 Conclusions

The objective behavioral paradigms described here were designed to measure a variety of different dimensions including state control, anxiety and behavioral inhibition, novelty seeking and impulsivity. They include measures that can be applied from infancy to old age, and can be used in field, semi-free ranging and captive settings. Each of these methods illustrates some of the properties of a well-defined objective behavioral test and the challenges of measuring temperament in nonhuman primates.

One point that emerges from this review is that the context of the test is as important as the stimulus. Novelty tests that involve removal from the home environment measure different aspects of temperament than tests that introduce novelty into a familiar environment where the subject can freely choose to approach and engage or not. An animal is in a different mental and emotional state when tested in a novel cage compared to a stable home environment. Inescapable novelty tests measure defensive and self-protective reactions that are likely to be derived from predator defense systems. Free-choice novelty tests measure active responses derived from appetitive systems related to acquiring food and sexual partners, and exploring new territory. The appropriate test will depend on the temperament construct being studied. Most of the tests for anxious temperament intentionally use inescapable novelty paradigms to study individual differences in behavioral response in a fear-invoking situation. The tests for novelty seeking and impulsivity are designed to measure voluntary response tendencies in a more neutral setting. The same stimulus, an intruder or a potentially threatening novel object, elicits a different response depending on the context (e.g., Spencer-Booth and Hinde 1969; Fairbanks et al. 2001; Kinnally et al. 2006). Interpretation of similarities and differences across studies must take these context differences into account.

Another important point that emerges from these studies is the need for more attention to defining the constructs being studied. This is particularly true for research on anxiety which is hindered by ambiguous terminology and shifting operational definitions. Kagan chose the term "behavioral inhibition" to describe the kind of child that shows negative affect when confronted with novel situations. Scoring for behavioral inhibition in children includes active behaviors such as "crying, sobbing and fretting" and "facial expression and vocalizations of distress" in addition to long latencies to approach and inhibition of play (Garcia Coll et al. 1984). Scoring systems for anxious temperament in nonhuman primates frequently include analogs of these responses along with a variety of other measures, such as movement, vocalization and threat behaviors that are not necessarily indicators of anxiety, although they may be in certain contexts. This has led to a relatively loose attribution of the term anxiety to a diverse array of behaviors. Studies suggest that objective behavioral tests of anxiety often reveal multiple response dimensions. Demonstration of convergent and discriminative validity of each of the outcome measures is essential if behavioral tests are going to advance our understanding of anxious temperament and anxiety disorders.

The third lesson from the research presented here is the value of establishing test–retest reliability in the measurement of behaviorally defined traits. It is important to remember that coded behaviors may reflect the state of the animal at the time of the test, but not all measures will represent underlying traits that define consistent individual differences over time. Test–retest reliability is an essential step to evaluate the ability of an individual measure or a composite index to measure trait-like individual differences.

The most useful paradigms for measuring temperament have developed a relatively small number of measures or composite scales that are independent, have high construct and discriminative validity, and high test–retest reliability. These measures are then consistently applied in subsequent research studies to identify causes and consequences of individual differences in traits. Greater attention to the steps outlined above will expand the value of research in this area. It will provide further insight into the mechanisms that influence individual differences in temperament, and into the role of those individual differences in predicting future health, social success and resilience to stress. The large datasets currently being generated for rhesus monkeys, baboons and vervets hold considerable promise for use in objective behavioral tests to identify the genetic, social and physiological causes and the social and biomedical consequences of variation in temperament. Following the guidelines proposed here to derive robust and reliable measures of temperament, with good construct and discriminative validity, will further these important goals.

Acknowledgments The authors would like to thank the staff of the Vervet Research Colony for making the research described here possible. Funding for animal care and for the development of objective behavioral tests for temperament at the Vervet Research Colony was provided by NIMH (R01-MH61852) and NCRR (P40-RR19963).

References

Aureli F, van Schaik CP (1991) Post-conflict behaviour in long-tailed macaques: II. Coping with uncertainty. Ethology 89:101–114

Bailey JN, Breidenthal SE, Jorgensen MJ et al. (2007) The association of DRD4 and novelty seeking is found in a nonhuman primate model. Psychiatric Genet 17:23–27

Bardi M, Bode A, Ramirez SM et al. (2005) Maternal care and development of stress responses in baboons. Am J Primatol 66:263–278

Bayley N (1969) Manual for the Bayley Scales of Infant Development. Psychological Corp, New York

Benjamin J, Li L, Patterson C et al. (1996) Population and familial association between the D4 dopamine receptor gene and measures of Novelty Seeking. Nat Genet 12:81–4

Bethea CL, Streicher JM, Coleman K et al. (2004) Anxious behavior and fenfluramine-induced prolactin secretion in young rhesus macaques with different alleles of the serotonin reuptake transporter polymorphism (5HTTLPR). Behav Genet 34:295–307

Brazelton TB (1973) Neonatal behavioral Assessment Scale. Lippincott, Philadelphia

Capitanio JP (1999) Personality dimensions in adult male rhesus macaques: Prediction of behaviors across time and situation. Am J Primatol 47:299–320

Capitanio JP, Mason WA, Mendoza SP et al. (2006) Nursery rearing and biobehavioral organization. In: Sackett GP, Ruppenthal G, Elias K (eds) Nursery rearing of nonhuman primates in the 21st century. Springer, New York

Capitanio JP, Mendoza SP, Baroncelli S (1999) The relationship of personality dimensions in adult male rhesus macaques to progression of simian immunodeficiency virus disease. Brain Behav Immun 13:138–154

Castles DL, Whiten A, Aureli F (1999) Social anxiety, relationships and self-directed behaviour among wild female olive baboons. Anim Behav 58:1207–1215

Champoux M, Hibbeln JR, Shannon C et al. (2002a) Fatty acid formula supplementation and neuromotor development in rhesus monkey neonates. Pediatr Res 51:273–281

Champoux M, Bennett A, Shannon C et al. (2002b) Serotonin transporter gene polymorphism, differential early rearing, and behavior in rhesus monkey neonates. Mol Psychiatry 7:1058–1063

Champoux M, Jaquish CE, Higley SB et al. (1999) Heritability of standardized biobehavioral assessment scores in rhesus monkey infants. Am J Primatol 49:42

Champoux M, Norcross J, Suomi SJ (2000) Rhesus monkeys with late-onset hydrocephalus differ from non-impaired animals during neonatal neurobehavioral assessments: Six-year retrospective analysis. Comp Med 50:218–224

Champoux M, Suomi SJ, Schneider ML (1994) Temperament differences between captive Indian and Chinese-Indian hybrid rhesus macaque neonates. Lab Anim Sci 44:351–357

Clarke AS, Boinski S (1995) Temperament in nonhuman primates. Am J Primatol 37:103–125

Cloninger CR (1987) A systematic method for clinical description and classification of personality. Arch Gen Psychiatry 44:573–588

Cronbach LJ (1951) Coefficient alpha and the internal structure of tests. Psychometrika 16:297–334

Diezinger F, Anderson JR (1986) Starting from scratch: A first look at a "displacement activity" in group-living rhesus monkeys. Am J Primatol 11:117–124

Ebstein RP, Novick O, Umansky R et al. (1996). Dopamine D4 receptor (D4DR) exon III polymorphism associated with the human personality trait of Novelty Seeking. Nat Genet 12:78–80

Fairbanks LA (2001) Individual differences in response to a stranger: Social impulsivity as a dimension of temperament in vervet monkeys. J Comp Psychol 115:22–28

Fairbanks LA, Bailey JN, Breidenthal SE et al. (2011) Environmental stress alters genetic regulation of novelty seeking in vervet monkeys. Genes Brain Behav June 1 [Epub ahead of print]

Fairbanks LA, Jorgensen MJ, Huff A et al. (2004b) Adolescent impulsivity predicts adult dominance attainment for male vervet monkeys. Am J Primatol 64:1–17

Fairbanks LA, McGuire MT (1988) Long-term effects of early mothering behavior on responsiveness to the environment in vervet monkeys. Dev Psychobiol 21:711–24

Fairbanks LA, McGuire MT (1993) Maternal protectiveness and response to the unfamiliar in vervet monkeys. Am J Primatol 30:119–129

Fairbanks LA, Melega WP, Jorgensen MJ et al. (2001) Social impulsivity inversely associated with CSF 5-HIAA and fluoxetine exposure in vervet monkeys. Neuropsychopharmacology 24:370–78

Fairbanks LA, Newman TK, Bailey JN et al. (2004a) Genetic contributions to social impulsivity and aggressiveness in vervet monkeys. Biol Psychiatry 55:642–647

Garcia Coll C, Kagan J et al. (1984) Behavioral inhibition in young children. Child Dev 55: 1005–1019

Gosling SD (2001) From mice to men: What can we learn about personality from animal research? Psychol Bull 127:45–86

Guadagnoli E, Velicer WF (1988) Relation of sample size to the stability of component patterns. Psychol Bull 103:265–275

He N, Bai J, Champoux M et al. (2004) Neurobehavioral deficits in neonatal rhesus monkeys exposed to cocaine in utero. Neurotoxicol Teratol 26:13–21

Higley JD, Mehlman PT, Higley SB et al. (1996a) Excessive mortality in young free-ranging male nonhuman primates with low cerebrospinal fluid 5-hydroxyindoleacetic acid concentrations. Arch Gen Psychiatry 53:537–543

Higley JD, Mehlman PT, Poland RE et al. (1996b) CSF testosterone and 5-HIAA correlate with different types of aggressive behaviors. Biol Psychiatry 40:1067–1082

Hinde RA, Spencer-Booth Y (1971) Effects of brief separation from mother on rhesus monkeys. Science 173:111–118

James AS, Groman SM, Seu E et al. (2007) Dimensions of impulsivity are associated with poor spatial working memory performance in monkeys. J Neurosci 27:14358–14364

Kagan J, Reznick JS, Snidman N (1988) Biological bases of childhood shyness. Science 240: 167–171

Kalin NH (2003) Nonhuman primate studies of fear, anxiety, and temperament and the role of benzodiazepine receptors and GABA systems. J Clin Psychiatry 64 (suppl 3):41–44

Kalin NH, Larson C, Shelton SE et al. (1998) Asymmetric frontal brain activity, cortisol, and behavior associated with fearful temperament in rhesus monkeys. Behav Neurosci 112: 286–292

Kalin NH, Shelton SE (1998) Ontogeny and stability of separation and threat-induced defensive behaviors in rhesus monkeys during the first year of life. Am J Primatol 44:125–135
Kalin NH, Shelton SE (2003) Nonhuman primate models to study anxiety, emotion regulation, and psychopathology. Ann N Y Acad Sci 1008:189–200
Kalin NH, Shelton SE, Davidson RJ (2007) Role of the primate orbitofrontal cortex in mediating anxious temperament. Biol Psychiatry 62:1134–1139
Kalin NH, Shelton SE, Davidson RJ et al. (2001) The primate amygdala mediates acute fear but not the behavioral and physiological components of anxious temperament. J Neurosci 21:2067–2074
Kalin NH, Shelton SE, Fox AS et al. (2005) Brain regions associated with the expression and contextual regulation of anxiety in primates. Biol Psychiatry 58:796–804
Kaplan JR, Fontenot MB, Berard J et al. (1995) Delayed dispersal and elevated monoaminergic activity in free-ranging rhesus monkeys. Am J Primatol 35:229–234
Kinnally E, Jensen HA, Ewing JH et al. (2006) Serotonin function is associated with behavioral response to a novel conspecific in marmosets. Am J Primatol 68:812–824
Laudenslager ML, Rasmussen KL, Berman CM et al. (1999) A preliminary description of responses of free-ranging rhesus monkeys to brief capture experiences: Behavior, endocrine, immune and health relationships. Brain Behav Immun 13:124–137
Linnoila M, Virkkunen M, George T et al. (1993): Impulse control disorders. Int Clin Psychopharmacol 8:53–56
Maestripieri D (1993a) Maternal anxiety in rhesus macaques (*Macaca mulatta*). I. Measurement of anxiety and identification of anxiety-eliciting situations. Ethology 95:19–31
Maestripieri D (1993b) Maternal anxiety in rhesus macaques (*Macaca mulatta*). II. Emotional bases of individual differences in mothering style. Ethology 95:32–42
Maestripieri D (2000) Measuring temperament in rhesus macaques: Consistency and change in emotionality over time. Behav Process 49:167–171
Maestripieri D, Martel FL, Nevison CM et al. (1992b) Anxiety in rhesus monkey infants in relation to interactions with their mother and other social companions. Dev Psychobiol 24:571–581
Maestripieri D, Schino G, Aureli F et al. (1992a) A modest proposal: Displacement activities as an indicator of emotions in primates. Anim Behav 44:967–979
Manuck SB, Flory JD, McCaffrey JM et al. (1998): Aggression, impulsivity and central nervous system serotonergic responsivity in a nonpatient sample. Neuropsychopharmacology 10:287–299
Manuck SB, Kaplan JR, Rymeski BA et al. (2003) Approach to a social stranger is associated with low central nervous system serotonergic responsively in female cynomolgus monkeys (*Macaca fascicularis*). Am J Primatol 61:187–194
Mason WA, Capitanio JP, Machado CJ et al. (2006) Amygdalectomy and responsiveness to novelty in rhesus monkeys (*Macaca mulatta*): Generality and individual consistency of effects. Emotion 6:73–81
Mehlman PT, Higley JD, Faucher I et al. (1994) Low CSF 5-HIAA concentrations and severe aggression and impaired impulse control in nonhuman primates. Am J Psychiatry 151:1485–1491
Mehlman PT, Higley JD, Faucher I et al. (1995) Correlation of CSF 5-HIAA concentration with sociality and the timing of emigration in free-ranging primates. Am J Psychiatry 152:907–913
Moeller FG, Barratt ES, Dougherty DM et al. (2001) Psychiatric aspects of impulsivity. Am J Psychiatry 158:1783–1793
Pavani S, Maestripieri D, Schino G et al. (1991) Factors influencing scratching behaviour in long-tailed macaques. Folia Primatol 57:34–38
Rogers J, Shelton SE, Shelledy W et al. (2008) Genetic influences on behavioral inhibition and anxiety in juvenile rhesus macaques. Genes Brain Behav 7:463–469
Santillan-Doherty AM, Munoz-Delgado J, Arenas R et al. (2006) Reliability of a method to measure novelty-seeking in nonhuman primates. Am J Primatol 68:1098–1113
Schino G, Perretta G, Taglioni AM et al. (1996) Primate displacement activities as an ethopharmacological model of anxiety. Anxiety 2:186–191

Schneider ML, Coe CL, Lubach GR (1992) Endocrine activation mimics the adverse effects of prenatal stress on the neuromotor development of the infant primate. Dev Psychobiol 25:427–439

Schneider ML, Moore CE, Kraemer GW (2001) Moderate alcohol during pregnancy: Learning and behavior in adolescent rhesus monkeys. Alcohol Clin Exp Res 25:1383–1392

Schneider ML, Moore CF, Suomi SJ et al. (1991) Laboratory assessment of temperament and environmental enrichment in rhesus monkey infants (*Macaca mulatta*). Am J Primatol 25:137–155

Schneider ML, Roughton EC, Lubach GR (1997) Moderate alcohol consumption and psychological stress during pregnancy induce attention and neuromotor impairments in primate infants. Child Dev 68:747–759

Schneider ML, Suomi SJ (1992) Neurobehavioral assessment in rhesus monkey neonates (*Macaca mulatta*): Developmental changes, behavioral stability and early experience. Inf Behav Dev 15:155–177

Shrout PE, Fleiss JL (1979) Intraclass correlations: Uses in assessing rater reliability. Psychol Bull 86:420–428

Sih A, Bell AM, Johnson JC (2004) Behavioral syndromes: An integrative overview. Q Rev Biol 79:241–277

Simpson MJA, Gore MA, Janus M et al. (1989) Prior experience of risk and individual differences in enterprise shown by rhesus monkey infants in the second half of their first year. Primates 30:493–509

Spencer-Booth Y, Hinde RA (1969) Tests of behavioural characteristics for rhesus monkeys. Behaviour 33:17–211

Spencer-Booth Y, Hinde RA (1971) Effects of brief separations from others during infancy on behavior of rhesus monkeys 6-24 months later. J Child Psychol Psychiatry 12:157–172

SPSS for Windows, Version 14.0 (2005) SPSS Incorporated, Chicago

Stevenson-Hinde J, Stillwell-Barnes R, Zunz M (1980a) Subjective assessment of rhesus monkeys over four successive years. Primates 21:66–82

Stevenson-Hinde J, Zunz M, Stillwell-Barnes R (1980b) Behaviour of one-year-old rhesus monkeys in a strange situation. Anim Behav 28:266–277

Troisi A, Schino G, D'Antoni M et al. (1991) Scratching as a behavioral index of anxiety in macaque mothers. Behav Neural Biol 56:307–313

Uher J, Asendorpf JB (2008) Personality assessment in the great apes: Comparing ecologically valid behavior measures, behavior ratings, and adjective rating. J Res Pers 42:821–838

Watson SL, Ward JP (1996) Temperament and problem solving in the small-eared bushbaby (*Otolemur garnettii*). J Comp Psychol 110:377–385

Weiss A, King JE, Perkins L (2006) Personality and subjective well-being in orangutans (*Pongo pygmaeus* and *Pongo abelii*). J Pers Soc Psychol 90:501–511

Williamson DE, Coleman K, Bacanu S-A et al. (2003) Heritability of fearful-anxious endophenotypes in infant rhesus macaques: A preliminary report. Biol Psychiatry 53:284–291

Wilson DS, Coleman K, Clark AB et al. (1993) Shy-bold continuum in pumpkinseed sunfish: An ecological study of a psychological trait. J Comp Psychol 107:250–256

Chapter 6
Predicting Primate Behavior from Personality Ratings

Lindsay Murray

Abstract An important reason for measuring personality in any species is the determination of the predictive power between personality ratings and manifest behavior. In this chapter, I review what is known to date about the relationships between personality traits and overt behaviors in primates. The majority of studies that have found such links have done so at the factor level, so I also present new data from my own research with chimpanzees, gorillas and bonobos, where I test for relationships between specific trait terms and observed behavior within the broader overall dimensions (factors) of personality, along with associations between these factor scores and behaviors. Principal components analysis of chimpanzee personality ratings yielded six components accounting for 77.5% of the variance in ratings. Demonstrating convergent and discriminant validity, I illustrate links between factor scores on the top three factors (each accounting for more than 15% of the variance) concerned with Confidence (C1), Sociability (C2) and Excitability (C3), and behaviors – e.g., Sociability (C2) scores correlate positively with several observed play indices. Examples of behaviors linked to discrete trait adjectives include significant correlations between rated playfulness and aggression and their respective observed behavioral frequencies. In addition, some relationships are specific to particular age or sex classes – e.g., the relationships between *confidence* and *effectiveness* and increased grooming are only evident among male chimpanzees, while immature individuals scoring higher on *intelligent* engage in more grooming. I also present preliminary findings of relationships between personality ratings and observed behaviors for the smaller samples of gorillas and bonobos. That the behavioral data here correlate significantly with both specific traits and with factor scores provides validation of the rating instrument and will benefit our understanding of nonhuman personality.

L. Murray (✉)
Department of Psychology, University of Chester,
Parkgate Road, Chester, CH2 1DH, UK
e-mail: l.murray@chester.ac.uk

A. Weiss et al. (eds.), *Personality and Temperament in Nonhuman Primates*, Developments in Primatology: Progress and Prospects, DOI 10.1007/978-1-4614-0176-6_6,

6.1 Introduction: Personality and Behavior

Personality is a fascinating aspect of the psychological life of any individual, whether human or nonhuman. Invariably, when we hear the term "personality," we want to gain an insight into the inner world of a particular individual and a big part of that involves how they behave. What are they likely to do? Will they be nice to me? Will I be better than them? Answering such questions depends largely on our ability to demonstrate clear and valid relationships between personality traits and the way individuals overtly behave.

A growing body of nonhuman research now helps us see interesting links between personality variables and behavior. In this chapter, I will review research validating personality trait ratings with manifest behavior among primates.

Personality traits can be defined as lasting dispositions which lead to characteristic behavior across situations, with different individuals reacting in different ways to the same situation (e.g., Cervone and Pervin 2008). If behavior is influenced by personality, the ability to measure personality should facilitate prediction of behavior. The early use of human personality tests as diagnostic tools (e.g., in personnel selection) did not, however, always allow such prediction, leading to criticism of the entire concept of traits (Mischel 1968). As the situation alone is equally wanting in allowing the prediction of behavior (Funder and Ozer 1983), the optimal approach is an interactionist one, taking account of both situational factors and personality traits. The more specifically defined the personality trait is, the more likely it is that behavior will be consistent across the relevant situations, although individuals also vary in the extent of consistency on particular traits, and in the extent of predictability across situations (Klirs and Revelle 1986).

6.2 Surveying the Literature

I began by carrying out systematic searches of the relevant databases (PsychINFO, Web of Science) using the terms "personality" and "temperament" (which are still often used interchangeably) and behavior, and searching for key names in the primate order as well as the more generic "ape," "monkey" and "primates" (after Gosling 2001; Jones and Gosling 2005). Studies that had subjective as well as behavioral measures of the same personality traits were targeted. Some article titles included both "temperament" and "behavior" but did not use any rating scale for example, individual differences in temperament do not just reflect behavioral and emotional characteristics, but have also been explored in relation to brain and physiological functioning (e.g., Kalin et al. 1998). A study of cortisol reactivity in tufted capuchins (Byrne and Suomi 2002) is an example of looking at personality and behavior in relation to one or more additional variables, often physiological or immunological measures.

Closer inspection of the abstracts and/or method and results sections showed that many studies included aspects of either personality or behavior – or both – but did not correlate separate personality ratings and behavioral codings. Examples of this include Fairbanks (2001), in which an index of social impulsivity was derived from a behavioral factor analysis, and Weiss et al. (2002) in which dominance was not measured separately as a behavioral coding but as an additional factor on the rating scale. I did not include dominance as a behavior as studies exploring links between personality and dominance often involve computation of other variables and/or subjective impressions of dominance rank. For example, Buirski et al. (1978) reported links between subjective dominance rank and Emotions Profile Index (EPI) personality traits such as *aggressive* and *distrustful* in chimpanzees. Following an iterative process, the references of relevant articles were scanned for new research not already identified. Results are shown in Table 6.1.

Table 6.1 shows that studies have focused entirely on apes and Old World monkeys, especially the Cercopithecines. Konečná et al.'s (2008) recent study was the first to examine personality in the Colobines and to use a Five-Factor Model (FFM)-based questionnaire on a monkey species. Sufficient detail about the sample is reported in most studies and some researchers have focused purely on males (Capitanio 1999, 2002; Kuhar et al. 2006; Konečná et al. 2008).

More studies report validation of factor scores (rather than traits) with observed behavior, while some report links between behaviors and both specific traits and factor scores (e.g., Stevenson-Hinde et al. 1980a; Murray 1995). Some researchers (Pederson et al. 2005; Konečná et al. 2008) do the opposite and present two types of behavior coding with factor scores: specific behaviors and those behaviors categorized into social contexts or components.

The focus of this review is on studies using trait rating scales of one type or another – arguably the only way to measure personality in nonhumans. The growing body of literature suggests that the trait paradigm to measure primate personality is psychometrically robust (e.g., Gosling 2001). However, interrater reliability is more often reported than is validity.

Another important methodological consideration is the independence of the trait ratings and behavioral observations and codings. This is not always reported clearly, but often these are done simultaneously by the same researcher(s), although if one researcher conducts the behavioral observations and also contributes to the personality ratings but other multiple raters are also involved and a high level of interrater reliability is demonstrated, the situation is improved. Arguably, it is optimal to have the ratings completed by one team of researchers and the behavioral codings by a separate team (e.g., Capitanio 1999; Pederson et al. 2005). In this way, if there is repetition in terms of behavioral definitions for trait terms, each set of raters and observers is only conceptualizing that description once and independently of each other. Capitanio's (1999) study did not only include independent ratings and observations but also incorporated a longitudinal element showing that personality factors predicted behavior up to 4.5 years later, as well as demonstrating that their predictive power is cross-situational.

Table 6.1 shows many examples of convergent validity between personality ratings and overt behaviors that would be predicted if they are both expressing the same latent variable; for example, sociability, playfulness, or aggression.

Table 6.1 Summary of research validating nonhuman primate personality ratings with observed behavior

Species/study	n[a]	Factor/trait/*other*	Behavior
Chimpanzees			
Buirski and Plutchik (1991)[b]	1	aggressive (+)	Infanticide, cannibalism
		depressed (+)	
		distrustful (+)	
		trustful (–)	
		timid (–)	
		controlled (–)	
		gregarious (–)	
Murray (1995)[d]	59	Confidence	Males' grooming rate (+)
			Males' rate being groomed (+)
			Males' % significant groomers (+)
			Males' % significant partners groomed (+)
			Immatures' responsibility for proximity (+)
		Sociability	Rate of play (+)
			% significant play partners (+)
			% time spent alone (–)
			Responsibility as groomee (+)
			Males' display rate (–)
			Males' responsibility as groomer (+)
			Immatures' rate being groomed (+)
			Immatures' % significant groomers (+)
			Immatures' aggression rate (–)
		Excitability	Rate of grooming (+)
			Median association time (–)
			Rate of aggression (+)
			Adults' rate being groomed (+)
			Adults' rate of play (+)
			Adults' % significant partners (–)
			Immatures' % time spent alone (+)
			Immatures' display rate (+)
Lilienfeld et al. (1999)[c]	34	*Chimpanzee Psychopathy Measure Score*	Play wrestling and tickling performed and received (+)
			Play chasing performed and received (+)
			Repetitive movement (–)
			Detachment (–)
			Gentle teasing received (–)
			Low-intensity agonism received (+)
King and Landau (2003)	49	Subjective well-being	Submissive Context Behaviors (–)
			Submissive avoidance (–)

(continued)

Table 6.1 (continued)

Species/study	n^a	Factor/trait/*other*	Behavior
Pederson et al. (2005)	49	Dominance	Submissive-watch (–)
			Aggressive display (+)
			Submissive Context Behaviors (–)
			Agonistic Context Behaviors (+)
		Extraversion	Idle (–)
			Public orientation-watch and greet (–)
			Gymnastics (+)
			Cling (+)
			Affiliative approach (+)
			Affinitive Context Behaviors (+)
			Public Orientation Context Behaviors (–)
		Dependability	Aggressive display (–)
			Agonistic Context Behaviors (–)
		Agreeableness	Aggressive display (–)
			Groom (+)
			Agonistic Context Behaviors (–)
		Emotionality	Aggressive display (+)
			Scratch (–)
			Agonistic Context Behaviors (+)
		Openness	Idle (–)
Vazire et al. (2007)[c]	52	aggressive	Contact-infant (–)
			Solitary play (–)
		belligerent, irritable	Threat (+)
		dominant	Contact-infant (–)
			Solitary play (–)
			Object manipulation (–)
			Self-groom (+)
		eccentric	Repetitive (+)
		depressed	Social play (–)
			Solitary play (–)
			Object manipulation (–)
		friendly, sociable	Social play (+)
			Contact-infant (+)
			Solitary play (+)
		solitary	Flee (–)
		solitary, sullen	Social play (–)
			Contact-infant (–)
			Solitary play (–)
			Object manipulation (–)
		active	Groom received (–)
			Contact-infant (+)
			Solitary play (+)
			Object manipulation (+)
		playful	Flee (+)
			Avoid (+)

(continued)

Table 6.1 (continued)

Species/study	n[a]	Factor/trait/*other*	Behavior
			Social play (+)
			Contact-infant (+)
			Solitary play (+)
			Object manipulation (+)
			Locomotion (+)
Bonobos			
Murray (1995)	4	active, excitable, fearful, insecure	Responsibility for play (–)
		playful, sociable	Responsibility for play (+)
		playful	Rate of play (+)
		curious	Rate of grooming (+)
		effective, popular	% significant partners groomed (+)
		deferential	% significant partners groomed (–)
		sociable	% time spent alone (–)
		fearful, insecure	% time spent alone (+)
		slow	Rate of aggression (+)
Gorilla			
Murray (1995)	13	tense	Rate of aggression (+)
		active	Females' display rate (+)
		confident, effective, irritable	Immatures' % significant groomers (+)
		deferential	Immatures' % significant groomers (–)
		slow, solitary	Immatures' rate being groomed (–)
		sociable	Immatures' rate of play (–)
Kuhar et al. (2006)	25 M	Extraverted	Initiate affiliative behaviors (+)
			Receive affiliative behaviors (+)
			Initiate contact aggression (+)
		Dominant	Initiate displacements (+)
			Receive displacements (–)
		Fearful	Initiate displacements (–)
		Understanding	Initiate noncontact aggression (–)
Great apes			
Uher and Asendorpf (2008)[e]	20 A	friendly to youngsters	Friendliness to youngsters
		persistent	Persistence
		physically active	Physical activity
		sexually active	Sexual activity
Olive baboons			
Buirski et al. (1973)[f]	7	*Protection* (less fearful)	Mean time being groomed (–)
		Deprivation (more sullen, jealous)	Mean time being groomed (+)
		Rejection (more defiant, belligerent)	Mean time being groomed (+)
		Destruction (more aggressive)	Mean time being groomed (+)

(continued)

Table 6.1 (continued)

Species/study	n^a	Factor/trait/*other*	Behavior
Vervet monkeys			
McGuire et al. (1994)[g]	86	Social competence	Solitary (–)
			Submit (–)
			Vigilance (+)
			Groom (+)
			Aggression (+)
		Playful/curious	Solitary (–)
			Active (+)
			Play (+)
			Explore (+)
		Opportunistic	Active (+)
			Aggressive (+)
Hanuman langurs			
Konečná et al. (2008)	27 AM	Agreeableness	Aggression (contacts vs. threats) (–)
			Displacement rate (–)
			Aggression rate (–)
			Touch penis (+)
			Grooming time (+)
			H'partners approaching (Shannon index) (+)
			Dominance component behaviors (–)
		Confidence	Aggression (close vs. distant) (–)
			Terminate groom (–)
			Locomotion (–)
			Grunt and grimace (+)
			Display (+)
			Displacement rate (+)
			Copulation (+)
			Passive affiliation time (+)
			Approach rate (+)
			Dominance component behaviors (+)
			Involvement component behaviors (+)
			Activity component behaviors (–)
		Extraversion	Synchronization/rest (–)
			Resting time (–)
			Terminate groom (+)
			Activity component behaviors (+)
Rhesus macaques			
Stevenson-Hinde et al. (1980a)	N/A	aggressive	Hits, threats and chases (+)
		effective	Displacement of/Avoidance by others (+)
		excitable	Displays/threats directed outside cage (+)
		fearful	Fear grins and avoidance of others (+)
		playful	Playful behavior (+)
		sociable	Total number of contact partners (+)

(continued)

Table 6.1 (continued)

Species/study	n[a]	Factor/trait/*other*	Behavior
		Confident	Mothers had infants that initiated contact and play with others, and did not maintain close proximity to their mothers
			One year olds maintain proximity to and initiate play with others and had experienced high rejection rate by mother at weaning age (16 weeks)
		Excitable	Mothers initiated close contact with infants, that did not maintain close proximity to others
			One year olds had high rate of mother-initiated close contact and restraint at 8 weeks
		Sociable	Mothers had infants that initiated contact with others
			One year olds had initiated contact with others at 8 weeks
Stevenson-Hinde et al. (1980b)	N/A	Confident	Infant males' latency to touch a banana in a Strange Situation test (+)
			Male response to novel mirror (gaze/lipsmacks) (–)
			Infant females' time to complete reinforcement test (+)
		Excitable	Infant males' time spent in test cage when novel ball present (+)
			Infant males' frequency of whoo calls in test (–)
			Infant males' time to complete reinforcement test (–)
			Infant males' time spent holding lever in slide test (+)
		Sociable	Infant males' activity in test (–)
			Infant females' time to complete reinforcement test (+)
			Infant males' time spent holding lever in slide test (+)
Capitanio (1999)[h]	33	Sociability	Approaches initiated and received in natal group (+)
	AM		Fear grimace initiate in natal group (–)
			Beginning of group formation:
			Threat initiate (–)
			Fear grimace receive (–)
			Approach receive (+)
			Proximity duration (+)

(continued)

Table 6.1 (continued)

Species/study	n[a]	Factor/trait/*other*	Behavior
			Nonsocial duration (–)
			Once group established:
			Fear grimace initiate (–)
			Approach initiate (+) and receive (+)
			Groom present initiate (+) and receive (+)
			Proximity duration (+)
			Nonsocial duration (–)
			Groom duration (+)
			Phase 1 dyad formation:
			Groom present initiate (+) and receive (+)
			Duration groom initiate (+)
			Lipsmack receive (+)
			Fear grimace receive (–)
			Sex present receive (+)
			Grunt receive (+)
			Proximity duration (+)
			Nonsocial duration (–)
			Phase 2 dyad formation:
			Sex present initiate (+) and receive (+)
			Grunt initiate (+) and receive (+)
			Duration groom initiate (+) and receive (+)
			Lipsmack receive (+)
			Fear grimace receive (–)
			Yawn receive (–)
			Groom present receive (+)
			Proximity duration (+)
			Nonsocial duration (–)
		Confidence	Groom receive in natal group (+)
			Fear grimace receive in natal group (+)
			Sex present receive in natal group (+)
			Beginning of group formation:
			Lipsmack initiate (–)
			Sex present receive (–)
			Fear grimace initiate (+)
			Once group established:
			Approach initiate (–) and receive (–)
			Phase 1 dyad formation:
			Lipsmack initiate (–) and receive (–)
			Grunt initiate (–) and receive (–)
			Sex present receive (–)

(continued)

Table 6.1 (continued)

Species/study	n^a	Factor/trait/*other*	Behavior
			Duration groom receive (–)
			Proximity duration (–)
			Nonsocial duration (+)
			Phase 2 dyad formation:
			Groom present initiate (+) and receive (–)
			Grunt initiate (–) and receive (–)
			Duration groom initiate (–) and receive (–)
			Lipsmack receive (–)
			Fear grimace receive (+)
			Sex present receive (–)
			Nonsocial duration (+)
		Excitability	Displacements initiated and received in natal group (+)
			Fear grimace initiate in natal group (+)
			Chase initiate in natal group (+)
			Sex present receive in natal group (+)
			Aggression initiate in natal group (–)
			Approach/contact receive in natal group (–)
			Groom receive in natal group (+)
			Response to threatening human:
			Back of cage location (–)
			Threat initiate (+)
			Beginning of group formation:
			Threat initiate (+)
			Lipsmack initiate (–)
			Fear grimace initiate (+) and receive (+)
			Sex present initiate (+) and receive (–)
			Groom present receive (–)
			Proximity duration (–)
			Nonsocial duration (+)
			Groom duration (–)
			Once group established:
			Fear grimace initiate (+) and receive (–)
			Approach receive (–)
			Nonsocial duration (+)
			Groom duration (–)
			Phase 1 dyad formation:
			Lipsmack initiate (+)

(continued)

Table 6.1 (continued)

Species/study	n[a]	Factor/trait/*other*	Behavior
			Fear grimace initiate (+) and receive (+)
			Sex present initiate (+)
			Duration groom receive (−)
			Phase 2 dyad formation:
			Lipsmack initiate (+)
			Grunt initiate (−)
			Fear grimace receive (+)
			Nonsocial duration (−)
		Equability	Approach initiate in natal group (−)
			Threat receive in natal group (+)
			Beginning of group formation:
			Approach initiate (−) and receive (−)
			Groom present initiate (−) and receive (−)
			Sex present receive (−)
			Proximity duration (−)
			Nonsocial duration (+)
			Groom duration (−)
			Once group established:
			Threat initiate (−)
			Groom present initiate (−) and receive (−)
			Phase 1 dyad formation:
			Fear grimace receive (−)
			Groom present receive (−)
			Grunt receive (+)
			Duration groom receive (−)
			Phase 2 dyad formation:
			Grunt initiate (−) and receive (+)
			Groom present receive (+)
			Duration groom receive (−)
			Proximity duration (+)
Capitanio (2002)	12 AM	Low sociable	"Sit and stare" response to video playback: yawn (+), watching (+), activity (−) during social stimulus; lipsmack (−) during nonsocial stimulus
			Higher latency to avert gaze from screen
Stump-tailed macaques			
Mondragón-Ceballos and Santillán-Doherty (1994)	27	Confident	Evasive (−)
			Copulatory (+)
			Feeding (−)
			Bizarre (−)

(continued)

Table 6.1 (continued)

Species/study	n^a	Factor/trait/*other*	Behavior
		Sociable	Submissive (+)
			Amiable (+)
			Self-grooming (–)
			Resting (–)
		Excitable	Threatening (+)
			Rambling (+)
			Bizarre (+)
Santillán-Doherty et al. (2006)	22	Novelty-seeking	Risk-taking
			Curiosity

[a] *N* is given only where clearly stated in the study. Samples are of mixed sex/age composition, unless indicated as *M* males; *F* females; *A* adults; *I* immatures

[b] Reports links between trait rating individual differences and subsequent unusual behavior, but behavior codings were not carried out concurrently with trait ratings

[c] Only correlations reaching $p<0.01$ level of significance are reported here

[d] Correlations shown here are between personality factors and behaviors. For correlations between personality traits and behaviors, see details in chapter

[e] Study pooled 5 of each of the following species into a Great Ape sample: orangutans, bonobos, chimpanzees, gorillas. Significant correlations ($p<0.01$) between adjective ratings and behavior measures are given here – see Uher and Asendorpf (2008) for correlations between behavior measures and "behavior ratings"

[f] Used Emotions Profile Index (EPI): protection, deprivation, rejection and destruction are emotions derived from scores on the traits defiant, assertive, submissive, obedient, belligerent, impulsive, alarmed, cautious, jealous and sullen; $n=7$; $p<0.05$

[g] Five of the eight behaviors observed in this study were not independent, as they were also included in the personality assessment as coded observed behaviors and were included in the factor analysis which generated the personality factors

[h] Only partial correlation coefficients greater than 0.25 are reported here. See Capitanio (1999) for descriptions of behavioral responses to specific video stimuli

6.2.1 Personality Trait Level Relationships with Behavior

In examining the relationships between behavior and specific personality traits, we can see that ratings on *playful* correlate positively with observed play behavior in chimpanzees (Vazire et al. 2007; Murray 1995 – see Table 6.2), bonobos (Murray 1995), and rhesus macaques (Stevenson-Hinde et al. 1980a). *Aggressive* ratings also correlate positively with overt aggression in chimpanzees (Murray 1995) and rhesus macaques (Stevenson-Hinde et al. 1980a), while rates of aggression correlate positively with ratings on *tense* in gorillas and *slow* in bonobos (Murray 1995). Higher ratings on the trait *sociable* are related to increased play behavior in chimpanzees (Vazire et al. 2007; Murray 1995). More *sociable* bonobos also initiate more play and spend less time alone (Murray 1995), while *sociable* rhesus macaques have more contact partners (Stevenson-Hinde et al. 1980a) and *sociable* chimpanzees have been found to increase contact with infants (Vazire et al. 2007).

The report by Buirski and Plutchik (1991), albeit involving only one adult female chimpanzee, illustrates a relationship between an unusual profile of personality

Table 6.2 Validated predictions of correlations between adjectives and behaviors among chimpanzees

Adjective	Behavior	All	*M*	*I*
Active	Rate of play	0.45		
Aggressive	Rate of aggression	0.42		
Confident	Rate being groomed[a]		0.72	
	Rate of grooming[a]		0.66	
Curious	Rate of play	0.44		
Deferential	% Significant groomers[a]		–0.68	
Effective	% Significant groomers[a]		0.69	
Fearful	% Significant partners groomed[a]		–0.66	
Irritable	Rate of grooming	0.44		
Playful	Rate of play	0.79		
	Responsibility for play	0.43		
	% Significant play partners	0.64		
Popular	% Time spent alone	–0.50		
	Rate of play	0.48		
	% Significant play partners	0.44		
	Rate being groomed[a]		0.65	
Protective	Rate being groomed	0.41		
	Rate of grooming	0.41		
Slow	Rate of play	–0.54		
Sociable	Rate of play	0.53		
	Responsibility for play	0.41		
	% Significant play partners[a]			–0.67
Solitary	% Time spent alone	0.44		
	Rate of play	–0.63		
	% Significant play partners	–0.56		
	Responsibility as groomee	–0.41		
Strong	Rate of displays	0.48		

Correlations significant at $p<0.001$ ($n=59$)
[a] Prediction validated among males or immatures only

ratings on the EPI and subsequent deviant behavior, involving the adult, Passion, and her daughter, Pom, who infamously killed and consumed several infants from their own community. There is, however, a retrospective aspect to this study as the authors were prompted to reexamine the ratings in light of the horrific behavior, rather than having made predictions at the time of the ratings; thus, the EPI instrument's diagnostic and predictive power may be overstated. The only other nonhuman primate study relating EPI ratings to behaviors showed a link in baboons between the amount of grooming received and individuals rated as less *fearful* (emotion dimension: Protection), and more *sullen, jealous* (Deprivation), *defiant, belligerent* (Rejection) and *aggressive* (Destruction) (Buirski et al. 1973).

Using an adapted version of the Stevenson-Hinde et al. (1980a) rating scale, I found that adult chimpanzees with higher ratings on *protective* received more grooming, while, using yet another instrument, Vazire et al. (2007) found an inverse relationship between grooming received and *sullen* in chimpanzees. These complicated findings illustrate some of the difficulties in interpreting and attempting to compare findings across studies. Not only are the species different, but so are the rating scales and sometimes the meanings of the same word: in the EPI, Protection is used as a label for Fearfulness (and presumably the need or desire for protection) whereas it would mean more or less the opposite in the scale I used.

With personality traits often being behaviorally defined, critics may argue that semantic similarity leads to somewhat obvious links between traits and behaviors, and there is a potential for circularity:

Q: Why is Boris playful?
A: Because he has a high rating on the adjective *playful.*
Q: How did he get that rating?
A: I filled out a questionnaire that defined *playful* as "plays a lot."

This is an extreme example (and not based on an actual definition from an existing questionnaire) but this objection could not be leveled at all adjectives as they do not have such clear-cut semantic similarity (e.g., *sullen, apprehensive, tense*). Even the adjective *sociable* is complex as it could potentially relate to a multitude of predictable behaviors, much as an underlying dimension of personality does. Pederson et al. (2005) suggest that such narrow trait ratings may be useful as they are easier to obtain than behavioral observations; however, to make such ratings, one needs to spend extended periods of time becoming acquainted with each rated individual. Nevertheless, once completed, a set of personality scores is an excellent way of gaining a quick snapshot impression of an individual and a potentially useful tool for decision-makers in charge of animal husbandry and welfare.

One way that researchers have tried to overcome some of the above criticisms is to use factor scores rather than (or in addition to) trait ratings. Objections due to semantic similarity and circularity cannot be so easily made if it is many or several of these traits that contribute in precise and meaningful ways to underlying global dimensions of personality.

6.2.2 Personality Factor Level Relationships with Behavior

Table 6.1 shows that Extraversion is positively linked to affinitive or affiliative behaviors in chimpanzees (Pederson et al. 2005) and gorillas (Kuhar et al. 2006), while Sociability relates positively to play behavior in chimpanzees (Murray 1995), affiliative behavior in rhesus macaques (Capitanio 1999), and amiable yet submissive behavior in stump-tailed macaques (Mondragón-Ceballos and Santillán-Doherty 1994). Excitability appears to be related to more grooming behavior in both rhesus macaques (Capitanio 1999) and chimpanzees (Murray 1995) but the relationship with aggressive behavior is different between these species, with more Excitable chimpanzees having a higher aggression rate while rhesus macaques higher on Excitability initiate less aggression. In stump-tailed macaques, Excitable scores correlate positively with behavior described as threatening, rambling, and bizarre (Mondragón-Ceballos and Santillán-Doherty 1994).

Personality ratings of rhesus macaque mothers have also been found to correlate with ratings of their infants throughout the infants' first year of life; for example, infants of highly Confident and Sociable mothers initiate more contact with other group members (Stevenson-Hinde et al. 1980a).

Confident scores correlate with increased grooming in chimpanzee (Murray 1995) and rhesus macaque (Capitanio 1999) males, and have also been found to relate to more copulatory behavior in langurs (Konečná et al. 2008) and rhesus (Capitanio 1999) and stump-tailed (Mondragón-Ceballos and Santillán-Doherty 1994) macaques. Some of the more surprising findings for more Confident primates include a decrease in aggression in langurs (Konečná et al. 2008) and a higher latency to touch a banana in a strange situation test in rhesus macaques (Stevenson-Hinde et al. 1980b).

Pederson et al. (2005) reported that chimpanzees' scores on personality dimensions are associated with behavior in particular contexts; for example, Dominance and Emotionality scores relate positively to agonistic behaviors. Scores on the Dominance factor correlate negatively with predictable submissive behaviors in chimpanzees (Pederson et al. 2005) and with displacements received by gorillas (Kuhar et al. 2006), and submissive context behaviors in turn are shown to relate inversely to chimpanzees' scores on a Subjective Well-Being scale (King and Landau 2003). In addition, scores on a comparable Social Competence factor relate negatively to submissive behavior in vervet monkeys (McGuire et al. 1994). A suite of behaviors constituting a Dominance behavioral component inversely relate to the Agreeableness factor in Hanuman langurs (Konečná et al. 2008). Individuals higher on Agreeableness spend more time grooming and this holds true for these langurs, as well as for chimpanzees (Pederson et al. 2005).

6.3 Elucidation of Personality-Behavior Relationships at Two Levels in an African Ape Study

Previously, I have combined analyses of individual chimpanzees' ratings on adjectives with the effects of situational influences such as grouping and rearing (Murray 1998), and of individual characteristics such as age and sex (Murray 2002), to determine the impact of these factors on personality. In this section, I will illustrate how both discrete personality trait ratings and personality factor scores relate to observed behavior in a sample of 59 chimpanzees from three UK zoos: a social group of 22 at Chester Zoo, another social group of 11 at London Zoo, and 26 individuals housed in groups, pairs or trios at Twycross Zoo. They ranged in age from 19 months to 45 years and contained 39 females and 20 males (see Murray 1995, 1998, 2002 for further details).

Relationships between traits and behaviors will also be examined for smaller samples of gorillas ($n=13$) and bonobos ($n=4$). The gorillas were housed at London and Twycross Zoos, were aged between 20 months and 30 years, and included five males and eight females. The bonobos at Twycross Zoo included one male and three females, aged 3–12 years.

6.3.1 Methods

6.3.1.1 Personality Ratings

I adapted the rating scale used by Stevenson-Hinde with rhesus macaques (Stevenson-Hinde and Zunz 1978; Stevenson-Hinde et al. 1980a), by omitting the item *understanding* and including the item *intelligent*. *Motherly* was changed to *maternal/paternal* and *subordinate* was changed to *deferential*. Zookeepers and I completed the 28-item scale, full details of which have been published previously (Murray 1998, 2002).

6.3.1.2 Reliability

Interrater correlations (Pearson) were examined for each ape individually, and raters were regarded as reliable when their ratings correlated significantly with at least one of the other raters ($p<0.01$). In fact, 91% of the reliable raters were reliable at the $p<0.001$ level of significance. Overall, a high percentage of significant interrater correlations were found, with 80% for male chimpanzees and 81% for females. Multiple regression analysis also showed that 83% of the variance in rating was accounted for by individual ape variation, with only 0.2% due to raters, and 16.8% due to raters and adjectives; a strong indication that raters were assessing individual subjects on adjectives differently as well as discriminating effectively between adjectives.

In addition, I computed the reliabilities of the adjectives: significant coefficients ranged from 0.40 for *curious* to 0.84 for *deferential*, with a mean of 0.57. Three of the 28 adjectives – *permissive, predictable*, and *sensitive* – were subsequently omitted from further analysis as they failed to reach the required level of significance ($p<0.001$) when correlation analyses were performed on the ratings of two raters per item across subjects.

6.3.1.3 Correlation and Principal Components Analysis

I calculated mean scores per adjective per individual and Pearson correlation analysis was performed on the data from each of the three species to identify how adjectives related to each other and clustered together. In addition, principal components analysis (PCA) was conducted on the chimpanzee data. Both varimax rotation and the unrotated solution yielded similar results, but I decided to use the unrotated solution, following Stevenson-Hinde and Zunz's (1978) method. Although six factors were extracted in the chimpanzee PCA, together accounting for 77.5% of the variance in chimpanzee personality ratings (see Appendix 1 for the six components which satisfied eigenvalue and scree plot tests), I will here – as previously (Murray 1998, 2002) – focus on the first three: Confidence, Sociability and Excitability, and relationships between these and other adjectives and behaviors, particularly those concerned with sociability, playfulness, and aggression.

To permit current and future cross-component comparisons, I calculated component (factor) scores for each individual and Spearman correlation analyses were carried out to measure the relationships between both factor scores and personality ratings with the behavioral measures.

As significant sex and age differences in both personality and behavior variables were already established (Murray 1995, 2002), here I conduct within-sex and within-age correlational analyses followed by tests for significant differences between sexes and age groups using Fisher's r to z transformation. Apes up to 107 months are classed as "Immature," and those of 108 months or more are classed as "Adults." Sex and age affect personality and behavior independently, but they will only be discussed here where they have a direct influence on the relationships between personality traits and/or component scores and observed behavior.

6.3.1.4 Behavioral Observations

Over 1,000 h of focal samples of behavior were collected (Altmann 1974), during which I recorded all occurrences of the following behaviors, along with partner preferences, initiator, and terminator where applicable: time spent alone, time spent in association, proximity, grooming, aggression, displays, and playing (see Murray 1995 for further details). Indices of responsibility for proximity, grooming, and play were calculated using the Hinde and Atkinson formula (1970). To allow meaningful comparison between apes, I also summarized the association, proximity, play, and grooming data by calculating percentages of significant partners using a median split.

I made predictions about correlations between particular behavioral measures and adjectives and here present correlations that were validated and those that were not. Additional correlations between ratings and behaviors which had not been predicted are also presented. Some pairings of adjectives are obviously descriptive opposites, such as *active* with *slow*, *confident* and *effective* with *apprehensive/deferential/fearful/insecure, aggressive* with *gentle* and *equable*, and, to an extent, *sociable* with *solitary*. Therefore, strong negative correlations were predicted accordingly. Some of the predictions reflect commonsense expectations; for example, that individuals rated more highly on *aggressive* and *playful* would have higher rates of observed aggression and play. The existing literature informed other predictions. Thus, as grooming is cited as a measure of sociability (e.g., Schino 2001, 2007), it was predicted that individuals rated more highly on *sociable* would engage in more grooming activities. Similarly, predictions about grooming were made for those adjectives relating to status and confidence, such that individuals of higher status may both give and receive more grooming (or groom/be groomed by more partners) (Simpson 1973; Goodall 1986), the giving of grooming by more dominant individuals perhaps serving a social bonding or alliance building function (Watts 2000) or acting to reduce tension (Simpson 1973).

6.3.2 Personality Trait Ratings and Behavior in Chimpanzees

Tables 6.2–6.4 present correlation coefficients between personality ratings and observed behavioral measures among chimpanzees showing convergent and discriminant validity. It is evident that there is a group of adjectives that strongly relate to a behavioral cluster concerned with play. Broadly, those chimpanzees rated as more playful, in terms of such measures as rates of play, number of play partners and responsibility for initiating play, are more likely to be *active, curious, playful, popular* and *sociable*, and less likely to be the *slower*, more *solitary* individuals. *Solitary* chimpanzees do indeed spend more time alone, while those individuals with higher scores on *strong* also have higher display rates. Similarly, chimpanzees who are rated as more *aggressive* display higher rates of aggression. Grooming rates are positively linked to chimpanzees who are rated as more *protective* and *irritable*.

Among male chimpanzees, there are associations between adjectives relating to *confidence* and *effectiveness* and grooming measures, such that more *confident* males both give and receive more grooming, and those who are rated as more *effective* are groomed by more partners. *Deferential* individuals, on the other hand, are groomed by fewer individuals, while those who are more *fearful*, *insecure* and *apprehensive* groom fewer conspecifics themselves. This could support the reassuring function of grooming whereby the more partners a male chimpanzee grooms, the less *insecure* he is. Similarly, the more minutes per hour a male spends being groomed by another, the more likely he is to be rated as more *confident.*

Grooming measures are often taken as indicators of sociability (e.g., Seyfarth 1977; Goosen 1987; Schino 2001) but, when I included sets of measures related to grooming and to play, I found that more measures relating to play than to grooming correlated highly with ratings on the adjective *sociable.* The only trend between *sociable* ratings and grooming measures is that more *sociable* chimpanzees tend to initiate more grooming (Table 6.3), but they did not give or receive more grooming, nor did they solicit it, and they did not have more grooming partners. Although such assumption of responsibility for grooming could be regarded as an indicator of the role of grooming in social bonding, the relationships between playfulness and sociability are stronger. Apes rated highly on *sociable* play more, have more play partners and, importantly, take more responsibility for initiating play.

One must remember that this could be a result of how the raters decided to rate the animal on that adjective. Despite being given distinct behavioral definitions for each adjective, raters will possibly, to an extent, have their own definition of such words as "sociable" as well as an implicit personality theory regarding which adjectives "go together." Being sociable may be seen as being fun-loving, active and happy, for instance, which are evident characteristics of playfulness, and therefore raters may be rating apes largely on "playful sociability." Adding some weight to this possibility, it is noteworthy that these two adjectives load highly positively together on the second component identified for explaining variation in chimpanzee personalities. An alternative explanation could question the reliability of the behavioral

Table 6.3 Predicted correlations not validated among chimpanzees

Adjective	Behavior	Correlation
Aggressive	Rate of displays	0.25
Apprehensive	Rate being groomed	–0.11
	% Significant groomers	–0.18
Confident	% Significant groomers	0.16
	% Significant partners groomed	0.12
	Rate of displays	–0.05
	Rate of aggression	0.20
Deferential	Rate being groomed	–0.11
Effective	**Rate being groomed**	**0.36**
	Rate of grooming (M)[a]	**0.50**
	% Significant partners groomed	0.24
	Rate of displays	0.12
	Rate of aggression	**0.27**
Equable	Rate of aggression	–0.15
	Rate of displays	–0.20
Excitable	**Rate of displays**	**0.37**
Fearful	Rate being groomed	–0.09
	% Significant groomers	**–0.27**
Gentle	**Rate of aggression**	**–0.26**
	Rate of displays	**–0.28**
Insecure	Rate being groomed	0.01
	% Significant groomers	–0.19
Irritable	**Rate of aggression**	**0.27**
Maternal/paternal	**Rate of aggression**	**0.68**
	Responsibility as groomer (I)[a]	0.26
	Rate of grooming (I)[a]	0.47
	Rate being groomed (M)[a]	**0.64**
	Responsibility for proximity (F)[a]	**–0.42**
	Rate of play (I)[a]	**0.65**
	Responsibility for play (M)[a]	**–0.65**
Playful	**% Time spent alone**	**–0.36**
Popular	Median association time	–0.23
	Resp. for proximity (A)[a]	**–0.42**
Protective	Rate of aggression	0.13
	% Significant groomers	0.21
Slow	Rate of grooming	0.25
	Rate being groomed	0.22
	Rate of displays	–0.11
Sociable	Median association time	–0.23
	% Time spent alone	–0.21
	Responsibility for proximity	0.13
	Responsibility as groomer	**0.37**
	Rate of grooming	0.22
	Rate being groomed	0.10
Solitary	Responsibility for proximity	0.01
	Responsibility for play	**–0.38**

Correlations not significant $p>0.001$ ($n=59$); *bold* denotes $p<0.05$

[a]Coefficient given for specific age/sex class, which was higher than for all chimpanzees together

Table 6.4 Other significant correlations between ratings and behaviors among chimpanzees not previously predicted

Adjective	Behavior	All	*M*	*A*	*I*
Active	% Significant play partners	0.41			
Aggressive	Responsibility for play	0.48			
	% Time spent alone[a]				0.63
Apprehensive	% Significant partners groomed[a]		−0.68		
Curious	% Significant play partners	0.46			
Excitable	% Time spent alone[a]				0.69
Gentle	Rate of play[a]				0.79
Insecure	% Significant partners groomed[a]		−0.70		
Intelligent	Rate of grooming[a]				0.69
Irritable	Median association time	−0.41			
	Responsibility for proximity[a]				0.64
Playful	Rate of displays[a]		−0.70		
Protective	Responsibility for play[a]			−0.70	
Slow	% Significant play partners	−0.53			
Solitary	Rate of aggression[a]				0.75
	Rate of displays[a]				0.68
Strong	% Time spent alone[a]				0.67
	% Significant play partners	−0.46			
	Rate of play	−0.41			
Tense	% Time spent alone[a]				0.79

Correlations significant at $p<0.001$ ($n=59$)
[a]Correlation significant among males, adults or immatures only

codings, although play and grooming are arguably two of the easiest behaviors to record, requiring little interpretation.

Table 6.3 shows correlations between adjectives and behaviors that had been predicted, but which were not validated by reaching the required level of significance ($p<0.001$). Only two of the relationship trends between ratings and behaviors, however, are correlated in a direction opposite to that predicted. In Table 6.4, significant correlations between ratings and behaviors for chimpanzees are shown which had not been predicted previously. Among immature chimpanzees, youngsters who spend more time alone are *stronger*, and more *excitable*, *aggressive* and *tense*, and those rated higher on *solitary* have higher rates of aggression and displays. Interestingly, more *intelligent* immature chimpanzees have higher rates of grooming, possibly emphasizing the ingratiating function of this activity (Goodall 1986).

6.3.3 Personality Trait Ratings and Behavior in Gorillas

Table 6.5 presents significant correlations between ratings and behaviors among gorillas, which are fewer than are present for chimpanzees. There is a strong correlation between being *tense* and a higher rate of aggression, and female gorillas scoring

Table 6.5 Significant correlations between ratings and behaviors among gorillas

Adjective	Behavior	All	*F*	*I*
Active	Rate of displays		0.96	
Confident	% Significant groomers[a]			1.00
Deferential	% Significant groomers[a]			−1.00
Effective	% Significant groomers[a]			1.00
Irritable	% Significant groomers[a]			1.00
Slow	Rate being groomed			−1.00
Sociable	Rate of play			−1.00
Solitary	Rate being groomed			−1.00
Tense	Rate of aggression	0.76		

Correlations significant at $p<0.001$ ($n=13$); *F* females; *I* immatures
[a]Correlations previously predicted

Table 6.6 Predicted correlations between ratings and behaviors among gorillas not significant

Adjective	Behavior	Correlation
Aggressive	**Rate of aggression**	**0.69**
Playful	**Rate of play**	**0.67**
Sociable	% Time spent alone	−0.24
Solitary	% Time spent alone	0.25

Correlations not significant $p>0.001$ ($n=13$); *bold* denotes $p<0.01$

higher on *active* have higher rates of displays. Confident immature gorillas are groomed by more partners, while their rate of play is inversely correlated with ratings on *sociable*; a surprising finding, and in marked contrast to the positive relationship existing among chimpanzees. Play behavior, like grooming, was relatively rare among gorillas and much of their play was of a more solitary nature, involving running and climbing or playing with objects in their enclosure.

Although not significant, the relationships among gorillas between ratings on *aggressive* and *playful* and their respective observed rates of behavior are nonetheless relatively strong (Table 6.6). Ratings of these apes on *sociable* and *solitary*, however, have no direct relationship with the amount of time spent alone by individuals.

6.3.4 Personality Trait Ratings and Behavior in Bonobos

Several significant correlations ($p<0.001$) between ratings and behaviors were identified among the small sample of four bonobos. Whereas in chimpanzees, *curiosity* correlates positively with rate of play, among bonobos, it correlates with rate of grooming. The status-related adjectives are, among bonobos, related to partners groomed by, rather than groomers of, the individual (as was the case for male chimpanzees). Thus, *deferential* bonobos groom fewer conspecifics, while *effective* and *popular* individuals groom more of their group members. While ratings on *solitary* do not correlate significantly with the amount of time spent alone, as predicted, there is the expected negative correlation between this behavioral measure and *sociable*. Bonobos who spend more time alone are also likely to be more *insecure* and *fearful*. While the predicted correlation between *aggressive* and rate of aggression was not significant, there was the expected correlation between ratings on *playful* and actual rates of play, and more *active*, *excitable, fearful* and *insecure* individuals initiate less play, while more *playful* and *sociable* bonobos initiate more play. A positive correlation also exists between ratings on *slow* and rate of aggression.

6.3.5 Summary of Trait–Behavior Correlations

From the above, it is evident that certain adjectives correlate with behavioral measures which one could anticipate, and which therefore confirms the validity of these ratings. These include *aggressive, playful*, and *solitary.* For the chimpanzees, Spearman correlation coefficients ranged between 0.41 and 0.79 ($p<0.001$), which is very similar to the range of trait–behavior correlation coefficients obtained in an earlier macaque study: 0.45–0.73 ($p<0.001$) (Stevenson-Hinde et al. 1980a).

1. *Active* chimpanzees have higher rates of play and more play partners, while active female gorillas have higher display rates. Active bonobos initiate less play.
2. *Aggressive* chimpanzees have higher rates of aggression and initiate more play. Immature chimpanzees with higher scores on aggressive spend more time alone. The positive correlation between rated and observed aggression is strong but not significant for gorillas and bonobos.
3. *Apprehensive* male chimpanzees groom fewer partners.
4. *Confident* male chimpanzees give and receive more grooming, and confident immature gorillas are groomed by more partners.
5. *Curious* chimpanzees have higher rates of play and more play partners, while curious bonobos engage in more grooming.
6. *Deferential* male chimpanzees and immature gorillas are groomed by fewer others, while deferential bonobos groom fewer partners themselves.
7. *Effective* male chimpanzees and immature gorillas are groomed by more partners, while effective bonobos groom more partners.
8. *Excitable* immature chimpanzees spend more time alone, while excitable bonobos initiate less play.
9. *Fearful* male chimpanzees groom fewer partners, while more fearful bonobos spend more time alone.
10. *Gentle* immature chimpanzees have higher rates of play.
11. *Insecure* male chimpanzees groom fewer partners, while insecure bonobos spend more time alone.

12. *Intelligent* immature chimpanzees engage in more grooming, perhaps as a strategy of ingratiation.
13. *Irritable* chimpanzees spend less time in association with each partner but have higher grooming rates. Irritable immature chimpanzees take more responsibility for proximity to others while irritable immature gorillas receive grooming from more partners.
14. *Playful* chimpanzees have higher rates of play, more play partners, and initiate more play. In addition, playful males have lower display rates. The positive correlation between rated and observed play is also significant for bonobos (who also initiate more play) and is relatively strong for gorillas too.
15. *Popular* chimpanzees spend less time alone and more time playing with more play partners, while popular males receive more grooming. Popular bonobos groom a greater number of partners.
16. *Protective* chimpanzees give and receive more grooming, with the more protective adults initiating less play.
17. *Slow* chimpanzees play less and have fewer play partners, while slower immature gorillas receive less grooming, and slow bonobos have higher rates of aggression.
18. *Sociable* chimpanzees play more and initiate more play, with more sociable immature chimpanzees having more play partners. Sociable immature gorillas, by contrast, play less, while sociable bonobos spend less time alone and initiate more play.
19. *Solitary* chimpanzees spend more time alone, play less, have fewer play partners, and solicit less grooming. Immature chimpanzees who are more solitary have higher rates of aggression and displays, while immature gorillas rated as more solitary receive less grooming. The positive correlation between ratings on Solitary and actually spending more time alone is strong for bonobos, but weak for gorillas.
20. *Strong* chimpanzees display more but play less and have fewer play partners. Immature chimpanzees rated as stronger spend more time alone.
21. *Tense* immature chimpanzees spend more time alone, while gorillas rated as more tense have higher rates of aggression.

6.3.6 Personality Factor Scores and Behavior in Chimpanzees

The principal components, or personality dimensions, through which we can better understand chimpanzees are similar to those of Confident to Fearful, Active to Slow and Sociable to Solitary found in rhesus macaques (Stevenson-Hinde and Zunz 1978), and from which Confident, Excitable and Sociable scores were also calculated (Stevenson-Hinde et al. 1980a), and to the Sociability, Confidence, Excitability and Equability components identified in Capitanio's (1999) study with the same species. Some similarities also exist to components labeled Sociable, Dominance, and Anxiety found in stump-tailed macaques (Mondragon-Ceballos et al. 1991), and to those of Socially Competent, Playful/Curious, and Opportunistic found in vervet monkeys (McGuire et al. 1994). They also bear marked resemblances to the behavioral factors identified for chimpanzees (van Hooff 1970) and rhesus macaques (Chamove et al. 1972). Separate Affinitive and Play motivational systems were identified for chimpanzees, as opposed to the all-encompassing Affectionate/Affiliative/Sociable/Play factor found among rhesus monkeys, but these can all be likened to the Sociability component (C2) in this study. The Fearful and Submissive factors identified for rhesus macaques and chimpanzees, respectively, can also be likened to the Confidence component (C1), and this may equate to the Neuroticism/Stability dimension of human personality. The Hostile/Aggressive factor found in rhesus is more akin to the Excitability component (C3), considered by Chamove et al. (1972) as Psychoticism, but a distinction is made by van Hooff between Aggressive and Excitement for chimpanzees in terms of behavioral counts.

Here, I present the following list of behavioral measures and how they correlate with scores on the components concerned with Confidence (C1), Sociability (C2), and Excitability (C3).

1. *Median association time* – Chimpanzees with higher scores on Excitability (C3) have lower median association times.
2. *% Time spent alone* – The scores of chimpanzees on Sociability (C2) are inversely correlated with the amount of time they spend alone. Among immature chimpanzees, high Excitability (C3) scores are positively associated with greater amounts of time spent alone.
3. *% Significant partners* – Adult chimpanzees with higher scores on Excitability (C3) associate with fewer significant partners.
4. *Responsibility for proximity* – Immature chimpanzees with higher Confidence (C1) scores take more responsibility for proximity.
5. *Rate of grooming* – Chimpanzees with higher scores on Excitability (C3) groom more. Male chimpanzees with high scores on Confidence (C1) have higher rates of grooming.
6. *Rate of being groomed* – Male chimpanzees with higher scores on Confidence (C1) receive more grooming. Immature chimpanzees with high Sociability (C2) scores are groomed at higher rates, while adults higher on Excitability (C3) receive more grooming.
7. *% Significant partners groomed* – Male chimpanzees who score highly on Confidence (C1) groom a greater number of partners.
8. *% Significant groomers* – Male chimpanzees with higher scores on Confidence (C1) also are groomed by more groomers. It is immature chimpanzees with higher scores on Sociability (C2), however, that receive grooming from a greater number of groomers.
9. *Responsibility as groomer* – Male chimpanzees with higher Sociability (C2) scores take more responsibility for initiating grooming.
10. *Responsibility as groomee* – Chimpanzees with higher Sociability (C2) scores solicit more grooming.
11. *Rate of play* – Higher rates of play are evident in chimpanzees who have higher Sociability (C2) scores. Adult chimpanzees with higher Excitability (C3) scores also have higher rates of play.

12. *% Significant play partners* – Chimpanzees with higher scores on Sociability (C2) have more play partners.
13. *Responsibility for play* – No significant findings.
14. *Rate of aggression* – Chimpanzees with higher scores on Excitability (C3) have higher rates of aggression. Immature chimpanzees with higher scores on Sociability (C2), however, have lower rates of aggression.
15. *Rate of displays* – Male chimpanzees with higher Sociability (C2) scores have lower display rates, whereas immature chimpanzees with higher scores on Excitability (C3) display at a higher rate.

These correlation coefficients are comparable to the range of partial correlation coefficients between personality dimensions and a variety of behavioral situations reported by Capitanio (1999), and are relatively higher than the personality dimension and behavior correlations reported for gorillas (Kuhar et al. 2006) and both the dimension-behavioral context and dimension-specific behavior correlations demonstrated by Pederson et al. (2005) for chimpanzees.

6.3.7 Multi-Level Relationships Between Personality and Behavior in Chimpanzees

That measurable individual personalities in captive chimpanzees exist has support from concordance among raters in the quantitative assessment of adjective ratings and from validation of these ratings with observed behavioral measures. This is inconsistent with accusations of ratings being subjective, anthropomorphic, and/or based on the raters' implicit personality theory (see Borkenau 1992 for a review of the controversy). Instead, our descriptions of individuals in their social context are enhanced at more than one level – from examining discrete trait rating patterns to utilizing a more global score on personality dimensions (for a classification of individuals according to personality types based on the pattern of scores representing levels of Confidence, Sociability and Excitability, see Murray 2002, 2005).

Such multi-level assessment of individuals provides information about personality to satisfy differing requirements. Discussing the rationale behind the development of the facet scales within each of the five factors in the NEO-PI-R (Costa and McCrae 1992), the authors state that, "Interpretation on the domain level yields a rapid understanding of the individual; [while] interpretation of specific facet scales gives a more detailed assessment." (Costa and McCrae 1995, p. 21). Nonhuman primate personality research would benefit from more consistency in methodology and instruments. Further development of chimpanzee personality instruments should include sufficient detail to provide scoring of facet-level equivalence.

At one level, each adjective used in the present study could be regarded as a trait. From the ratings on the adjectives, therefore, the highest five or seven trait scores could be used to describe each individual (these numbers suggested as providing definitive pictures of a human individual, e.g., Digman 1990). Thus, to contrast two adult male chimpanzees, for example, we could describe Boris as *effective, popular, confident, curious, strong, paternal* and *intelligent*, whereas we would see that William is *deferential, fearful, gentle, apprehensive, strong, eccentric and excitable.* Beyond this level, however, it is evident that certain adjectives are highly intercorrelated and form trait clusters. Among chimpanzees, there are clusters relating to (1) confidence and status, (2) sociability and activity, (3) intelligence and opportunism, and (4) aggression and excitability (Murray 1995, 2002). Cluster (1) is evident for gorillas and bonobos as well, while cluster (2) is represented among bonobos as sociability and among gorillas as activity. Cluster (4) is very similar to the adjectives relating to aggression in gorillas and to protection and aggression in bonobos, while cluster (3) is matched in gorillas, but is not present among bonobos (Murray 1995). Of course, these clusters give a strong indication of likely underlying dimensions and, because the clusters are comparable across species, it is possible to calculate scores for the gorillas and bonobos based on standardized ratings of the individuals on selected adjectives from each of the components identified for the chimpanzees (Murray, in prep.).

The meaning of factors, or dimensions of personality, can be interpreted in different ways. Thus, the chimpanzees' factor scores taken as representing levels of Confidence, Sociability and Excitability may be regarded as central (Allport 1937) or source (Cattell 1946) traits of personality, which can characterize an individual.

A criticism of placing too much emphasis on factor scores is that specificity is sacrificed for the sake of generality (Buss 1989; John 1989; McAdams 1992). However, here I offer an elucidation of the relationships between overt behaviors and both overall personality factors and specific trait adjectives. Examination of how behaviors link to factor scores offers a quicker and more general "snapshot" picture while the relationships of behaviors to specific adjective ratings likely have greater power for providing richer detail because they are more discrete measurements. Therefore, Figs. 6.1–6.3 are diagrammatic representations of the predictive relationships between adjectives and behaviors, according to the underlying factors of personality. They provide a readily understood characterization of the chimpanzee personalities as they show relationships between the adjectives (narrower trait terms) and behaviors, within the overall dimensions of personality, rather than just between factor scores and behaviors.

Because all adjectives which had factor loadings above 0.317 are shown, some appear on more than one component; however, they are all presented so that each component and its relationships with the behavioral measures may be considered independently.

6.3.7.1 C1: Confident/Apprehensive

Figure 6.1 shows a clear delineation between adjectives loading highly positively on C1 and a subset of behaviors that correlate with those adjectives, and between adjectives loading highly negatively on this component and another distinct group of behaviors correlating with them. The predictive value of these relationships between

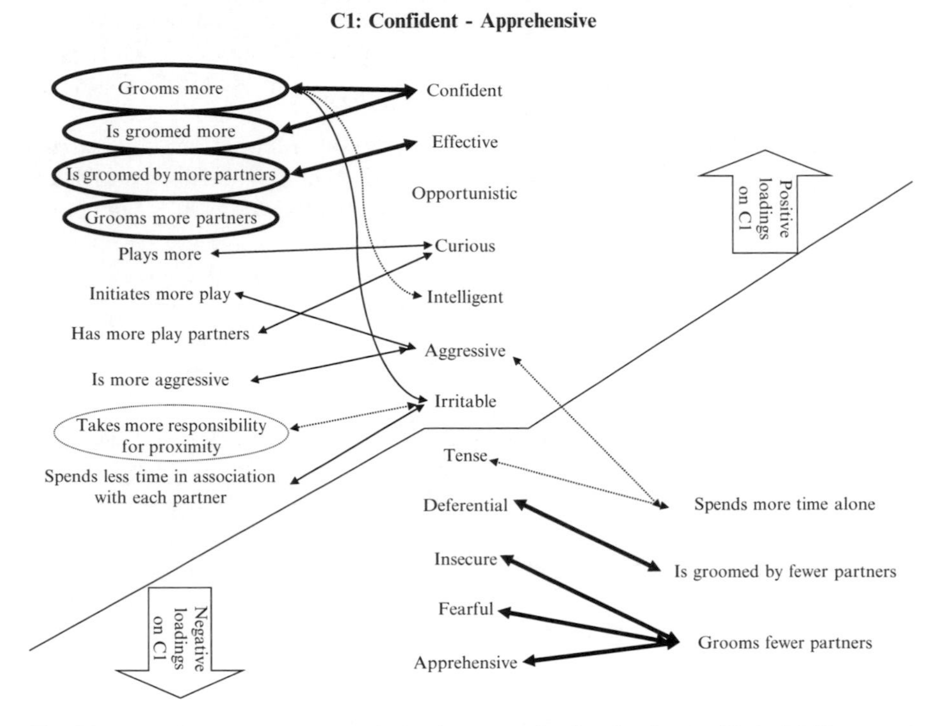

Fig. 6.1 Predictive relationships ($p<0.001$) between adjectives loading on C1 (at 0.317 or more) and behaviors in chimpanzees. *Normal lines* = all chimpanzees; *bold* = relationships present only in males; *dotted* = relationships present only in immatures; *ovals* = behavior correlates with high scores on C1 ($p<0.01$) in males (*bold*) or immatures (*dotted*) only

adjectives and behaviors is excellent as, with the exception of one relationship among immature chimpanzees, there is no overlap, i.e., none of the adjectives loading positively are positively related to what could be termed the "negative" behaviors (e.g., spending more time alone and/or not engaged in social behaviors). From Fig. 6.1, a high/low distinction can be made, whereby high ratings on the adjectives above the line, and/or high scores on C1 (which reflect high ratings on the positively loading adjectives and low ratings on the negatively loading adjectives) are related to more grooming, play and aggression, while high ratings on the adjectives below the line are related to less grooming and spending more time alone.

While such predictive power is encouraging, the relationships illustrated may not necessarily always hold. Caution should be exercised particularly when generalizing to other groups of individuals because, not only will the age or sex ratio of other groups vary, but so may other circumstances such as grouping, rearing history, and the experiences of the individuals. However, these diagrams also highlight which relationships are differentially found within a specific sex or age class. The relationships between increased grooming and the traits *confidence* and *effectiveness*, for example, are only evident among male chimpanzees. This is important because it enables prediction among a specific group of individuals. Similarly, the finding that, among immature chimpanzees, more *tense* individuals spend more time alone, could be useful for animal caretakers when taken as an indication of tension within the group.

A relationship between grooming frequency and dominance rank in primates including chimpanzees and baboons has frequently been identified (Sparks 1969; Buirski et al. 1973; Simpson 1973; Seyfarth 1980; Schino 2001, 2007), with dominant animals grooming others less frequently, but receiving more grooming and deference themselves. However, correlations between grooming measures and Component 1, relating to confidence and effectiveness, are not entirely in accordance with this generalization. Specifically, male chimpanzees with high C1 scores groom more partners and are groomed by more partners, but they have higher rates of grooming. Vervet monkeys scoring highly on a status-linked factor termed Social Competence also engage in more grooming (McGuire et al. 1994). Male chimpanzees rated highly on *effective* are groomed by more partners, while more *confident* males receive more grooming.

The finding that immature chimpanzees rated as more *intelligent* engage in more grooming than their peers can be interpreted as supporting Goodall's (1986) view that lower status animals groom those with a higher rank than their own, using grooming as a strategy of manipulation and ingratiation. It can also be used as a method of appeasement, perhaps by an immature chimpanzee who has been punished by the alpha male for being too rough in play. It may also reflect an interchange relationship (Seyfarth and Cheney 1984; Cheney et al. 1986; Hemelrijk and Ek 1991), in which grooming is exchanged for coalitionary support in antagonistic encounters. These explanations in which grooming and support during aggression can be seen as reciprocally altruistic (Trivers 1971; Seyfarth and Cheney 1984; Hemelrijk 1994; Schino 2007) also attribute to individuals, including those that are young, the capability for future planning, in that they plan to develop affiliative relationships to obtain, maintain, or increase their social status.

6.3.7.2 C2: Sociable/Solitary

A clear high/low distinction is also evident between adjectives loading highly positively on C2 and an associated group of behaviors, and adjectives loading highly negatively on this component and another group of behaviors (Fig. 6.2). Individuals with high ratings on the adjectives above the line, or high scores on C2, are more likely to play more with more partners, solicit more grooming and spend less time alone. Those rated more highly on adjectives below the line, conversely, play less, spend more time alone, are aggressive, and display more. They also engage in more grooming. Males who are more playful and/or who have higher C2 scores are less likely to display frequently, while immatures scoring highly on C2 are less aggressive and receive more grooming from more partners.

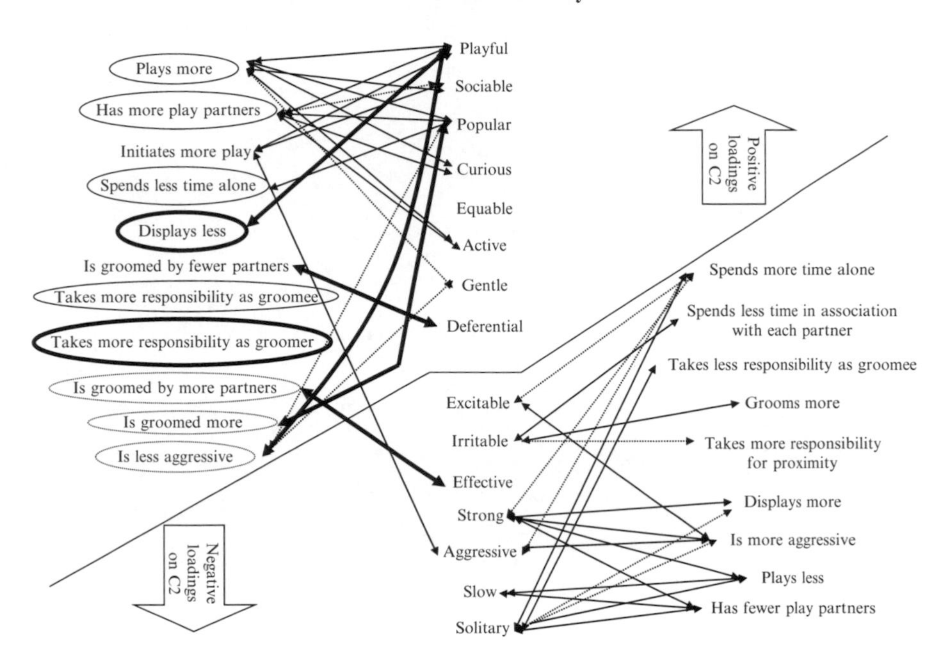

Fig. 6.2 Predictive relationships ($p<0.001$) between adjectives loading on C2 (at 0.317 or more) and behaviors in chimpanzees. *Normal lines* = all chimpanzees; *bold* = relationships present only in males; *dotted* = relationships present only in immatures; *ovals* = behavior correlates with high scores on C2 ($p<0.01$) in all chimpanzees (*normal lines*), in males (*bold*) or in immatures (*dotted*)

The findings that chimpanzees scoring highly on C2 spend less time alone and play more with more partners are in accordance with previous findings that vervet monkeys' scores on a Playful/Curious factor were positively related to play and negatively related to being solitary (McGuire et al. 1994). They also support the labeling of this C2 component as Sociable/Playful to Solitary (Murray 1995), since it has been argued that, rather than using factor loadings, factors should be identified from their correlations with external criteria (Kline 1992, 1993). In revisiting my earlier finding that more measures relating to play rather than to grooming correlate with sociability (Murray 2002, 2005), I categorized the chimpanzees as "unsociable," "sociable" or "very sociable" according to their scores on C2, and I also classified them in terms of levels of grooming and play behavior using sample median splits. I found that all unsociable chimpanzees are nonplayful, but nevertheless groom at medium and high rates; and most very sociable individuals are low groomers but classed as either playful or very playful. Binomial tests revealed no significant relationship between sociability categories and grooming, but nonplayful apes are more likely to be unsociable ($p<0.001$) and playful apes significantly more likely to be very sociable ($p<0.05$). Perhaps the best indicator for generalizing about how sociable individuals are therefore is their C2 score.

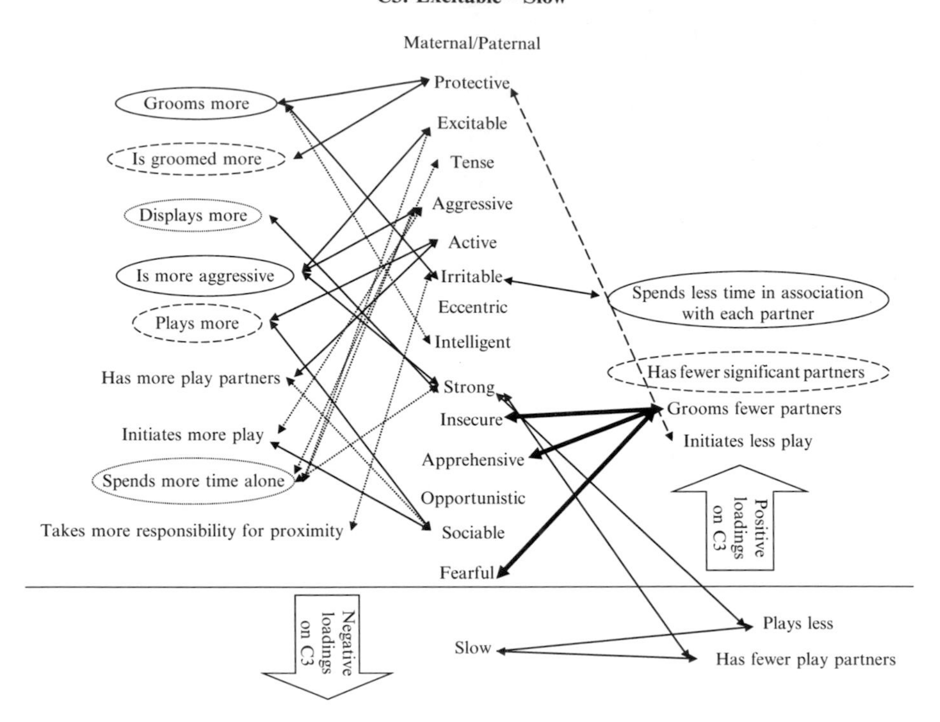

Fig. 6.3 Predictive relationships ($p<0.001$) between adjectives loading on C3 (at 0.317 or more) and behaviors in chimpanzees. *Normal lines* = all chimpanzees; *bold* = relationships present only in males; *dotted* = relationships present only in immatures; *dashed* = relationships present only in adults; *ovals* = behavior correlates with high scores on C3 ($p<0.01$) in males (*bold*) or immatures (*dotted*) only

6.3.7.3 C3: Excitable/Slow

Figure 6.3 shows that the "high/low" distinction on the third component is less clear-cut. All the adjectives above the line have high positive loadings on C3, while only one – *slow* – has a high negative loading. Chimpanzees with high ratings on *slow* play less. However, making predictions about the behavior of chimpanzees with high ratings on the other adjectives, or higher scores on C3, is more problematic. The behaviors relating to these adjectives can broadly be encompassed within the realms of more play, more grooming, and more aggression. Chimpanzees with high C3 scores groom more, but spend less time in association with each partner. Adults high on C3 play more and receive more grooming, although they have fewer partners overall. Immature chimpanzees scoring highly on C3 tend to spend more time alone and display more.

The trait *aggressive* loads most highly on C3, and individuals with high scores on this component also have higher rates of aggression. Ratings on the adjective *aggressive* are positively correlated with actual rates of aggression, and chimpanzees who display more are rated as *stronger*. Immature chimpanzees rated as more *aggressive* take more responsibility for initiating play, while immature chimpanzees rated as more *solitary* have higher rates of aggression and displays. Overt aggression is relatively rare in many groups of primates, e.g., olive baboons (*Papio anubis:* Buirski et al. 1973), although rivalry among male chimpanzees in captivity has resulted in the death of more than one male in multi-male groups (Wolfe, quoted in Goodall 1986, p. 419; de Waal 1986). In my study, such extreme aggression was not witnessed, and display rates were generally higher than rates of aggression. Nor were there many obvious cases of coalitions or alliances between individuals. Long-term stability in a group may reduce competition between males for females because the extended exposure to the same females by captive males may also reduce their sexual interest (Lambeth et al. 1990). The difference, therefore, between the Chester and Arnhem multi-male groups (see above: de Waal 1986) is hard to explain, but could be related to the stability of dominance relationships, which may act to constrain sexual and aggressive behavior. It could be that our increasing understanding of the relationships between personality and behavior will help us with just these sorts of questions, and can have implications for the welfare and well-being of animals.

6.4 Conclusions

The thrust of traditional trait theories (Allport 1937; Cattell 1973) is that traits determine behavior, with the reciprocal influence of behavior and learning back to traits being minimal. According to this directionality, *confident* male chimpanzees, for example, will groom more than those who are less confident; but the fact that an individual male grooms a great deal does not necessarily mean that he is more confident than others. This should be kept in mind when interpreting the information contained in Figs. 6.1–6.3, although their utility is likely to be in the direction of predicting from trait to behavior.

Measuring personality and being cognizant of how it can affect behavior has important applications for animal welfare. Certain personality profiles may be more suited for particular roles, for example, adopting an orphan (van Wulfften Palthe and van Hooff 1975; Holzer Blersch and Schmidt 1992), whereas others may warn of potential friction within a group by identifying potential negative reactions to the proposed introduction of a newcomer. Watts (1994), for example, found that female gorillas who were long-term residents within a stable group were relatively intolerant of immigrants, and that this intolerance persisted even after 2–3 years. The Confident, Sociable and Excitable factor scores of individuals, their personality types, or mean ratings on particular adjectives, can be used for selection of individuals for different purposes. To select the best first candidate to introduce to a hand-reared infant about to be integrated into an established group, for example, one might look for an individual with a relatively high Sociable score, but perhaps a mid-range score on Confident and Excitable, and with high ratings on the adjectives *maternal/paternal, equable*, and *gentle*. Selection according to personality profiles has also been suggested for managers of the Gorilla Species Survival Plan (Gold and Maple 1994). Desiring a "good candidate" for a bachelor group, for instance, Gold and Maple recommend looking for a male with a relatively high Extraverted score and a comparatively low Dominant score. Kuhar et al. (2006) also report that in the two known cases thus far of a juvenile male gorilla being introduced to a silverback for the purposes of forming a multi-male group, those silverbacks scored almost two standard deviations above the mean on the factor Understanding, which Kuhar et al. have identified as the most relevant (thus far) in giving an indication of individuals less likely to engage in noncontact aggression. Some of the links found here between personality and behavior might also be useful for captive management and conservation; for example, more tense immature chimpanzees spend more time alone and therefore groups in which chimpanzees spend more time alone may indicate group tension.

Many areas of psychological life are encompassed by the term *personality* in addition to interactions in the social arena. Varied aspects of the individual, such as emotion, motivation, attitudes, cognition, learning and arguably intelligence are incorporated in a description of personality (Cervone and Pervin 2008). We will probably always struggle to know for certain what an animal subjectively or consciously experiences just by observation and psychological researchers may wish for some way to ask subjects about themselves. Perhaps the language apes may yet provide another dimension to our information on such subjective topics. It is possible, however, that the taxonomy provided by models such as the FFM may be inadequate for personality description by these apes themselves, and their vocabulary and context usage will obviously be influenced by the humans who teach them.

An alternative methodology for studying personality in nonhumans is not likely to readily become available. The trait approach offers us many insights into understanding closely related species, but caveats should always be borne in mind. One of the most extreme criticisms of factor models of personality argues that they offer little more than an envisaged very broad description of how human beings might be described to aliens (McAdams 1992). It is possible that critics would extend this view to the utilization of factor scores from similar instruments used to measure personality in nonhumans. Ratings are similarly criticized as limited by the omission of any context, such that the whole exercise deteriorates to the point where the aim is merely to "…get a general, superficial, and virtually nonconditional picture of … personality." (McAdams 1992, p. 350). However, the rating scale method, as applied to the apes in this study, can provide both a description of the overall pattern of an individual's behavior in terms of factor scores (in contrast to a possible myriad of pure measurements of more discrete events, which might not enable a clear "picture" of individuality), yet also yield more detailed information at the adjective trait level. Furthermore, although the factors that emerged as quantifiable for these apes are similar to those found for other ape and monkey species, the range of variation or strength of manifestation of those personality domains may not be, and the traits are not necessarily semantically identical across species.

How individuals resemble and differ from each other in terms of their personalities and overt behaviors are some of the most interesting problems that can be tackled. In this chapter, I have reviewed studies providing validation of personality trait ratings and factors. A picture is steadily building to inform us about links between the behaviors we observe in primates and their "inner world" of individuality. In addition to relationships between discrete trait ratings and overt behaviors, I have demonstrated links between more global scores on major underlying personality factors and behaviors, and I have shown that some behaviors are differentially related in males/females and immatures/adults to personality traits and factors.

From the phylogenetic perspective, the net should be cast wider now to include New World monkeys and to consolidate and broaden findings on the Old World monkeys and apes, especially in relation to personality–behavior relationships on multi-faceted levels. This may yield important information on the evolution of personality structure and its relation to behavior. We also need studies in the wild to confirm the constancy and species-specificity of personality and its links to behavior. While there is growing support for the generality of chimpanzee personality across different captive settings, including zoos and a sanctuary (King et al. 2005), it is possible that the relationships between personality variables and behavioral manifestations may yet be quite different given such diverse situational and experiential contexts. King and Weiss (2011) point out that personality data from the wild is virtually nonexistent, despite many long-term behavioral studies which could easily incorporate it as a variable. One would suspect the case to be very different with captive studies; yet the difference is in fact not so profound, with a relative paucity of studies addressing how personality and behavior relate. I echo the call of others to include personality measurement alongside other ongoing research and to standardize the methodology for greater comparability – and more intriguing insights into primate personality.

Acknowledgments This research was supported by SERC funding while the author was at Cambridge University, was approved by each of the zoos and by the SERC, and adhered to all UK legal requirements. I would like to thank Chester NEZS, London ZSL and Twycross Zoos for access to the ape collections and I am indebted to the keepers for completing rating scales. My thanks also to Colleen Schaffner for comments on an earlier draft of this chapter.

Appendix 1 Chimpanzee personality factors derived via principal-components analysis

C1	23.70%	C2	18.40%	C3	15.40%	C4	9.30%	C5	5.70%	C6	5.10%
Confident	0.924	Playful	0.829	Mat./Paternal	0.689	Gentle	0.822	Intelligent	0.637	Solitary	0.392
Effective	0.730	Sociable	0.737	Protective	0.604	Equable	0.645	Opportunistic	0.412	Strong	0.327
Opportunistic	0.588	Popular	0.692	Excitable	0.598	Slow	0.554	Slow	0.296	Eccentric	0.325
Curious	0.521	Curious	0.535	Tense	0.567	Protective	0.484	Curious	0.224	Curious	0.293
Intelligent	0.470	Equable	0.421	Aggressive	0.516	Mat./Paternal	0.462	Deferential	0.158	Equable	0.259
Aggressive	0.440	Active	0.374	Active	0.495	Popular	0.214	Mat./Paternal	0.157	Deferential	0.228
Irritable	0.337	Gentle	0.355	Irritable	0.486	Effective	0.184	Apprehensive	0.138	Active	0.210
Popular	0.292	Deferential	0.349	Eccentric	0.470	Solitary	0.168	Protective	0.109	Opportunistic	0.166
Active	0.279	Fearful	0.264	Intelligent	0.405	Eccentric	0.154	Tense	0.102	Intelligent	0.151
Sociable	0.237	Mat./Paternal	0.139	Strong	0.384	Deferential	0.119	Fearful	0.067	Tense	0.142
Strong	0.233	Insecure	0.094	Insecure	0.364	Strong	0.102	Irritable	0.062	Apprehensive	0.129
Protective	0.215	Opportunistic	0.082	Apprehensive	0.363	Confident	0.044	Insecure	0.022	Effective	0.073
Mat./Paternal	0.196	Intelligent	−0.010	Opportunistic	0.353	Intelligent	0.043	Sociable	0.020	Playful	0.023
Equable	0.194	Confident	−0.017	Sociable	0.351	Apprehensive	0.037	Solitary	0.000	Protective	0.021
Playful	0.147	Apprehensive	−0.041	Fearful	0.335	Excitable	0.003	Aggressive	−0.073	Confident	−0.062
Slow	0.067	Eccentric	−0.146	Playful	0.276	Curious	−0.010	Confident	−0.096	Aggressive	−0.066
Solitary	−0.202	Protective	−0.172	Deferential	0.240	Insecure	−0.023	Playful	−0.098	Sociable	−0.095
Excitable	−0.223	Tense	−0.224	Effective	0.233	Sociable	−0.029	Equable	−0.124	Insecure	−0.119
Gentle	−0.225	Excitable	−0.358	Curious	0.212	Fearful	−0.052	Gentle	−0.154	Gentle	−0.124
Eccentric	−0.228	Irritable	−0.366	Popular	0.161	Tense	−0.100	Effective	−0.209	Popular	−0.134
Tense	−0.633	Effective	−0.458	Solitary	0.073	Irritable	−0.103	Excitable	−0.276	Slow	−0.157
Deferential	−0.721	Strong	−0.487	Gentle	0.021	Playful	−0.156	Active	−0.288	Fearful	−0.173
Insecure	−0.752	Aggressive	−0.504	Confident	−0.046	Opportunistic	−0.165	Popular	−0.296	Mat./Paternal	−0.229
Fearful	−0.759	Slow	−0.506	Equable	−0.200	Aggressive	−0.237	Eccentric	−0.325	Excitable	−0.355
Apprehensive	−0.850	Solitary	−0.784	Slow	−0.332	Active	−0.416	Strong	−0.375	Irritable	−0.516

Personality Rating Scale (Murray, 1995, adapted from Stevenson-Hinde & Zunz, 1978)

1. **Active** - moves about a lot.
2. **Aggressive** - causes harm or potential harm.
3. **Apprehensive** - seems to be anxious about everything; fears and avoids any kind of risk.
4. **Confident** - behaves in a positive, assured manner; not restrained or tentative.
5. **Curious** - readily explores new situations.
6. **Deferential** - gives in readily to others.
7. **Eccentric** - shows stereotypes or unusual mannerisms.
8. **Effective** - gets own way; can control others; leads.
9. **Equable** - reacts to others in an even calm way; is not easily disturbed.
10. **Excitable** - over-reacts to any change.
11. **Fearful** - fear-grins; retreats readily from others or from outside disturbances.
12. **Gentle** - responds to others in an easy kind manner.
13. **Insecure** - hesitates to act alone; seeks reassurance from others.
14. **Intelligent** - clever; tactical; manipulative; inventive.
15. **Irritable** - reacts negatively with little provocation.
16. **Maternal/Paternal** - warm and receptive to infants.
17. **Opportunistic** - seizes a chance as soon as it arises.
18. **Permissive*** - could, but does not, interfere with the behaviour of others.
19. **Playful** - initiates play and joins in when play is solicited.
20. **Popular** - is sought out as a companion by others.
21. **Predictable*** - consistent in behaviour.
22. **Protective** - prevents harm or possible harm to others.
23. **Sensitive*** - responds or reacts with minimal cues to the behaviour of others.
24. **Slow** - moves and sits in a relaxed manner; moves slowly and deliberately; not easily hurried.
25. **Sociable** - seeks companionship of others.
26. **Solitary** - spends time alone.
27. **Strong** - depends upon sturdiness and muscular strength.
28. **Tense** - shows restraint in posture and movement.

*Adjectives subsequently omitted from analysis due to poor intra-item reliability.

Instructions to raters for completing the personality ratings

Ratings must be made independently without discussion with other raters. The ratings should be made on the basis of your knowledge of the individual animals within the last month and should be completed as soon as possible. If there is any reason why the behaviour of the animals is not typical of any other period of time, please feel free to comment on this on a separate sheet of paper. For each animal, please rate each item in turn, independently, and according to its definition. For each item, place a circle around the number which you feel best illustrates the strength of the particular behaviour in each individual:

1. None of the behaviour
2. Traces of the behaviour present
3. Behaviour distinctly present, but a little less than average
4. Average, or slightly above average
5. Strong, but not outstanding, evidence of the behaviour
6. Strong and conspicuous behaviour, approaching the extreme
7. Extreme manifestation of the behaviour

References

Allport GW (1937) Personality: a psychological interpretation. Holt Rinehart & Winston, New York
Altmann J (1974) Observational study of behavior: Sampling methods. Behaviour 69:227–263
Borkenau P (1992) Implicit personality theory and the five-factor model. J Pers 60:295–327
Buirski P, Plutchik R (1991) Measurement of deviant behavior in a Gombe chimpanzee: Relation to later behavior. Primates 32:207–211
Buirski P, Kellerman H, Plutchik R et al. (1973) A field study of emotions, dominance, and social behaviour in a group of baboons (*Papio anubis*). Primates 14:67–78
Buirski P, Plutchik R, Kellerman H (1978) Sex differences, dominance, and personality in the chimpanzee. Anim Behav 26:123–129
Buss AH (1989) Personality as traits. Am Psychol 44:1378–1388
Byrne G, Suomi SJ (2002) Cortisol reactivity and its relation to homecage behavior and personality ratings in tufted capuchin (*Cebus apella*) juveniles from birth to six years of age. Psychoneuroendocrinology 27:139–154
Capitanio J (1999) Personality dimensions in adult male rhesus macaques: Prediction of behaviors across time and situation. Am J Primatol 47:299–320
Capitanio J (2002) Sociability and responses to video playbacks in adult male rhesus monkeys (*Macaca mulatta*). Primates 43:169–177
Cattell, RB (1946) The description and measurement of personality. World Books, Yonkers
Cattell, RB (1973) Personality pinned down. Psychol Today 7:40–46
Cervone D, Pervin LA (2008) Personality theory and research. Wiley, Hoboken
Chamove AS, Eysenck HJ, Harlow HF (1972) Personality in monkeys: factor analyses of rhesus social behavior. Q J Exp Psychol 24:496–504
Cheney DL, Seyfarth, RM, Smuts BB (1986) Social relationships and social cognition in nonhuman primates. Science 234:1361–1366
Costa, PT Jr, McCrae RR (1992) NEO-PI-R professional manual. Psychological Assessment Resources Inc, Odessa
Costa, PT Jr, McCrae RR (1995) Domains and facets: Hierarchical personality assessment using the Revised NEO Personality Inventory. J Pers Assess 64:21–50
de Waal FBM (1986) The brutal elimination of a rival among captive male chimpanzees. Ethol Sociobiol 7:237–251
Digman JM (1990) Personality structure: Emergence of the five-factor model. Annu Rev Psychol 41:417–440
Fairbanks LA (2001) Individual differences in response to a stranger: Social impulsivity as a dimension of temperament in vervet monkeys (*Cercopithecus aethiops sabaeus*). J Comp Psychol 115:22–28

Funder DC, Ozer DJ (1983) Behavior as a function of the situation. J Pers Soc Psychol 44:107–112
Gold KC, Maple TL (1994) Personality assessment in the gorilla and its utility as a management tool. Zoo Biol 13:509–522
Goodall J (1986) The chimpanzees of Gombe: Patterns of behavior. Belknap, Cambridge
Goosen C (1987) Social grooming in primates. In: Mitchell G, Erwin J (eds) Comparative primate biology, Vol. 2b: Behavior, cognition and motivation. Alan R. Liss, New York
Gosling SD (2001) From mice to men: What can we learn about personality from animal research? Psychol Bull 127:45–86
Hemelrijk CK (1994) Support for being groomed in long-tailed macaques, *Macaca fascicularis*. Anim Behav 48:479–481
Hemelrijk CK, Ek A (1991) Reciprocity and interchange of grooming and 'support' in captive chimpanzees. Anim Behav 41:923–935
Hinde RA, Atkinson S (1970) Assessing the roles of social partners in maintaining mutual proximity, as exemplified by mother-infant relations in rhesus monkeys. Anim Behav 18:169–176
Holzer BB, Schmidt CR (1992) Adoption of an additional infant by a western lowland gorilla (*Gorilla gorilla gorilla*). Folia Primatol 58:190–196
John OP (1989) Towards a taxonomy of personality descriptors. In: Buss DM, Cantor N (eds) Personality psychology: Recent trends and emerging directions. Springer-Verlag, New York
Jones AC, Gosling SD (2005) Temperament and personality in dogs (*Canis familiaris*): A review and evaluation of past research. Appl Anim Behav Sci 95:1–53
Kalin NH, Larson C, Shelton SE et al. (1998) Asymmetric frontal brain activity, cortisol, and behavior associated with fearful temperaments in rhesus monkeys. Behav Neurosci 112:286–292
King JE, Landau VI (2003) Can chimpanzee (*Pan troglodytes*) happiness be estimated by human raters? J Res Pers 37:1–15
King JE, Weiss A (2011) Personality from the perspective of a primatologist. In: Weiss A, King JE, Murray L (eds) Personality and temperament in nonhuman primates. Springer, New York
King JE, Weiss A, Farmer KH (2005) A chimpanzee (*Pan troglodytes*) analogue of cross-national generalization of personality structure: Zoological parks and an African sanctuary. J Pers 73:389–410
Kline P (1992) Handbook of psychological testing. Routledge, London
Kline P (1993) Personality: The psychometric view. Routledge, London
Klirs EG, Revelle W (1986) Predicting variability from perceived situational similarity. J Res Pers 20:34–50
Konečná M, Lhota S, Weiss A et al. (2008) Personality in free-ranging Hanuman langur (*Semnopithecus entellus*) males: Subjective ratings and recorded behavior. J Comp Psychol 122:379–389
Kuhar CW, Stoinski TS, Lukas KE et al. (2006) Gorilla Behavior Index revisited: Age, housing and behavior. Appl Anim Behav Sci 96:315–326
Lambeth SP, Bloomsmith MA, Alford PL (1990) The effects of estrous cycling on agonism and wounding in multi-male chimpanzee groups. Am J Primatol 20:206
Lilienfeld SO, Gershon J, Duke M et al. (1999) A preliminary investigation of the construct of psychopathic personality (Psychopathy) in chimpanzees (*Pan troglodytes*). J Comp Psychol 113:365–375
McAdams DP (1992) The five-factor model in personality: A critical appraisal. J Pers 60: 329–361
McGuire MT, Raleigh MJ, Pollack DB (1994) Personality features in vervet monkeys: the effects of sex, age, social status, and group composition. Am J Primatol 33:1–13
Mischel W (1968) Personality and assessment. Wiley, New York
Mondragon-Ceballos R, Santillan-Doherty AM (1994) The relationship between personality and age, sex and rank in captive stumptail macaques. In: Roeder JJ, Thierry B, Anderson JR,

Herrenschmidt N (eds) Current primatology, Vol. II: Social development, learning and behavior. University of Louis Pasteur, Strasbourg
Mondragon-Ceballos R, Santillan-Doherty AM, Chiappa P (1991) Correlation between subjective assessment of individuality traits and ethological sampling of behaviour in stumptail macaques. In: Ehara A, Kimura T, Takenaka O, Iwamoto M (eds) Primatology today: Proceedings of the XIIIth Congress of the International Primatological Society, Nagoya and Kyoto, 18–24 July 1990. Elsevier Science, Amsterdam
Murray LE (1995) Personality and individual differences in captive African apes. PhD Thesis, University of Cambridge
Murray, LE (1998). The effects of group structure and rearing strategy on personality in Chimpanzees *Pan troglodytes* at Chester, London and Twycross Zoos. Int Zoo Yearb 36: 97–108
Murray LE (2002) Individual differences in chimpanzee *Pan troglodytes* personality and their implications for the evolution of mind. In: Harcourt C, Sherwood B (eds) New perspectives in primate evolution and behavior. Westbury, Otley, UK
Murray LE (2005) Personality types in chimpanzees (*Pan troglodytes*): Another 'Big 5'. Am J Primatol 66:53–54
Pederson AK, King JE, Landau VI (2005) Chimpanzee (*Pan troglodytes*) personality predicts behavior. J Res Pers 39:534–549
Santillán-Doherty AM, Munoz-Delgado J, Arenas R et al. (2006) Reliability of a method to measure novelty-seeking in nonhuman primates. Am J Primatol 68:1098–1113
Schino G (2001) Grooming, competition and social rank among female primates: a meta-analysis. Anim Behav 62:265–271
Schino G (2007) Grooming and agonistic support: A meta-analysis of primate reciprocal altruism. Behav Ecol 18:115–120
Seyfarth RM (1977) A model of social grooming among adult female monkeys. J Theor Biol 65:671–698
Seyfarth RM (1980) The distribution of grooming and related behaviors among adult female vervet monkeys. Anim Behav 28:798–813
Seyfarth RM, Cheney DL (1984) Grooming, alliances and reciprocal altruism in vervet monkeys. Nature 308:541–543
Simpson MJA (1973) The social grooming of male chimpanzees. In: Michael RP, Crook JH (eds) Comparative ecology and behavior of primates. Academic Press, London, pp. 411–505
Sparks J (1969) Allogrooming in primates: A review. In: Morris D (ed) Primate ethology. Anchor Books, Garden City
Stevenson-Hinde J, Stillwell-Barnes R, Zunz M (1980a) Subjective assessment of rhesus monkeys over four successive years. Primates 21:66–82
Stevenson-Hinde J, Stillwell-Barnes R, Zunz M (1980b) Individual differences in young rhesus monkeys: Consistency and change. Primates 21:498–509
Stevenson-Hinde J, Zunz M (1978) Subjective assessment of individual rhesus monkeys. Primates 19:473–482
Trivers RL (1971) The evolution of reciprocal altruism. Q Rev Biol 46:35–57
Uher J, Asendorpf JB (2008) Personality assessment in the Great Apes: Comparing ecologically valid behavior measures, behavior ratings, and adjective ratings. J Res Pers 42:821–838
van Hooff, JA (1970) A component analysis of the structure of the social behavior of a semi-captive chimpanzee group. Experientia 26:549–550
van Wulfften Palthe T, van Hooff, JARAM (1975) A case of adoption of an infant chimpanzee by a suckling foster chimpanzee. Primates 16:231–234
Vazire S, Gosling SD, Dickey AS et al. (2007) Measuring personality in nonhuman animals. In: Robins RW, Fraley RC, Krueger RF (eds) Handbook of research methods in personality psychology. Guilford Press, New York

Watts DP (1994) Social relationships of immigrant and resident female mountain gorillas, II: Relatedness, residence, and relationships between females. Am J Primatol 32:13–30

Watts DP (2000) Grooming between male chimpanzees at Ngogo, Kibale National Park. II. Influence of male rank and possible competition for partners. Int J Primatol 21:211–238

Weiss A, King JE, Enns RM (2002) Subjective well-being is heritable and genetically correlated with Dominance in chimpanzees (*Pan troglodytes*). J Pers Soc Psychol 83:1141–1149

Chapter 7
Primate Personality and Behavioral Endocrinology

Stephanie F. Anestis

Abstract Although hormones are best known for their physiological functions, elegant studies in a variety of species have also demonstrated important effects of hormones on behavior (and vice versa: behavior's effects on hormone levels). Behavioral endocrinology is an exciting field because the relationship between hormones and behavior is complex and in many ways still poorly defined. Initial studies in primates focused primarily on associations between specific behaviors such as aggression or mating and hormones such as testosterone, cortisol, and progesterone. However, as primatologists began to recognize the importance of behavioral style variation and the influence of personality on all aspects of behavior, the physiological correlates of this variable also began gaining attention. In this review, I briefly discuss the mechanisms by which hormones affect behavior before reviewing important research on the role of hormones in maternal style, dominance relationships, and personality. I also discuss the practical and theoretical implications of the relationship between primate personalities and hormone levels. I suggest that this field could benefit from more research in two primary areas: first, the hormonally mediated costs and benefits of certain behavioral styles, and second, personality variation in wild primates and its endocrine correlates.

7.1 Introduction: Endocrinology in Behavior Research

Behavioral endocrinology is the study of the interaction between hormones and behavior. Hormones are chemical messengers that travel from one cell to cause an effect in another. Most hormones circulate in the bloodstream until they reach

S.F. Anestis (✉)
Department of Anthropology, Yale University, New Haven, CT 06520-8277, USA
e-mail: stephanie.anestis@yale.edu

A. Weiss et al. (eds.), *Personality and Temperament in Nonhuman Primates*,
Developments in Primatology: Progress and Prospects, DOI 10.1007/978-1-4614-0176-6_7,

appropriate receptors on or in target cells (an exception is some neurohormones, which can also be released into the port connecting the hypothalamus and pituitary gland, two important endocrine structures in the brain). In response to hormone-receptor binding, cells may react by increasing or decreasing the transcription rate of specific genes, or by initiating intracellular chemical reactions (e.g., turning on inactive enzymes or receptors). Hormones play a role in almost every physiological function, and are therefore central to an organism's survival and reproduction.

The behavioral effects of hormones have been recognized for centuries as a result of early experiments in castration. People noticed that removing the testes of domesticated male animals not only prevented them from reproducing, but also (depending on the timing of castration) resulted in often dramatic behavioral effects. The personalities of the animals could actually be changed by an alteration of their physiological state. In particular, male animals such as horses and bulls became tamer, less aggressive, and more easily managed.

The field of behavioral endocrinology includes experimental and observational studies on a huge range of animal species. In the simplest cases, behaviors cease in the absence of a hormone and resume when the hormone is replaced. A classic example is mounting behavior by male rats, which disappears in the absence of testosterone, despite appropriate behavioral cues from females (Meisel and Sachs 1994). However, the presence of testosterone itself is not enough to elicit mounting behavior in the absence of behavioral cues (e.g., an estrous female), demonstrating hormones' typically indirect effects on behavior. More commonly, the relationship between hormones and behavior is quite complicated, often involving multiple hormones and a range of variation in behavior that can be quite subtle. Importantly, hormones affect behavior patterns by increasing the *probability* they will occur in the presence of other behavioral triggers, not by directly causing them.

Hormones can be grouped into a variety of classes based on their biological structure. Although all hormones can potentially affect behavior, the peptide and steroid hormones have been the most extensively studied in behavioral endocrinology. The mechanisms involved in hormonal influences on behavior are varied and complex. Brain cells, or neurons, in all regions and in all structures are equipped with a variety of hormonal receptors for both peptide and steroid hormones. When a peptide hormone binds to a receptor on a nerve cell, it activates the signal transduction pathway, or a series of steps that convey a message to the cell nucleus (Nelson 2005). Steroid hormones, on the other hand, enter cells and interact directly with the cell nucleus. In both cases, hormonal messages are typically responsible for starting, stopping, increasing, or decreasing transcription of specific genes; these are termed genomic effects, and can result in changes in the numbers of enzymes, neurotransmitters, and even receptors on cells, which in turn result in behavioral changes. However, hormones can also affect behavior through mechanisms that do not directly involve transcription. For example, estrogen can increase the excitability of a neuron (Teyler et al. 1980), or regulate serotonin transporters (Rehavi et al. 1987). There are probably also many ways in which hormones affect behavior that have not yet been described.

Since behavior can also affect hormone levels, patterns of causality are not easily inferred from correlational studies. Testosterone responses to competitive events in human males are one example: testosterone levels typically rise in the anticipation of a competitive encounter, then increase further in winners and decrease in losers (Mazur et al. 1992; Bernhardt et al. 1998; Gonzales-Bono et al. 2000). Cortisol, known as the "stress" hormone, is also responsive to behavior and environment in primates and other animals, increasing as a result of acute stressful events such as disorientation and capture (baboons: Sapolsky 1985; squirrel monkeys: Steklis et al. 1986; chimpanzees: Anestis et al. 2006) and separation from conspecifics (squirrel monkeys: Vogt et al. 1980; Goeldi's monkeys: Dettling et al. 1998). High cortisol can also be chronic, or associated with long-term behavioral patterns and outcomes; for example, increased cortisol levels have been associated with periods of unstable dominance relationships (baboons: Sapolsky 1983; Bergman et al. 2005; rhesus monkeys: Gust et al. 1991) and either low or high dominance rank in a social group (reviewed in Abbott et al. 2003).

7.2 Why Are Hormones Important to the Study of Primate Personality?

Relatively few studies have described animal personality in a quantitative, rather than a qualitative, way. Ultimately, animal personalities are interesting not just because they illuminate one more way in which animals exhibit social and cognitive complexity, but also because personality variation can have direct effects on individual differences in fitness, for example by affecting immunity and health (Capitanio 2011). Isolating and attempting to specify the components of such a complex variable is necessary to ascertain how it impacts reproductive success. One way to do this is to consider attributes that cluster together, or in other words to identify specific "behavioral styles" using direct behavioral observations of the animals' behavior. The identification and exploration of primate behavioral styles, the focus of this book, is therefore crucial to the understanding of primate social behavior from an evolutionary perspective.

If personality is influenced by natural or sexual selection, we should be able to find individual differences in variables that affect reproductive success that vary according to particular traits or characteristics, including behavioral styles. What we ultimately want to know as evolutionary theorists is: how much does personality affect natural selection? This is where endocrinology can go far in illuminating the significance of personality variation. Measuring reproductive success directly in long-lived animals is rarely possible, yet reproductive success is a primary variable of interest. To circumvent this difficulty, researchers can study variables assumed to be correlated with fitness. Variation in hormone levels is one such measure. Because hormones affect every functional system in the organism, their influence on growth, maintenance, and reproduction – the central components of individual life history – is particularly relevant to discussions of evolutionary fitness.

Most studies of hormone–behavior relationships have focused on short-term associations between a behavioral event and hormone levels. For example, the timing of ovulation can be estimated using measures of female gonadal hormones, and these can be related to female mating behavior (Brockman and Whitten 1996; Wallis 2002). Experimental studies can reveal the hormonal response to a specific event, for example, the stress of sedation (Sapolsky 1982; Anestis et al. 2006). These types of questions are interesting and illuminating because they give us insight into the specific actions of particular hormones, how behavior is influenced by hormone levels, or how individuals vary in their endocrine responsivity. However, another type of hormone measurement – namely estimations of baseline levels – is particularly important if we are interested in personality-related endocrine variation, because it may reveal an individual's experience over the long term. It can reveal the consequences of occupying high or low rank, of residing with a despotic or lenient alpha male, and, importantly, of having a specific behavioral style. In the following sections, I will review the literature on (1) maternal style and hormones and (2) dominance rank, personality, and hormones to illustrate the ways in which hormonal data can illuminate our investigations into individual variation.

7.3 Hormones and Maternal Style

A vast literature on the biology of parenting exists, particularly in rodents. These studies have demonstrated the role of prenatal maternal hormones on parenting behavior (Rosenblatt et al. 1988), as well as the effects of maternal behavioral differences on the endocrine and behavioral systems of young animals (Sapolsky 2004). For example, infants of rat mothers who lick, groom, and arched-back nurse them more than average display reduced reactivity to stress (Weaver et al. 2004). Pryce (1992) presents an integrative discussion of the factors motivating maternal behavior in mammals and underscores the importance of hormones in regulating maternal arousal, which in turn affects how mothers respond to offspring. Clearly, maternal style has a major effect on the social, behavioral, and physiological development of young (Hrdy 1999).

In primates, the relationship between mothering style and hormone levels in both mothers and infants has been studied in only a few species, partially because maternal (and paternal) behavior involves a variety of often complex behaviors and because mothering is extended in primates due to a long juvenile period. Nevertheless, differences in mothering styles were one of the first descriptors of behavioral style variation in primates. For example, Altmann (1980) noted two distinct mothering styles in her long-term studies of baboons; some mothers she termed "laissez-faire" because they typically allowed older infants to wander a short distance away, whereas the "restrictive" mothers allowed their infants little freedom and were much quicker to retrieve them upon separation. Jane Goodall's studies of chimpanzees revealed mostly anecdotal, but richly described variation in personalities, and she used mothering behavior as an important descriptor of female chimpanzee personalities (Goodall 1986).

7.3.1 Cortisol

Stress, defined as the disruption of homeostasis in an organism, has diverse effects on an animal's physiology, affecting almost every major system (Sapolsky 1993, 1994; Honess and Marin 2006). Cortisol, a steroid hormone secreted by the adrenal glands in response to activation of the hypothalamic–pituitary–adrenal (stress) axis, is a useful physiological marker indicating both acute and chronic stress in primates. Cortisol secretion under stress is adaptive because it prepares the organism to deal with the crisis at hand, diverting energy away from long-term processes and toward those essential to immediate survival. It mobilizes glucose, and promotes the breakdown of lipids and proteins. It also increases heart rate and blood pressure, effectively increasing the rate of delivery of oxygen and glucose to those areas that need it for overcoming the stressor (e.g. as Sapolsky describes, allowing a zebra to outrun a predator; 1994). However, chronically high levels of cortisol can have negative effects on a wide range of physiological systems, including the immune system (Riley 1981; Goujon et al. 1997; Norbiato et al. 1997), the reproductive system (Attardi et al. 1997; Loucks and Redman 2004), and the brain (Sapolsky and Pulsinelli 1985; Fuchs and Flugge 2003; Buwalda et al. 2005). In humans and other social animals, psychosocial stress, as opposed to physical stress such as exhaustion, extreme temperatures, etc., is prominent and can be chronic. Dominance status clearly has important effects on the degree of stress experienced by animals ranging from rodents to primates (see next section; Sapolsky 1982; Creel et al. 1996; Creel 2001; Bartolomucci et al. 2005), but individual differences in the perception of stress and the type of coping mechanism (variables often associated with personality) are also associated with circulating and excreted levels of cortisol (Pruessner et al. 1997; Vollrath 2001; Carere et al. 2003; Reimers et al. 2003; Wommack and Delville 2003; Roy 2004). Ultimately, cortisol is an important hormone because it potentially affects a wide range of physiological processes and therefore directly affects reproductive success.

Some degree of hypercortisolism is typical of primate pregnancies (McLean and Smith 1999), but few authors have studied how variation in prepartum or postpartum cortisol levels affects maternal behavior or maternal style. Bardi and colleagues, in a series of studies on maternal style and hormone levels in baboons (Bardi et al. 2004, 2005), found that mothers with higher prepartum cortisol levels exhibited more maternal affiliative behaviors with offspring (e.g., watching, cuddling, and grooming infants) than those with lower levels. On the other hand, mothers with higher levels of postpartum cortisol scored higher in maternal stress behaviors (scratch, startle, muzzle wipe, and brow wipe), and generally spent less time in contact with infants (Bardi et al. 2004). The authors suggest that while prepartum cortisol levels prime mothers for the demanding experience of motherhood, possibly by increasing the female's attention toward the infant or her general arousal, high postpartum levels may be more indicative of anxiety, which can have a negative effect on parenting.

Maternal style in baboons also affects *offspring* reactivity to stress, even beyond infancy. Bardi et al. (2005) found that juveniles raised by mothers who showed more positive interactions with them – specifically maintaining contact and directing affiliative behavior toward their infant – exhibited high levels of vigilance but less active stress behaviors (as in the previous study, scratch, startle, muzzle wipe, and brow wipe) during a stress test in which the animal was briefly isolated from the social group. Infants of mothers with a less affiliative style spent more time locomoting in the test environment (perhaps exhibiting a fight/flight response), showing a more active behavioral response to stress and a greater cortisol response than those infants raised by the mothers with a more affiliative style. Bardi and coauthors suggest that their results are similar to studies of rodents showing a profound effect of mothering style and infant handling on stress responsivity of offspring later in life (Francis and Meaney 1999).

Macaque mothers also show variation in a variety of behavioral dimensions related to behavioral style, particularly rejection and protectiveness (Schino et al. 1995). Indicators of infant anxiety such as scratching and crying are positively associated with maternal postpartum cortisol levels in both Japanese and rhesus macaques (Bardi and Huffman 2006). In rhesus only, infant anxiety is also related to prepartum maternal cortisol: infants of mothers with *lower* prepartum cortisol were more anxious. These data are consistent with data from baboons, again suggesting that high prepartum cortisol levels facilitate caregiving, while high postpartum levels reflect anxiety and may be maladaptive. High prepartum cortisol levels may serve as a primer for maternal behavior, increasing mothers' attentiveness to their infants in the immediate postpartum period, whereas high postpartum cortisol reflects high levels of stress that may have negative effects on parenting behavior (Bardi et al. 2004).

7.3.2 Sex Steroids: Estrogen and Progesterone

Studies of parenting behavior and its endocrine correlates have not been limited to Old World monkeys; marmosets and tamarins are excellent model species for the study of parenting behavior because they typically exhibit extensive maternal and paternal involvement in raising young. These studies have focused less on cortisol and more on female ovarian hormones, specifically estrogen and progesterone. Exactly how these hormones affect maternal behavior – and how relative levels during pregnancy, parturition, and lactation relate to behavioral style differences – is still being explored. However, the estrogen/progesterone ratio (which allows a relative comparison of these two hormones) and the relationship between prepartum levels in late pregnancy and postpartum levels (similar to studies of cortisol discussed above) are likely to be important (Bardi et al. 2004).

Pryce et al. (1988) found that prepartum levels of urinary estradiol, a form of estrogen, were higher in red-bellied tamarin mothers who displayed adequate maternal behavior than in those displaying inadequate maternal behavior. In addition, marmoset females who experienced the greatest increase in estrogen relative to progesterone in late pregnancy also showed the most maternal behavior as measured by variables such as the percentage of time spent nursing/carrying infants, and response latency to neonate crying. However, the authors note that in the overall sample of eight mother–infant pairs, there was no relationship between prepartum sex steroid levels and maternal behavior (Pryce et al. 1995). In a study of tufted-ear marmosets that incorporated a within-subjects design, Fite and French (2000) found that higher prepartum levels of estradiol were associated with *higher* infant mortality. They suggest that a within-subjects design, where an individual female's gonadal steroid excretion is compared across successful and unsuccessful reproductive events, may more accurately reflect the association between hormones and maternal behavior than between-subjects designs because of a large degree of inter-subject variability. These inconsistent results – where depending on the study, both higher and lower levels of prepartum estradiol are associated with poor infant caregiving and infant mortality – reveal the complex nature of the hormone–behavior relationship in regulating maternal style.

Clearly, assessment of pregnancy and peripartum ovarian steroid levels in callitrichids cannot be used as a sole predictor of infant outcome. These studies raise the interesting question of whether variation in maternal behavior in callitrichids really constitutes individual maternal style, or if maternal behavior is more closely tied to experiential factors (e.g., primiparity vs. multiparity or dominance rank). To date, behavioral style has not been quantified or explicitly studied in marmosets and tamarins.

A series of studies in Old World monkeys has also suggested a role for sex steroids in the regulation of maternal behavior. Rhesus and Japanese macaque mothers with a maternal style characterized by high rejection rates of infants, including a variety of behaviors involving the mother breaking contact from the infant, exhibited lower levels of excreted estrogen metabolites both pre- and postpartum than mothers with a less rejecting style (Bardi et al. 2001). The authors found a similar trend between estrogen metabolites and time in maternal contact for hamadryas baboons, though in this species a more robust relationship was found between high prepartum PdG (pregnanediol-3-glucoronide, a progesterone metabolite), low postpartum PdG, and high maternal stress, indicating that considering levels of multiple hormones (e.g., both cortisol and ovarian hormones) can help elucidate the clearly complex relationship between maternal behavior and endocrinology (Bardi et al. 2004). No single hormone is responsible for the onset or maintenance of parenting behavior, a result consistent with the variety of elements involved in maternal–infant interactions.

However, not all Old World monkey studies have found individual variation in mothering associated with estrogen or progesterone levels. Maestripieri and

Megna (2000) did not find robust differences in estrogen or progesterone, or the degree of decline in these hormones postpartum, between abusive and nonabusive rhesus macaque mothers. Maternal abuse of infants might be considered one extreme end on a maternal style spectrum, though it may also be a pathological response, and therefore not indicative of the normal relationship between hormones and parenting behavior.

7.3.3 Prolactin

Neuropeptides, particularly oxytocin, prolactin, and arginine vasopressin, are positively associated with social and parenting behavior in males and females in a variety of mammals (Bridges 1994; Young 1999; Curley and Keverne 2005). Prolactin level variation has been studied in several primate species, though maternal style variation was typically not the focus of these studies. For example, Soltis et al. (2005) found that levels of prolactin increased with amount of caregiving given to infant squirrel monkeys, even when allo-mothers were included in the analysis. In addition, groups with resident infants exhibited overall higher mean prolactin levels than groups without infants. They hypothesize that physical contact, and not just lactation and nursing, may be the mediator for prolactin secretion.

Despite the potentially important relationship between peptide hormones such as prolactin and behavior, the study of primate inter-individual differences in these hormones is still in its infancy. One reason is the uncertainty about how peripheral (e.g., circulating and excreted) levels of these hormones are related to central levels (e.g., levels in the brain), where behavior is initiated and maintained. Measuring levels in the brain is almost impossible in socially living animals without disrupting normal behavioral processes, so we are typically limited to studies of excreted metabolites present in the cerebral spinal fluid (CSF) or blood. However, CSF and blood sampling is also invasive. Therefore, until assays are developed that can accurately detect individual differences in these hormones in urine or other more easily collected biological media (e.g., oxytocin in urine: Fries et al. 2005; Seltzer and Ziegler 2007), this field will be limited.

On the other hand, those peptide hormones most closely associated with the physiological functions of maternal care, especially lactation, may not be good candidates for behavioral variation, because variation is better explained by physical differences including milk production. Unlike cortisol, for example, prolactin levels in nursing mammalian mothers predictably rise and fall with a physical stimulus (nipple stimulation; Grattan 2002). In common marmosets, prolactin elevations in fathers and helpers may be related to the physical effort of carrying offspring, as prolactin was higher when two offspring were carried rather than one, but elevations did not occur in response to infant cues alone (e.g., birth of infants; Mota and Sousa 2000; Mota et al. 2006). Nevertheless, prolactin's association with behavior at a general level is unmistakable (Ziegler 2000), and the hypothesis that prolactin and other peptides are associated with differences in maternal style is worth investigating.

Future studies should focus on developing assays to measure peptides in urine and feces, and resolving the relationship between central and circulating levels so that we can decide which types of hypotheses can be addressed with peptide hormone data.

7.4 Dominance Relationships, Behavioral Style, and Hormones

Dominance relationships are a salient feature of most primate societies. In many species, dominance rank affects feeding efficiency (e.g., brown capuchins, *Cebus apella*: Janson 1985 or baboons, *Papio anubis*: Barton et al. 1996) and male mating success (Cowlishaw and Dunbar 1991), although rank effects on reproductive success have been inferred more often than documented (but see Gouzoules et al. 1982; Altmann et al. 1995; Packer et al. 1995; Pusey et al. 1997; Di Fiore 2003). Rank can also have physiological effects. For example, in a wide variety of primate species low-ranking animals are characterized by faster heart rates and higher cortisol levels than in dominant peers (baboons: Sapolsky 1995; marmosets: Saltzman et al. 1998; macaques: Aureli et al. 1999); these differences may reflect differential levels of stress. Investigations of rank effects are thus important because they shed light on how daily life can be dramatically different for two individuals living in the same social group, even when they belong to the same age and sex class.

However, a narrow focus on dominance risks ignoring ways in which individuals navigate their social environment independently of rank. Low-ranking individuals may have varied coping strategies, resulting in large individual differences in stress levels as measured by glucocorticoids, immune parameters, and/or behavioral indicators of anxiety (Davidson et al. 1993; Laudenslager and Boccia 1996; Virgin and Sapolsky 1997), and high-ranking individuals may exhibit different responses to perceived threats to their status (Sapolsky and Ray 1989). Indeed, the primate literature is replete with anecdotal evidence of such individual variation (Altmann 1980; de Waal 1982; Goodall 1986; Sapolsky and Ray 1989; Boyson 1994; Ron 1994; Laudenslager and Boccia 1996; Nishida and Hosaka 1996; Virgin and Sapolsky 1997). For example, adult male chimpanzees use multiple approaches to try to obtain high rank, and their concern with rank varies (de Waal 1982; Goodall 1986; Nishida and Hosaka 1996). Although few authors have tried to quantify such personality differences, they can have an impact on the relationship between rank and hormone levels.

7.4.1 Cortisol

Robert Sapolsky, in his studies of rank effects on cortisol and testosterone levels in wild baboons (*Papio anubis*) at Amboseli, found that dominance rank alone was not enough to explain individual variation in hormone levels. He and student

Justina Ray (1989) therefore assessed the impact of behavioral style on physiological variables, particularly endocrine parameters. They recorded behavioral data on social interactions using standard ethological techniques and then used these data to identify behavioral styles; for example, males differed in how likely they were to distinguish between a threatening and neutral interaction with a rival (measured by the probability of return to the pre-interaction activity after the interaction). Although low-ranking males exhibited higher mean serum cortisol levels than high-ranking males, large differences also occurred *within* the high-ranking cohort. A subset of the high-ranking individuals – including those good at distinguishing between threatening and neutral interactions, for example, and those who spent a considerable time with adult females – showed significantly lower cortisol levels than high-ranking individuals who did not exhibit similar styles. These less adept, high-ranking males actually exhibited cortisol levels similar to those of low-ranking males. Additional studies revealed that differences in cortisol levels among low-ranking animals were also influenced by behavioral style (Virgin and Sapolsky 1997).

Chimpanzees are an excellent species for investigating behavioral style because of their behavioral complexity, and because personality differences clearly exist among individuals (de Waal 1982; Goodall 1986; Boyson 1994; Nishida and Hosaka 1996). In my studies of young, captive chimpanzees, I have followed Sapolsky's method of using ethological observation methods to assess behavioral styles. I created behavioral indices intended to cover the variety of behaviors exhibited by the chimpanzees, including affiliation, agonism, and neutral interactions (see Table 7.1). A principal components analysis applied to these data, covering longitudinal data on over 30 chimpanzees in multiple captive groups, has identified six behavioral style components, which I labeled Smart, Affiliative, Playful, Aggressive, Friendly, and Mellow based on the behavioral indices driving each (Anestis 2005; Table 7.2).

Although dominance rank, assessed using data on pant-grunts and supplants, was not associated with baseline urinary cortisol in this sample, score in the Smart component was: the individuals who scored highest in this component showed higher levels of cortisol than the individuals who scored lowest (Fig. 7.1). Although this result may initially seem counterintuitive – shouldn't animals with the most social awareness in their study groups be the least stressed? – it may actually be that being socially skilled carries some costs, namely psychosocial stress. Spending a lot of time and energy investing in alliances, monitoring other individuals' interactions, and honing other social skills, though likely advantageous in mating success, may impose some survival costs if elevated cortisol levels are sustained long-term.

In this sample of captive chimpanzees, four of the six behavioral style components were associated with dominance rank: aggressive, mellow, smart, and playful (see Table 7.2 for indices driving these components). This suggests that, not surprisingly, behavioral style and dominance rank are not entirely independent measures. In those species in which dominance rank plays a relatively important role in social interactions, an individual's patterns of behavior in a social group will always be influenced by his or her relative place in the hierarchy.

Table 7.1 Chimpanzee behavioral style indices and how they are calculated (Anestis 2005)

Index	Calculation
Chooses safe interactions	Approaches lower-ranking (LR) individuals more often than expected based on the no. of available LR individuals
Initiates winnable agonistic interactions	% of agonistic interactions initiated & won (opponent pant grunts, runs, moves, screams, cries, and/or gives fear face)
Uses coalitions	% of aggressive interactions (attack, hit, threaten, and/or chase) in which individual formed a coalition either as initiator or by joining a third party
Coalition partners	No. of different coalition partners used in the study period, corrected for the number of individuals in the group
Is aggressive	% of initiated social interactions that are aggressive (attack, hit, threaten and/or chase)
Is mellow	% of "no reaction" responses to approaches by conspecifics
Does not react	% of "no reaction" responses to aggression (attack, hit, threaten, and/or chase) against self
Participates in grooming	% of total no. of group grooming sessions in which individual was a participant
Gets groomed	% of individual's total grooming sessions as groomee
Play offers accepted	% of play offers to other group members that are accepted (result in a play session)
Play partners	No. of different play partners over the study period, corrected for total number of individuals in group
Participates in play	% of total no. of group play sessions in which individual was a participant
Has friends	No. of different friends (defined in text) divided by the total number of individuals in the group
Spreads friendship	% of affiliation (groom, hug, and/or touch) with primary partner
Is affiliative	% of initiated social interactions that are affiliative (groom, hug, and/or touch)

Table 7.2 Six principal components derived from the 18 indices listed in Table 7.1

Component	Driving behavioral indices (factor loadings >0.6 or <–0.6); see Table 7.1 for behavioral indices
Smart	Uses coalitions, gets groomed, play offers accepted
Affiliative	Participates in grooming, is affiliative
Playful	Play partners, participates in play
Aggressive	Coalition partners, is aggressive
Friendly	Has friends, spreads friendship
Mellow	Is mellow, does not react

7.4.2 Testosterone

Testosterone is a steroid hormone secreted in both male and female primates (though in much higher concentrations in males) as the end product of the hypothalamic–pituitary–gonadal axis. High testosterone levels impose energetic and other physiological costs because the hormone increases investment in energetically expensive

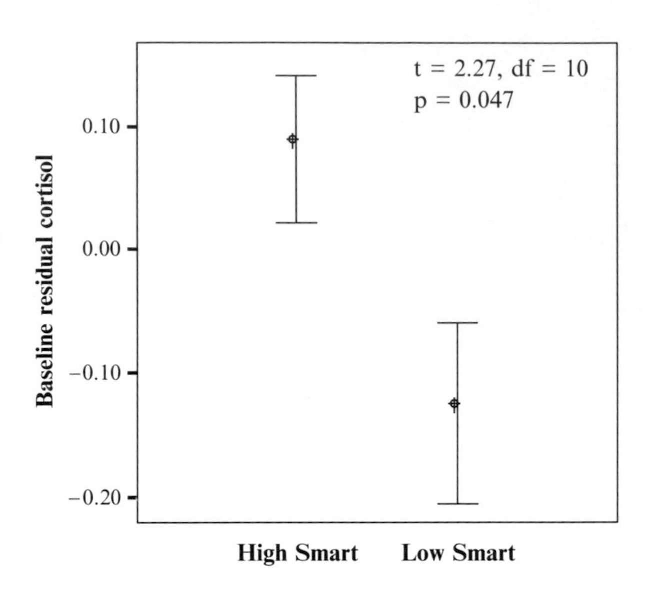

Fig. 7.1 Significant differences between mean baseline cortisol levels (±standard error) in individuals scoring high and low in the Smart component of behavioral style (Anestis 2005). Data shown are means after correction for time of day

muscle tissue (Elia 1992; Bardin 1996); it therefore plays a role in an organism's energy allocation decisions, mediating the trade-off between investment in reproductive effort and survival (Bribiescas 1996, 2001a, b; Ellison 2003). Testosterone supplementation in several different species has demonstrated this trade off: testosterone-augmented males experience increased mortality as a result of increased metabolic costs (lizards: Marler and Moore 1988; dark-eyed juncos: Ketterson and Nolan 1992). In addition to energetic costs, high testosterone levels increase susceptibility to disease (Zuk and McKean 1996; Moore and Wilson 2002; Tanriverdi et al. 2003), which also negatively impacts survivorship.

Testosterone also has important behavioral effects primarily in males, including activational effects on a variety of behaviors such as mating and aggression (Bardin and Catterall 1981; Bermond et al. 1982; Griffin 1996; Albertsson-Wikland et al. 1997). Since maintaining high testosterone levels is costly, yet may increase fitness in males who can afford it, baseline testosterone differences among males are potentially interesting. For example, testosterone is positively associated with dominance rank or status in some studies (rhesus macaques: Rose et al. 1971; squirrel monkeys: Coe et al. 1983; gray mouse lemurs, *Microcebus murinus*: Perret 1992; mountain gorillas: *Gorilla gorilla beringei*: Robbins and Czekala 1997, sifaka, *Propithecus verreauxi*: Brockman et al. 2001, mandrills, *Mandrillus sphinx*: Setchell and Dixson 2001, 2002, chimpanzees: Muehlenbein et al. 2004; Muller and Wrangham 2004), but not in others (Japanese macaques, *Macaca fuscata*: Eaton and Resko 1974; Barrett et al. 2002; squirrel monkeys: Steklis et al. 1986, stumptail macaques, *Macaca arctoides*: Nieuwenhuijsen et al. 1987, muriquis, *Brachyteles arachnoids*: Strier et al. 1999, bonobos, *Pan paniscus*:

Sannen et al. 2004). The absence of a consistent relationship between dominance rank and testosterone, and the fact that dominance status is not always simply correlated with aggression, indicate that the relationship between testosterone and behavioral variables is complex (Anestis 2010). The behavioral variables that are associated with testosterone therefore remain to be elucidated, and variation in personality is a prime candidate as a mediator of the relationship between testosterone and dominance. It is worth noting that there is also some debate as to the appropriateness of the methods used to measure testosterone in excreted samples (Hauser et al. 2008), and that therefore methodological issues – such as species variation in the amount of testosterone metabolite excreted in urine – may be contributing to the clouded relationship between testosterone and behavioral variables.

In a study using blood samples, Sapolsky found that rank and testosterone are not related under typical circumstances in savanna baboons (Sapolsky 1991). However, during periods of social instability, when rank is contested and solidified, high-ranking males show a different pattern of response to the physiological challenge of sedation than low-ranking males: while testosterone eventually decreases in all males, in dominant males testosterone levels actually show a transient increase before falling, whereas subordinates do not show a similar increase. Sapolsky has also considered the important role of behavioral style in predicting testosterone differences within groups of similar-ranking males. Most notably, low-ranking male baboons who initiated aggressive interactions or displaced aggression after losing an agonistic interaction exhibited higher basal testosterone concentrations than those males who never did (Virgin and Sapolsky 1997). This effect was independent of overall participation in aggressive interactions.

In my studies of young chimpanzees, I have found that some behavioral styles are related to dominance, while others are not. For example, while individuals who score highly on the Aggressive component are typically high-ranking, there is no relationship between an individual's score in the Friendly component and rank (Anestis 2006). In other words, some high-scoring Friendly individuals have alpha status, while others are at the bottom of their social hierarchy. This means that behavioral style can tell us something new and interesting about an individual's social relationships, something that cannot be learned by assessing dominance rank alone. However, the relationship between testosterone and behavioral style in chimpanzees is less clear. I did not find significant correlations between testosterone and behavioral styles independent of dominance rank and levels of aggression. There are several possible reasons for this, including the fact that my study involved juveniles and adolescents; testosterone may be more closely tied to rates of aggression than to aggressive style at young ages. It is also possible that testosterone is more related to states (current environments or situational variables that affect behavior, for example, dominance relationships) than traits (relatively stable characteristics of an individual) in chimpanzees. Current studies are investigating the relationship between behavioral styles and other hormones, especially neurohormones, in both adolescent and adult chimpanzees and using longitudinal data to determine how stable behavioral styles are over time.

7.4.3 Monamine Hormones

Serotonin, dopamine, and norepinephrine are neither steroid nor peptide hormones, but rather belong to a class of hormones called monamines. Monamines are primarily secreted by the adrenal glands, located on top of the kidneys, and the pineal gland in the brain. They can function as neurotransmitters, and therefore have more direct effects on behavior than steroid or peptide hormones. Brain monamine levels are typically not measured directly, but are estimated based on levels of their metabolites present in the CSF. Studies on humans have demonstrated a relationship between monamine hormone levels and impulsivity, aggression, and sensation-seeking, among other traits (Netter et al. 1996; Zuckerman 1996; Manuck et al. 1998), but contrasting results from studies attempting to study the same series of traits should serve as a warning that the relationships between monamines and personality dimensions in humans are not yet fully elucidated.

A variety of personality dimensions in humans has been studied in relation to monamine levels. Dopamine is positively associated with reward/novelty-seeking and possibly exploratory behavior (Depue 1995; Zuckerman 1995). Epinephrine and norepinephrine are secreted by the adrenal glands in response to stress, but their baseline levels may display trait-like characteristics that could be related to personality traits. For example, lower levels of norepinephrine metabolites in humans scoring high in Neuroticism (Ballenger et al. 1983). Serotonin has received much attention because of its potential role in depression (Rubinow et al. 1998; van Praag 2004), but it is also related to impulsivity and aggression in nondepressive subjects (Zuckerman 1995) and can also affect sexual behavior (Baum 2002).

Although measures of CSF levels are probably accurate, it is important to keep their limitations in mind, particularly the fact that they represent general indices of neuronal activity and cannot tell us about monamine action in specific brain regions (Kaplan et al. 2002). One solution to this problem is the use of brain imaging studies that can highlight personality-related variation in the use of particular brain regions associated with the activity of specific monamines. For example, Sugiura et al. (2000) found increased cerebral blood flow to the paralimbic cortex, where dopamine receptor density is elevated, in association with high scores on the novelty-seeking personality dimension, supporting previous studies citing a link between dopamine and this trait. In addition, studying genetic polymorphisms in monamine receptor expression is as, if not more, important than studying actual circulating levels, as individual variation in gene expression has been associated with differences in personality traits, especially sensation seeking and anxiety/neuroticism (see Ebstein et al. 2002; Ebstein 2006 for review).

Variation in serotonin and dopamine levels and these hormones' relationship to aggression and dominance behaviors have been the subject of extensive study in monkeys, but the picture is not entirely clear. For example, while in some cases elevated serotonin levels are associated with increased aggression and other behaviors typical of dominant animals, in other cases the opposite is true. Kaplan et al. (2002) found that levels of the cerebrospinal fluid serotonin metabolite 5-hydroxyindoleacetic acid (5-HIAA) were inversely correlated with rank in male cynomolgus monkeys, but only after controlling for levels of the dopamine metabolite homovanillic acid (HVA). HVA, on the other hand, was positively associated with dominance status in both male and female monkeys. These authors suggest that increased dopaminergic activity and decreased serotonin activity may represent two different pathways to dominance, one based on reward potentiation (dopamine) and the other on increased aggression and impulsivity (serotonin). Since monamine levels are stable in individuals over time (Higley et al. 1996), they may be trait-like biomarkers that can affect an individual monkey's potential for attaining high status (Kaplan et al. 2002).

Several researchers have demonstrated correlations between behavioral traits and monamine metabolite levels in *Cercopithecus aethiops* (grivet and vervet monkeys). Fairbanks et al. (1999) measured levels of CSF monamines in free-ranging grivet monkeys in Awash National Park, Ethiopia by trapping and sedating the animals. They compared monamine metabolite levels in those monkeys that reentered traps and those that did not as a measure of impulsivity or risk-taking. Monkeys that reentered traps had lower levels of the norepinephrine metabolite 3-methoxy-4-hydroxyphenylglycol (MHPG) than those that did not. These differences were unrelated to age, which also had been related to monamine levels. The authors suggest that those animals that entered the traps a second time were less afraid of them as a result of decreased hypothalamic–pituitary–adrenal (stress) axis activation and therefore reduced MHPG.

In some species, a hormone–personality combination associated with risk-taking and impulsivity can increase a male's chances of attaining high rank after emigration from the natal group (Fairbanks et al. 2004; Fairbanks and Jorgenson 2011). An adolescent male vervet's chances of attaining alpha status can be predicted using data on his body weight, adolescent impulsivity (measured by an intruder test), and levels of serotonin and dopamine metabolites with remarkable reliability (Fairbanks et al. 2004). Specifically, animals that showed less fear toward a strange intruder approached more quickly, exhibited lower levels of 5-HIAA, and were more likely to become high ranking.

Manuck et al. (2003) extended these results to captive female cynomolgus monkeys showing similar associations between serotonergic activity and response to a social stranger, and suggested that these hormone–behavior relationships are not limited to one sex (Manuck et al. 2003). Westergaard et al. (2003) also found an association between low levels of 5-HIAA and impulsivity in free-ranging rhesus females, mirroring results from males. These authors suggest that impulsive behavior may lead to different outcomes in males and females in matrifocal societies; females, because they are surrounded by kin allies, are "buffered" from the negative effects of impulsivity, which can lead to social isolation and severe aggression among males, who are not similarly surrounded by kin allies. Data on the relationship between dominance, impulsivity, and monamine levels in a broader range of species are clearly needed to elucidate the differences between the sexes.

The serotonin transporter gene is also polymorphic in monkeys, and the length of the polymorphic region is associated with variation in anxiety, aggression, and other traits (Trefilov et al. 2000; Higley et al. 2011; Suomi et al. 2011). In a study focused on a specific behavioral outcome in rhesus monkeys – age at dispersal in males – one allelic variant of the serotonin transporter gene resulted in a significantly younger age at dispersal than the other (57.1 vs. 67.5 months), which may indicate differences in risk-taking behavior (Trefilov et al. 2000). The homozygous individual may have experienced reduced serotonin levels, consistent with the association between low serotonin, aggression, impulsivity, and other behavioral traits (see above, and also Mehlman et al. 1994; Higley et al. 1996).

7.5 Future Directions

Psychologists have repeatedly demonstrated that trait-like differences exist among humans, and we describe people based on their unique personalities. In recent years, interest has increased in the role of personality in disease risk, and many such relationships between physiology and personality are mediated or at least influenced by hormones. Nevertheless, although most primate researchers refer to the personalities of the animals they study, few studies have quantitatively assessed the relationship between behavioral style and hormone levels in nonhuman primates. This has probably stemmed from perceived methodological limitations, but in fact these limitations are not real; relatively simple ways to measure both behavioral style and hormone levels exist. An investigation into how hormones vary in nonhuman primates as a result of variation in personality is likely to yield both practical and theoretical insights. Practical implications include captive management, where knowledge of how different personality types respond physiologically to grouping arrangements, physical aspects of enclosures, and change could affect the health and welfare of the animals. It could also hold value for breeding programs if personality variation is related to successful outcome; for example, individuals could be chosen for breeding based on their behavioral styles, which could affect (for example) their success at mothering.

The theoretical implications of quantifying personality also make this field worthy of study. Hormones are one proximate mechanism through which natural selection has shaped behavior (Clarke and Boinski 1995). When hormones affect personality, they affect an individual's life course. On the flip side, when personality affects hormones, it has the potential to affect an individual's reproductive success by influencing the reproductive axis. Though not the subject of this chapter (Sih 2011), these investigations lead to questions such as why natural selection preserves such variation at all, or, alternatively, why it has not shaped individuals to exhibit limitless possibilities for behavioral flexibility (McElreath and Strimling 2006). More studies are clearly needed on the hormonally mediated costs and benefits of certain behavioral styles in primate species.

This field would benefit from more rigorous attention to the construct of behavioral style in primates. Recently, studies using subjective personality measures have contributed to the advancement of this field (Weiss et al. 2000; Itoh 2002; de Waal 2002; Weiss et al. 2002; Gosling et al. 2003; King and Landau 2003; Maninger et al. 2003;

Capitanio et al. 2004; Capitanio and Widaman 2005; King et al. 2005; Martin 2005; Pederson et al. 2005), but quantitative measures of behavioral style using ethological data are still almost completely absent from the literature. While authors will report on differing rates of maternal behaviors or may comment on dominance style variation, in many cases these differences are not formally defined. In some cases, researchers may be hesitant to claim personality differences if their consistency and stability over time has not been demonstrated. We therefore need a greater number of longitudinal studies that can assess which differences are temporally stable. For example, do baboon or macaque mothers typically display the same maternal styles to each successive infant? Do female chimpanzees maintain a similar behavioral style when they emigrate to a new community? The answers to these questions will help us determine whether or not nonhuman primate personalities are akin to human personalities, which are generally considered to be inherent to individuals and therefore consistent over time and across environments (Finn 1986; Furnham and Heaven 1999; Caspi et al. 2003; Shiner et al. 2003).

Researchers are also no longer limited to experimental manipulations or data from animals in captive settings because of the expansion in the types of techniques available for biological sample collection and storage, as well as advances in hormone analysis (Whitten et al. 1998; Knott 2005). This has made possible the study of how hormones affect *social* behavior and an increased interest in personality as a biological variable, but we need to do more. The interaction of hormones and behavior in typical social environments and natural settings is rarely studied. In addition, studies of personality have focused almost exclusively on Old World monkeys and apes, but there is no theoretical reason for this bias. Of course, all primate endocrine studies must adhere rigorously to assay validation standards in order to ensure that hormone measurements are accurate, reliable, and repeatable.

Finally, because behavioral style involves the assessment of a variety of complex and sometimes subtle behaviors, and because hormones do not directly cause behavior, studies of hormone/style relationships are not likely to ever result in simple correlations (Zuckerman 1995; Anestis 2010). The variance in hormone levels that can be explained by differences in behavioral style is never going to be very high, and multiple hormones (and neurotransmitters) affect the expression of all personality traits. Hormones also interact with each other and with neurotransmitters, making the picture even more complicated. However, this does not make this a futile question; ultimately, biology is interesting partially because of its complexity, and much can be learned from trying to tease apart the intricate connections between hormones and behavior.

7.6 Conclusions

The study of primate behavioral endocrinology has increased our understanding of the biological bases of primate behavior, including social behavior. Primate personality is one aspect of social behavior, one that is gaining attention as an important factor in inter-individual variation. Finding out how particular behavioral styles are related to hormone levels is not just an interesting exercise that further demonstrates the similarities between humans and other primates, but rather is a process that has vast practical and theoretical implications.

Acknowledgments Funding for this project was provided by the Schwartz Family Foundation, the National Science Foundation under grant no. 0120175, The L.S.B. Leakey Foundation, and the Yale University Williams Fund.

References

Abbott DH, Keverne EB, Bercovitch FB et al. (2003) Are subordinates always stressed? A comparative analysis of rank differences in cortisol levels among primates. Horm Behav 43:67–82

Albertsson-Wikland K, Rosberg S, Lannering B et al. (1997) Twenty-four-hour profiles of luteinizing hormone, follicle-stimulating hormone, testosterone, and estradiol levels: a semilongitudinal study throughout puberty in healthy boys. J Clin Endocrinol Metab 82:541–549

Altmann J (1980) Baboon mothers and infants. Harvard University Press, Cambridge

Altmann J, Sapolsky R, Licht P (1995) Baboon fertility and social status. Nature 377:688–689

Anestis SF (2005) Behavioral style, dominance rank, and urinary cortisol in young chimpanzees (*Pan troglodytes*). Behaviour 142:1245–1268

Anestis SF (2006) Testosterone in juvenile and adolescent male chimpanzees (*Pan troglodytes*): Effects of dominance rank, aggression, and behavioral style. Am J Phys Anthropol 130:536–545

Anestis SF, Bribiescas RG, Hasselschwert DL (2006) Age, rank, and personality effects on the cortisol sedation stress response in young chimpanzees. Physiol Behav 89:287–294

Anestis S. 2010. Hormones and social behavior in primates. Evol Anthropol 19:66–78.

Attardi B, Tsujii T, Friedman R et al. (1997) Glucocorticoid repression of gonadotropin-releasing hormone gene expression and secretion in morphologically distinct subpopulations of GT1-7 cells. Mol Cell Endocrinol 131:241–255

Aureli F, Preston SD, de Waal FBM (1999) Heart rate responses to social interactions in free-moving rhesus macaques (*Macaca mulatta*): A pilot study. J Comp Psychol 113:59–65

Ballenger JC, Post RM, Jimerson DC et al. (1983) Biochemical correlates of personality traits in normals: An exploratory study. Pers Individ Dif 4:615–625

Bardi M, Shimizu K, Fujita S et al. (2001) Hormonal correlates of maternal style in captive macaques (*Macaca fuscata* and *M. mulatta*). Int J Primatol 22:647–662

Bardi M, French JA, Ramirez SM et al. (2004) The role of the endocrine system in baboon maternal behavior. Biol Psychiatry 55:724–732

Bardi M, Bode AE, Ramirez SM et al. (2005) Maternal care and development of stress responses in baboons. Am J Primatol 66:263–278

Bardi M, Huffman MA (2006) Maternal behavior and maternal stress are associated with infant behavioral development in macaques. Dev Psychobiol 48:1–9

Bardin CW (1996) The anabolic action of testosterone. New Eng J Med 335:52–53

Bardin CW, Catterall JF (1981) Testosterone: A major determinant of extragenital sexual dimorphism. Science 211:1285–1294

Barrett GM, Shimizu K, Bardi M et al. (2002) Endocrine correlates of rank, reproduction, and female-directed aggression in male Japanese macaques (*Macaca fuscata*). Horm Behav 42:85–96

Bartolomucci A, Palanza P, Sacerdote P et al. (2005) Social factors and individual vulnerability to chronic stress exposure. Neurosci Biobehav Rev 29:67–81

Barton RA, Byrne RW, Whiten A (1996) Ecology, feeding competition and social structure in baboons. Behav Ecol Sociobiol 38:321–329

Baum MJ (2002) Neuroendocrinology of sexual behavior in the male. In: Becker J, Breedlove M, Crews D, McCarthy M (eds) Behavioral endocrinology. MIT Press, Cambridge

Bergman TJ, Beehner JC, Cheney DL et al. (2005) Correlates of stress in free-ranging male chacma baboons, *Papio hamadryas ursinus*. Anim Behav 70:703–713

Bermond B, Mos J, Meelis W et al. (1982) Aggression induced by stimulation of the hypothalamus: effects of androgens. Pharm Biochem Behav 16:41–45

Bernhardt PC, Dabbs JM, Jr., Fielden JA et al. (1998) Testosterone changes during vicarious experiences of winning and losing among fans at sporting events. Physiol Behav 65:59–62

Boyson S (1994) Individual differences in the cognitive abilities of chimpanzees. In: Wrangham RW, McGrew WC, de Waal FBM, Heltne PG (eds) Chimpanzee cultures. Harvard University Press, Cambridge

Bribiescas RG (1996) Testosterone levels among Ache hunter-gatherer men: a functional interpretation of population variation among adult males. Hum Nat 7:163–188

Bribiescas RG (2001) Reproductive ecology and life history of the human male. Yearb Phys Anthropol 44:148–176

Bribiescas RG (2001) Reproductive physiology of the human male: An evolutionary and life history perspective. In: Ellison PT (ed) Reproductive ecology and human evolution. Aldine de Gruyter, New York

Bridges RS (1994) The role of lactogenic hormones in maternal behavior in female rats. Acta Paediatr Suppl 397:33–39

Brockman DK, Whitten PL (1996) Reproduction in free-ranging *Propithecus verreauxi*: Estrus and the relationship between multiple partner matings and fertilization. Am J Phys Anthropol 100:57–69

Brockman DK, Whitten PL, Richard AF et al. (2001) Birth season testosterone levels in male Verreaux's sifaka, *Propithecus verreauxi*: Insights into socio-demographic factors mediating seasonal testicular function. Behav Ecol Sociobiol 49:117–127

Buwalda B, Kole MHP, Veenema AH et al. (2005) Long-term effects of social stress on brain and behavior: A focus on hippocampal functioning. Neurosci Biobehav Rev 29:83–97

Capitanio JP (2011) Nonhuman primate personality and immunity: Mechanisms of health and disease. In: Weiss A, King JE, Murray L (eds) Personality and temperament in nonhuman primates. Springer, New York

Capitanio JP, Widaman KF (2005) Confirmatory factor analysis of personality structure in adult male rhesus monkeys (*Macaca mulatta*). Am J Primatol 65:289–294

Capitanio JP, Mendoza SP, Bentson KL (2004) Personality characteristics and basal cortisol concentrations in adult male rhesus macaques (*Macaca mulatta*). Psychoneuroendocrinology 29:1300–1308

Carcre C, Groothuis TGG, Mostl E et al. (2003) Fecal corticosteroids in a territorial bird selected for different personalities: Daily rhythm and the response to social stress. Horm Behav 43: 540–548

Caspi A, Harrington H, Milne B et al. (2003) Children's behavioral styles at age 3 are linked to their adult personality traits at age 26. J Pers 71:495–514

Clarke AS, Boinski S (1995) Temperament in nonhuman primates. Am J Primatol 37:103–125

Coe CL, Smith ER, Mendoza SP et al. (1983) Varying influence of social status on hormone levels in male squirrel monkeys. In: Steklis HD, Kling AS (eds) Hormones, drugs, and social behavior in primates. Spectrum Publications, New York

Cowlishaw G, Dunbar RIM (1991) Dominance rank and mating success in male primates. Anim Behav 41:1045–1056

Creel S (2001) Social dominance and stress hormones. Trends Ecol Evol 16:491–497

Creel S, Creel NM, Monfort SL (1996) Social stress and dominance. Nature 379:212–214.

Curley JP, Keverne EB (2005) Genes, brains and mammalian social bonds. Trends Ecol Evol 20:561–567

Davidson RJ, Kalin NH, Shelton SE (1993) Lateralized response to diazepam predicts temperamental style in rhesus monkeys. Behav Neurosci 107:1106–1110

Depue RA (1995) Neurobiological factors in personality and depression. Eur J Pers 9:413–439
Dettling A, Pryce CR, Martin RD et al. (1998) Physiological responses to parental separation and a strange situation are related to parental care received in juvenile Goeldi's monkeys (*Callimico goeldii*). Dev Psychobiol 33:21–31
Di Fiore A (2003) Molecular genetic approaches to the study of primate behavior, social organization, and reproduction. Yearb Phys Anthropol 122:62–99
Eaton GG, Resko JA (1974) Plasma testosterone and male dominance in a Japanese macaque troop compared with repeated measures of testosterone in laboratory males. Horm Behav 5:251–259
Ebstein RP (2006) The molecular genetic architecture of human personality: Beyond self-report questionnaires. Mol Psychiatry 11:427–445
Ebstein RP, Zohar AH, Benjamin J et al. (2002) An update on molecular genetic studies of human personality traits. Appl Bioinformatics 1:57–68
Elia M (1992) Organ and tissue contribution to metabolic rate. In: Kinney JM and Tucker HN (eds) Energy metabolism: Tissue determinants and cellular corollaries. Raven Press, New York
Ellison PT (2003) Energetics and reproductive effort. Am J Hum Biol 15:342–351
Fairbanks LA, Fontenot MB, Phillips-Conroy JE et al. (1999) CSF Monoamines, age and impulsivity in wild grivet monkeys (*Cercopithecus aethiops aethiops*). Brain Behav Evol 53: 305–312
Fairbanks LA, Jorgenson MJ (2011) Objective behavioral tests of temperament in nonhuman primates. In: Weiss A, King JE, Murray L (eds) Personality and temperament in nonhuman primates. Springer, New York
Fairbanks LA, Jorgensen MJ, Huff A et al. (2004) Adolescent impulsivity predicts adult dominance attainment in male vervet monkeys. Am J Primatol 64:1–17
Finn SE (1986) Stability of personality self-ratings over 30 years: Evidence for an age/cohort interaction. J Pers Soc Psychol 50:813–818
Fite JE, French JA (2000) Pre- and postpartum sex steroids in female marmosets (*Callithrix kuhlii*): Is there a link with infant survivorship and maternal behavior? Horm Behav 38:1–12
Francis DD, Meaney MJ (1999) Maternal care and the development of stress responses. Curr Opin Neurobiol 9:128–134
Fries ABW, Ziegler TE, Kurian JR et al. (2005) Early experience in humans is associated with changes in neuropeptides critical for regulating social behavior. Proc Nat Acad Sci USA 102:17237–17240
Fuchs E, Flugge G (2003) Chronic social stress: Effects on limbic brain structures. Physiol Behav 79:417–427
Furnham A, Heaven PCL (1999) Personality and social behaviour. Arnold, London
Gonzales-Bono E, Salvador A, Ricarte J et al. (2000) Testosterone and attribution of successful competition. Aggress Behav 26:235–240
Goodall J (1986) The chimpanzees of Gombe: Patterns of behavior. Bellknap Press, Boston
Gosling SD, Lilienfeld SO, Marino L (2003) Personality. In: D Maestripieri (ed) Primate psychology. Harvard University Press, Cambridge
Goujon E, Laye S, Parnet P et al. (1997) Regulation of cytokine gene expression in the central nervous system by glucocorticoids: mechanisms and functional consequences. Psychoneuroendocrinology 22 (suppl 2):S75–S80
Gouzoules H, Gouzoules S, Fedigan L (1982) Behavioural dominance and reproductive success in female Japanese macaques (*Macaca fuscata*). Anim Behav 30:1138–1150
Grattan D (2002) Behavioural significance of prolactin signalling in the central nervous system during pregnancy and lactation. Reproduction 123:497–506
Griffin JE (1996) Male reproductive function. In: Griffin JE, Ojeda SR (eds) Textbook of endocrine physiology. Oxford University Press, New York
Gust DA, Gordon TP, Wilson ME et al. (1991) Formation of a new social group of unfamiliar female rhesus monkeys affects the immune and pituitary adrenocortical systems. Brain Behav Immun 5:296–307
Hauser B, Deschner T, Boesch C (2008) Development of a liquid chromatography-tandem mass spectrometry method for the determination of 23 endogenous steroids in small quantities in primate urine. J Chromatogr B 862:100–112

Higley JD, King STJ, Hasert MF et al. (1996) Stability of interindividual differences in serotonin function and its relationship to severe aggression and competent social behavior in rhesus macaque females. Neuropsychopharmacology 14:67–76

Higley JD, Suomi SJ, Chaffin AC (2011) Impulsivity and aggression as personality traits in non-human primates. In: Weiss A, King JE, Murray L (eds) Personality and temperament in nonhuman primates. Springer, New York

Honess PE, Marin CM (2006) Behavioural and physiological aspects of stress and aggression in nonhuman primates. Neurosci Biobehav Rev 30:390–412

Hrdy S (1999) Mother nature: A history of mothers, infants, and natural selection. Ballantine, New York

Itoh K (2002) Personality research with non-human primates: Theoretical formulation and methods. Primates 4:249–261

Janson C (1985) Aggressive competition and individual food consumption in wild brown capuchin monkeys, *Cebus apella*. Behav Ecol Sociobiol 18:125–138

Kaplan JR, Manuck SB, Fontenot MB et al. (2002) Central nervous system monamine correlates of social dominance in cynomolgus monkeys (*Macaca fascicularis*). Neuropsychopharmacology 26:431–443.

Ketterson ED, Nolan VJ (1992) Hormones and life histories: an integrative approach. Am Nat 140:S33-S62

King JE, Landau V (2003) Can chimpanzee (*Pan troglodytes*) happiness be estimated by human raters? J Res Pers 37:1–15

King JE, Weiss A, Farmer KH (2005) A chimpanzee (*Pan troglodytes*) analogue of cross-national generalization of personality structure: Zoological parks and an African sanctuary. J Pers 73: 389–410

Knott C (2005) Radioimmunoassay of estrone conjugates from urine dried on filter paper. Am J Primatol 67:121–135

Laudenslager ML, Boccia ML (1996) Some observations on psychosocial stressors, immunity, and individual differences in nonhuman primates. Am J Primatol 39:205–221

Loucks AB, Redman LM (2004) The effect of stress on menstrual function. Trends Endocrinol Metab 15:466–471

Maestripieri D, Megna NL (2000) Hormones and behavior in rhesus macaque abusive and nonabusive mothers: 2. Mother-infant interactions. Physiol Behav 71:43–49

Maninger N, Capitanio JP, Mendoza SP et al. (2003) Personality influences tetanus-specific antibody response in adult male rhesus macaques after removal from natal group and housing relocation. Am J Primatol 61:73–83

Manuck SB, Flory JD, McCaffrey JM ct al. (1998) Aggression, impulsivity and central nervous system serotonergic responsivity in a non-patient sample. Neuropsychopharmacology 19:287–299

Manuck SB, Kaplan JR, Rymeski BA et al. (2003) Approach to a social stranger is associated with low central nervous system serotonergic responsivity in female cynomolgus monkeys (*Macaca fascicularis*). Am J Primatol 61:187–194

Marler CA, Moore MC (1988) Evolutionary costs of aggression revealed by testosterone manipulations in free-living male lizards. Behav Ecol Sociobiol 23:21–26

Martin JE (2005) The influence of rearing on personality ratings of captive chimpanzees (*Pan troglodytes*). Appl Anim Behav Sci 90:167–181

Mazur A, Booth A, Dabbs JMJ (1992) Testosterone and chess competition. Soc Psychol Quart 55:70–77

McElreath R, Strimling P (2006) How noisy information and individual asymmetries can make 'personality' an adaptation: A simple model. Anim Behav 72:1135–1139

McLean M, Smith R (1999) Corticotropin-releasing hormone in human pregnancy and parturition. Trends Endocrinol Metab 10:174–178.

Mehlman PT, Higley JD, Faucher I et al. (1994) Low CSF 5-HIAA concentrations and severe aggression and impaired impulse control in nonhuman primates. Am J Psychiatry 151:1485–1491

Meisel RL, Sachs BD (1994) The physiology of male sexual behavior. In: Knobil E, Neill JD (eds) The physiology of reproduction. Raven Press, New York

Moore SL, Wilson K (2002) Parasites as a viability cost of sexual selection in natural populations of mammals. Science 297:2015–2018.

Mota MT, Sousa MBC (2000) Prolactin levels of fathers and helpers related to alloparental care in common marmosets, *Callithrix jacchus*. Folia Primatol 71:22–26

Mota MT, Franci CR, Sousa MBC (2006) Hormonal changes related to paternal and alloparental care in common marmosets (*Callithrix jacchus*). Horm Behav 49:293–302

Muehlenbein M, Watts DP, Whitten PL (2004) Dominance rank and fecal testosterone levels in adult male chimpanzees (*Pan troglodytes schweinfurthii*) at Ngogo, Kibale National Park, Uganda. Am J Primatol 64:71–82

Muller MN, Wrangham RW (2004) Dominance, aggression and testosterone in wild chimpanzees: A test of the 'challenge hypothesis'. Anim Behav 67:113–123

Nelson RJ (2005) An introduction to behavioral endocrinology. Sinauer Associates, Sunderland

Netter P, Hennig J, Roed IS (1996) Serotonin and dopamine as mediators of sensation seeking behavior. Neuropsychobiology 34:155–165

Nieuwenhuijsen K, de Neef KJ, van der Werff ten Bosch JJ et al. (1987) Testosterone, testis size, seasonality, and behavior in group-living stumptail macaques (*Macaca arctoides*). Horm Behav 21:153–169

Nishida T, Hosaka K (1996) Coalition strategies among adult male chimpanzees of the Mahale Mountains. In: McGrew WC, Marchant LF, Nishida T (eds) Great ape societies. Cambridge University Press, Cambridge

Norbiato G, Bevilacqua M, Vago T (1997) Glucocorticoids and the immune system in AIDS. Psychoneuroendocrinology 22 (suppl 1):S19–S25

Packer C, Collins DA, Sindimwo A et al. (1995) Reproductive constraints on aggressive competition in female baboons. Nature 373:60–63

Pederson A, King JE, Landau VI (2005) Chimpanzee (*Pan troglodytes*) personality predicts behavior. J Res Pers 39:534–549

Perret M (1992) Environmental and social determinants of sexual function in the male lesser mouse lemur (*Microcebus murinus*). Folia Primatol 59:1–25

Pruessner JC, Gaab J, Hellhammer DH et al. (1997) Increasing correlations between personality traits and cortisol stress responses obtained by data aggregation. Psychoneuroendocrinology 22:615–625.

Pryce CR (1992) A comparative systems model of the regulation of maternal motivation in mammals. Anim Behav 43:417–441

Pryce CR, Abbott DH, Hodges JK et al. (1988) Maternal behavior is related to prepartum urinary estradiol levels in red-bellied tamarin monkeys. Physiol Behav 44:717–726

Pryce CR, Mutschler T, Dobeli M et al. (1995) Prepartum sex steroid hormones and infant directed behaviour in primiparous marmoset mothers (*Callithrix jacchus*). In: Pryce CR, Martin D, Skuse D (eds) Motherhood in human and nonhuman primates. Karger, Basel

Pusey A, Williams J, Goodall J (1997) The influence of dominance rank on the reproductive success of female chimpanzees. Science 277:828–831

Rehavi M, Sepcuti H, Weizman A (1987) Upregulation of imipramine binding and serotonin uptake by estradiol in female rat brain. Brain Res 410:135–139

Reimers M, Schwarzenberger F, Preuschoft S (2003) Personality as a predictor for stress susceptibility and stress coping mechanisms in former laboratory chimpanzees (*Pan troglodtyes*). Folia Primatol 74:216–217

Riley V (1981) Psychoneuroendocrine influences on immunocompetence and neoplasia. Science 212:1100–1109

Robbins MM, Czekala NM (1997) A preliminary investigation of urinary testosterone and cortisol levels in wild male mountain gorillas. Am J Primatol 43:51–64

Ron T (1994) An ambitious baboon: Independent rise in rank by a single female. Folia Primatol 63:71–74

Rose RM, Holaday JW, Bernstein IS (1971) Plasma testosterone, dominance rank and aggressive behaviour in male rhesus monkeys. Nature 231:366–368

Rosenblatt JS, Mayer AD, Giordano AL (1988) Hormonal basis during pregnancy for the onset of maternal behavior in the rat. Pyschoneuroendocrinology 13:29–46
Roy MP (2004) Patterns of cortisol reactivity to laboratory stress. Horm Behav 46:618–627
Rubinow DR, Schmidt PJ, Roca CA (1998) Estrogen-serotonin interactions: Implications for affective regulation. Biol Psychiatry 44:839–850
Saltzman W, Schultz-Darken NJ, Wegner FH et al. (1998) Suppression of cortisol levels in subordinate female marmosets: reproductive and social contributions. Horm Behav 33:58–74
Sannen A, van Elsacker L, Heistermann M et al. (2004) Urinary testosterone-metabolite levels and dominance rank in male and female bonobos (*Pan paniscus*). Primates 45:89–96
Sapolsky RM (1982) The endocrine stress-response and social status in the wild baboon. Horm Behav 16:279–292
Sapolsky RM (1983) Endocrine aspects of social instability in the olive baboon (*Papio anubis*). Am J Primatol 5:365–379
Sapolsky RM (1985) Stress-induced suppression of testicular function in the wild baboon: Role of glucocorticoids. Endocrinology 116:2273–2278
Sapolsky RM (1991) Testicular function, social rank and personality among wild baboons. Psychoneuroendocrinology 16:281–293
Sapolsky RM (1993) Neuroendocrinology of the stress-response. In: JB Becker, SM Breedlove D Crews (eds) Behavioral endocrinology. MIT Press, Cambridge
Sapolsky RM (1994) Why zebras don't get ulcers: A guide to stress, stress-related diseases, and coping. W. H. Freeman, New York
Sapolsky RM (1995) Social subordinance as a marker of hypercortisolism: Some unexpected subtleties. In: Chrousos G, McCarty R, Pacak K, Cizza G, Sternberg E, Gold P, Kvetnansky R (eds) Stress: Basic mechanisms and clinical implications. New York Academy of Sciences, New York
Sapolsky RM (2004) Mothering style and methylation. Nature Neurosci 7:791–792
Sapolsky RM, Pulsinelli WA (1985) Glucocorticoids potentiate ischemic injury to neurons: Therapeutic implications. Science 229:1397–1400
Sapolsky RM, Ray JC (1989) Styles of dominance and their endocrine correlates among wild olive baboons (*Papio anubis*). Am J Primatol 18:1–13
Schino G, D'Amato FR, Troisi A (1995) Mother-infant relationships in Japanese macaques: Sources of inter-individual variation. Anim Behav 49:151–158
Seltzer LJ, Ziegler TE (2007) Non-invasive measurement of small peptides in the common marmoset (*Callithrix jacchus*): A radiolabeled clearance study and endogenous excretion under varying social conditions. Horm Behav 51:436–442
Setchell JM, Dixson AF (2001) Changes in the secondary sexual adornments of male mandrills (*Mandrillus sphinx*) are associated with gain and loss of alpha status. Horm Behav 39: 177–184
Setchell JM, Dixson AF (2002) Developmental variables and dominance rank in adolescent male mandrills (*Mandrillus sphinx*). Am J Primatol 56:9–25
Shiner RL, Masten AS, Roberts JM (2003) Childhood personality foreshadows adult personality and life outcomes two decades later. J Pers 71:1145–1170
Soltis J, Wegner FH, Newman JD (2005) Urinary prolactin is correlated with mothering and allomothering in squirrel monkeys. Physiol Behav 84:295–301
Steklis HD, Raleigh MJ, Kling AS et al. (1986) Biochemical and hormonal correlates of dominance and social behavior in all-male groups of squirrel monkeys (*Saimiri sciureus*). Am J Primatol 11:133–145
Strier KB, Ziegler TE, Wittwer DJ (1999) Seasonal and social correlates of fecal testosterone and cortisol levels in wild male muriquis (*Brachyteles arachnoides*). Horm Behav 35:125–134
Sugiura M, Kawashima R, Nakagawa M et al. (2000) Correlation between human personality and neural activity in cerebral cortex. NeuroImage 11:541–546
Suomi SJ, Chaffin AC, Higley JD (2011) Reactivity and behavioral inhibition as personality traits in nonhuman primates. In: Weiss A, King JE, Murray L (eds) Personality and temperament in nonhuman primates. Springer, New York

Tanriverdi F, Silveira L, MacColl G et al. (2003) The hypothalamic-pituitary-gonadal axis: Immune function and autoimmunity. J Endocrinol 176:293–304

Teyler TJ, Vardaris RM, Lewis D et al. (1980) Gonadal steroids: Effects on excitability of hippocampal pyramidal cells. Science 209:1017–1019

Trefilov A, Berard J, Krawczak M et al. (2000) Natal dispersal in rhesus macaques is related to serotonin transporter gene promoter variation. Behav Genet 30:295–301

van Praag HM (2004) Can stress cause depression? Prog Neuropsychopharmacol Biol Psychiatry 28:891–907

Virgin CEJ, Sapolsky RM (1997) Styles of male social behavior and their endocrine correlates among low-ranking baboons. Am J Primatol 42:25–39

Vogt JL, Coe CL, Lowe E et al. (1980) Behavioral and pituitary-adrenal response of adult squirrel monkeys to mother-infant separation. Psychoneuroendocrinology 5:181–190

Vollrath M (2001) Personality and stress. Scand J Psychol 42:335–347

de Waal FBM (1982) Chimpanzee Politics. Johns Hopkins University Press, Baltimore

de Waal FBM (2002) Social roles, alternative strategies, personalities, and other sources of individual variation in monkeys and apes. J Res Pers 36:541–542

Wallis J (2002) Seasonal aspects of reproduction and sexual behavior in two chimpanzee populations: A comparison of Gombe (Tanzania) and Budongo (Uganda). In: Boesch C, Hohmann G, Marchant LF (eds) Behavioural diversity in chimpanzees and bonobos. Cambridge University Press, Cambridge

Weaver ICG, Cervoni N, Champagne FA et al. (2004) Epigenetic programming by maternal behavior. Nature Neurosci 7:847–854

Weiss A, King JE, Figueredo AJ (2000) The heritability of personality factors in chimpanzees (*Pan troglodytes*). Behav Genet 30:213–221

Weiss A, King JE, Enns R (2002) Subjective well-being is heritable and genetically correlated with Dominance in chimpanzees (*Pan troglodytes*). J Pers Soc Psychol 83:1141–1149

Westergaard GC, Suomi SJ, Chavanne TJ et al. (2003) Physiological correlates of aggression and impulsivity in free-ranging female primates. Neuropsychopharmacology 28:1045–1055

Whitten PL, Brockman DK, Stavisky RC (1998) Recent advances in noninvasive techniques to monitor hormone-behavior interactions. Am J Phys Anthropol 41:1–23

Wommack JC, Delville Y (2003) Repeated social stress and the development of agonistic behavior: Individual differences in coping responses in male golden hamsters. Physiol Behav 80:303–30

Young LJ (1999) Oxytocin and vasopressin receptors and species-typical social behaviors. Horm Behav 36:212–221

Ziegler TE (2000) Hormones associated with non-maternal infant care: A review of mammalian and avian studies. Folia Primatol 71:6–21

Zuckerman M (1995) Good and bad humors: Biochemical bases of personality and its disorders. Psychol Sci 6:325–332

Zuckerman M (1996) The psychobiological model for impulsive sensation seeking: A comparative approach. Neuropsychobiology 34:125–129

Zuk M, McKean KA (1996) Sex differences in parasite infections: Patterns and processes. Int J Parasitol 26:1009–1024

Chapter 8
Chimpanzee Faces Under the Magnifying Glass: Emerging Methods Reveal Cross-Species Similarities and Individuality

Kim A. Bard, Augusta D. Gaspar, and Sarah-Jane Vick

Abstract Independently, we created descriptive systems to characterize chimpanzee facial behavior, responding to a common need to have an objective, standardized coding system to ask questions about primate facial behaviors. Even with slightly different systems, we arrive at similar outcomes, with convergent conclusions about chimpanzee facial mobility. This convergence is a validation of the importance of the approach, and provides support for the future use of a facial action coding system for chimpanzees, *ChimpFACS*. Chimpanzees share many facial behaviors with those of humans. Therefore, processes and mechanisms that explain individual differences in facial activity can be compared with the use of a standardized systems such as *ChimpFACS* and *FACS*. In this chapter we describe our independent methodological approaches, comparing how we arrived at our facial coding categories. We present some Action Descriptors (ADs) from Gaspar's initial studies, especially focusing on an ethogram of chimpanzee and bonobo facial behavior, based on studies conducted between 1997 and 2004 at three chimpanzee colonies (The Detroit Zoo; Cleveland Metroparks Zoo; and Burger's Zoo) and two bonobo colonies (The Columbus Zoo and Aquarium; The Milwaukee County Zoo). We discuss the potential significance of arising issues, the minor qualitative species differences that were found, and the larger quantitative differences in particular facial behaviors observed between species, e.g., bonobos expressed more movements containing particular action units (Brow Lowerer, Lip Raiser, Lip Corner Puller) compared with chimpanzees. The substantial interindividual variation in facial behavior within each species was most striking. Considering individual differences and the impact of development, we highlight the flexibility in facial activity of chimpanzees. We discuss the meaning of facial behaviors in nonhuman primates, addressing specifically

K.A. Bard (✉)
Psychology Department, University of Portsmouth, King Henry Building,
Portsmouth, PO1 2DY, UK
e-mail: Kim.Bard@port.ac.uk

A. Weiss et al. (eds.), *Personality and Temperament in Nonhuman Primates*,
Developments in Primatology: Progress and Prospects, DOI 10.1007/978-1-4614-0176-6_8,

individual attributes of Social Attraction, facial expressivity, and the connection of facial behavior to emotion. We do not rule out the communicative function of facial behavior, in which case an individual's properties of facial behavior are seen as influencing his or her social life, but provide strong arguments in support of the role of facial behavior in the expression of internal states.

8.1 Introduction: Chimpanzee and Bonobo Facial Behavior

We come to this chapter with different backgrounds: Bard applies principles of developmental psychology to the study of chimpanzees (Bard 2005, 2007; van IJzendoorn et al. 2009; Bard et al. 2006); Gaspar applies principles of ethology to the study of facial behavior across species, speaking to issues in evolution (Gaspar 2001, 2006); and Vick applies principles of psychology to study primate communication (e.g., gaze and facial expression: Bethell et al. 2007; Vick and Anderson 2003; Vick et al. 2006). We collected evidence of individuality (Bard and Gardner 1996; Bard 1998; Gaspar 1996; Gaspar et al. 2004), including group differences (Bard 2003; Bard et al. 2005), and individual idiosyncrasies with facial behaviors (Gaspar 2006).

In this chapter, we ask questions about variation in facial expression from an evolutionary perspective: Why should extensive individual variation occur within the communication repertoire of a species? Should signals be unambiguous, standard, and species typical? We discuss this further, because there are indeed good biologically, developmentally, and evolutionarily based reasons to explore interindividual differences in communication signals and sequences, although this type of study continues to be rare in the literature.

8.1.1 From Universal Expressions to Individuality

The phylogenetic approach to understanding how facial expressions evolved is based on the first scientific study of facial expressions by Darwin (1872/1965). The earliest reports of nonhuman primate facial repertoires refer to facial communication in the chimpanzee (van Hooff 1967; van Lawick-Goodall 1968) and in the gorilla (Schaller 1964; Fossey 1983), and describe prototypic facial expressions, associated with emotion and/or communication: e.g., a "pout" face when individuals are feeling or communicating distress; a "play face" when individuals are happy or signaling playful intent. Facial expressions were conceived as facial displays (Andrew 1963), even in humans (Ekman and Friesen 1975), with universality in both the sender and in the receiver (Ekman et al. 1969). One assumption underlying this approach is that (at least some) facial expressions have an innate basis, with stereotypical appearance across individuals. Therefore, comparative studies could reconstruct the phylogenic history of facial displays (Pollick and de Waal 2007). Yet, as we will demonstrate in this chapter, there are large individual differences among facial displays in humans and chimpanzees. The magnitude of differences among individuals has overwhelming implications for adhering exclusively to a phylogenetic approach based on universal facial displays. In this chapter, we discuss these issues and advocate for the importance of studying many noninnate variables to explain universal and idiosyncratic aspects of facial behaviors.

The search for species-typical inventories of facial displays has resulted in reports of repertoires varying in size, between 6 and 51 different facial expressions in chimpanzees (van Hooff 1962, 1967, 1972; van Lawick-Goodall 1968; Berdecio and Nash 1981; Chevalier-Skolnikoff 1982; Gaspar 2001; Parr et al. 2005; Pollick and de Waal 2007), and between 5 and 46 facial expressions in bonobos (de Waal 1988; Gaspar 2001; Pollick and de Waal 2007). Extensive idiosyncrasy occurs in communicative repertoires (Hopkins et al. 2007), with idiosyncratic gestures, individual differences in the frequency of gestures, group-specific gestures in gorillas and bonobos (Pika et al. 2003, 2005), and group differences between chimpanzee and bonobos in flexible use of gestures (Pollick and de Waal 2007). In human studies, a distinction is made between expressive and unexpressive individuals, particularly in children (Underwood 1997). Moreover, expressivity, transparency, and other properties of the facial communication of humans have important social interaction correlates (Boyatzis and Satyaprazad 1994; Underwood 1997; Murphy and Faulkner 2000).

Many studies providing an inventory for each species will discount interindividual differences and ignore the potential anthropocentric biases inherent in the human perception of faces (Waller et al. 2007). The reported number of distinct facial expressions (1) are categorized by human observers who have spent years viewing human faces with a speedy configural processing system that often ignores or misperceives incongruent features (for details see Waller et al. 2007); (2) falls below the high range of distinct *gestalten* that chimpanzees are capable of doing, as a result of their nearly identical-to-human facial musculature (Burrows et al. 2006; Waller et al. 2006); and (3) is highly influenced by the particular coding systems chosen by different observers. Coding systems vary based on the questions addressed (Bakeman and Gottman 1997). For example, if asking about the social value of a smile then a system that codes for happiness might be ideal, whereas if asking about the quality of emotion expressed in a smile, then coding for particular components of the smile is essential. These types of questions reflect the theoretical debate about whether facial behaviors are an index of emotion or of other internal states, such as moods or feelings, or are signals in social interactions (Hinde 1985; Fridlund 1994; Preuschoft and van Hooff 1997; Parkinson 2005; Gaspar 2006). Additionally, we urge caution in creating inventories of facial expression types, as these often rely exclusively on prototypic categories, ignoring the dynamic qualities of facial expressions.

8.1.2 FACS Approaches to Nonhuman Primate Facial Displays

At the basic level of description of facial expression, research has been limited by the lack of a standardized, objective descriptive system that can be applied equivalently to human and to chimpanzee faces. The Facial Action Coding System

(FACS: Ekman and Friesen 1978) is anatomically based and describes surface appearance changes related to muscle action. The majority of Action Units (AUs) describes the action of a single underlying muscle. The FACS approach allows us to answer questions about the structure of primate facial displays. By identifying component movements, the FACS codes from the bottom-up, rather than describing how appearance may fit onto gestalt expression templates thereby allowing variation between similar facial configurations to be detected. Importantly, if the specific muscle causing facial movement cannot be determined, these actions are still described in detail but labeled as Action Descriptors (AD) (Ekman et al. 2002).

The top-down approach predominates in chimpanzee facial expression studies; researchers start with the overall appearance of commonly observed expressions and then dismantle these by describing the appearance of the component features, some much more consistently (e.g., van Hooff 1967) than others. While providing more detail on expressions and their variation than most approaches, coding is nonetheless selective and focused upon specific combinations. Reliance on describing only particular landmark features (e.g., a distinctive mouth shape) means that individual variation in facial morphology may result in the expression being difficult to recognize across individuals (Oster and Ekman 1978).

A few studies have used FACS to describe facial behavior in nonhuman primates. These initial attempts to employ a FACS approach to primate facial expressions revealed that such detailed coding of facial expressions is possible in primates (Preuschoft and van Hooff 1995; Steiner et al. 2001; Ueno et al. 2004). However, these studies were selective in only using the most intuitive AUs. Moreover, the manner of translating AUs across species is often understated: how one translates the human FACS for use with other primates needs to be carefully specified because primates differ in facial morphology, and the appearance of facial movements differs, sometimes dramatically, with facial morphology. Most relevant here, these initial FACS-based primate studies took a top-down approach: they documented facial inventories for species, and then applied a FACS approach to describe species-specific facial displays while individual differences in facial displays were not reported.

Past research has largely focused on peak expressions rather than movement of the face per se. However, a bottom-up approach starting with how specific movements may alter individual features is a more useful means for studying a species' facial repertoire. Studying individual movements, rather than looking only for configurations at their apex, broadens the field of communication studies allowing for specification of smaller and more subtle facial movements (Dawkins 1986 cited in Dawkins and Guilford 1991). Without a methodology to address subtle facial movements, systematic exploration of their potential signal value is not feasible.

The need to specify how AUs are translated from adult humans for use in different study populations is demonstrated by the FACS for human infants (BabyFACS) created to take differences between infant and adult facial morphology into account (Oster and Ekman 1978; Oster 2005). The FACS approach advocates using a particular methodological process in developing the coding system, specifically grounding facial movements in the underlying muscle structure. While such a discussion may seem pedantic, the standardization value of FACS is that it provides a common language, based in musculature, for describing changes in facial appearance. To apply FACS to distinct groups (whether babies or other species) means to fully consider underlying muscular architecture in coding facial movements with established AUs, or to identify that detailed facial ADs are being used that are not directly related to muscle action (as for AUs).

Thus, in this chapter we depend on the independently created systems that code facial movements (Gaspar 2001; Vick et al. 2007) to protect against the bias of top-down classifications (Waller et al. 2007). Both systems allow delineation of features that comprise particular facial expressions rather than assuming all facial expressions of a category contain the same features for all individual cases (a bottom-up approach). Both coding systems are atheoretical, protecting the descriptions of facial behavior from conceptual bias. Component features (AUs or ADs, translated into FACS equivalents when applicable) are used to describe facial expressions similar to the processes used by Ekman and Friesen (1978) for human expressions and van Hooff (1962, 1967) for nonhuman primate expressions. Facial expression, for us, refers to a *gestalt* (or configuration) of AUs or ADs.

As this review indicates, there is little continuity among studies of chimpanzee facial behavior; coding systems are usually designed to answer the specific research question and therefore may have very limited applications or use in comparing individuals both between and within species. Thus, comparisons among studies are hindered by differences in both methodology and definitions of facial behaviors. It is possible that not only is there discordance in terms of the labels that these expressions are given but that there may also be variation in terms of the facial display classifications themselves, as reported for some manual gestures (e.g., Whiten et al. 1999; Pollick and de Waal 2007).

8.1.3 Describing Chimpanzee and Bonobo Facial Events "From Scratch": Creating Descriptor Systems

In 1997, Gaspar began a study of bonobo (*Pan paniscus*) and chimpanzee (*Pan troglodytes*) facial behavior with three main goals: (1) to study the diversity of facial expressions in each species; (2) to investigate whether expressions and their contexts differed across the two *Pan* species; and (3) to determine how similar in form and function these were to human facial expressions as described by Ekman and colleagues (Ekman and Friesen 1975; Ekman et al. 1987). In 2001, Bard and her ChimpFACS team, Marcia Smith Pasqualini, Lisa Parr, Bridget Waller, and Sarah-Jane Vick, with no knowledge of Gaspar's previous work, set out to develop and subsequently disseminate a facial action coding system for chimpanzees that paralleled the FACS for humans.

Gaspar's project began with a decision to describe facial events "from scratch," in other words to develop a coding system for chimpanzees and bonobos without previous input of AUs from FACS. Three months during the summer of 1997 were devoted to ad libitum description of unitary actions and other descriptors of bonobo facial behavior, and another month was dedicated to ad libitum observation of chimpanzee facial behaviors. Gaspar created a detailed coding system that would allow for the composing of configurations in terms of facial actions and details of appearance changes – thus forming *gestalten*, full face configurations of various descriptive units.

Bard's ChimpFACS project began with 3 months of FACS training and certification. Subsequently, the team reviewed existing videotapes of chimpanzee facial expressions concluding that they did not provide sufficiently detailed views of the chimpanzee face movements to create and illustrate a detailed coding manual to compare to FACS. Therefore, additional months were spent collecting new videotaped records. The ChimpFACS team also consulted with Paul Ekman on the process of developing ChimpFACS from naturalistic observations and with Harriet Oster on the process of modifying and adapting FACS to develop ChimpFACS. The team decision was to begin with a comparison of the facial musculature of the chimpanzee (Burrows et al. 2006; see figure in Waller et al. 2006) in conjunction with a comparison of facial morphology (see Fig. 1, p. 7 in Vick et al. 2007). ChimpFACS can now be learned by anyone. The manual with video clips, practice coding, and certification test are available at the website www.chimpfacs.com. Attaining reliability with experts allows for all users to achieve standardization. Due to this process, observations of facial expressions can be confidently coded using ChimpFACS (Parr et al. 2007; Vick and Paukner 2010).

Gaspar (2001) is the only study to date to use a detailed FACS-based approach to study interindividual variation in facial repertoire. Gaspar used a random sampling method to build a chimpanzee and bonobo *facial expression* repertoire from the bottom up. In this method, she randomly selected the same amount of facial configuration samples from all individuals. A minimum of 50 1-min video focal samples of continuous facial behavior for each individual bonobo ($N=15$) and each chimpanzee ($N=21$) were obtained from the video database. The video database included proximate interactions so that the context of behavior was observable. In addition, during the recording, verbal commentary clarified the context. The facial behavior coding procedure from these videos comprised two stages: (1) extracting two random still frames (hereafter snapshots) from each 1 min video focal; and (2) analyzing each snapshot for the constituent facial descriptors (full list in Table 8.1). Coding began with the still frame, but each AD was confirmed by analyzing the original video motion transition to the extracted frame. With this sampling method, 2,100 chimpanzee snapshots were gathered (100 facial pictures each for 21 individuals) and 1,500 bonobo facial snapshots (15 bonobos). Context analysis was conducted using a list of 32 possible contexts that had been elaborated from previous ad lib observations and video scans of the colonies. For these analyses, intra-rater longitudinal reliability was calculated: 10% of snapshots were recoded after a 3-month interval and the index of concordance (Martin and Bateson 1994) was 0.86 for facial coding and 0.80 for context.

Table 8.1 Table of action units (AUs), action descriptors (ADs), and musculature for different facial action coding systems (FACS)

Visual description (Gaspar)	FACS equivalence (Ekman et al. 2002)	Muscle (Burrows et al. 2006) or AD (Gaspar 2005)	Muscle action (Waller et al. 2006)	ChimpFACS (Vick et al. 2007)	Facial mobility (Dobson 2008)
1. Eyelids widely opened	AU 5 (D and E intensities)	Yes	–	No	–
2. Eyelids slightly opened	Pan descriptor	Yes	–	–	–
3. Eyelids moderately opened	Pan descriptor	Yes	–	–	–
4. Eyebrows raised	AU1 + 2 brow raise	Yes	Yes	Yes	Yes
5. Eyebrows drawn together + inner corners raised	AU1 + 4 inner brow raised and drawn together	Yes	–	No	No
6. Eyebrows drawn together and downwards	AU4 brows lowered	Yes	No	No	No
7. Eyebrows relaxed	Baseline brows	–	–	–	–
8. Upper lip protruded	AU T18A (pucker upper lip)	Yes	No	No	Yes
9. Lower lip protruded	AU B18A (pucker lower lip)	–	No	No	Yes
10. Upper lip subducted	AU T28 (suck upper lip)	AD	–	Yes	–
11. Lower lip subducted	AU B28 (suck lower lip)	AD	–	Yes	–
12. Upper teeth exposed	Pan descriptor – not exclusive	AD	–	–	–
13. Lower teeth exposed	Pan descriptor – not exclusive	AD	–	–	–
14. Mouth slightly open (lips + teeth part)	AU25 + 26	AD	–	Yes	Yes
15. Mouth moderately open (lips + teeth)	AU25 + 26 (lips part and jaw dropped)	AD	–	Yes	Yes
16. Mouth widely open	AU27 jaw stretch	AD	–	Yes	Yes
17. Upper lip lifted	AU10 upper lip raise	Yes	Yes	Yes	Yes
18. Upper lip folded up and outward	AU T22 (upper lip) lip funnel	Yes	Yes	Yes	Yes
19. Lower lip lifted	AU17 (chin raise)	Yes	Yes	Yes	Yes

(continued)

Table 8.1 (continued)

Visual description (Gaspar)	FACS equivalence (Ekman et al. 2002)	Muscle (Burrows et al. 2006) or AD (Gaspar 2005)	Muscle action (Waller et al. 2006)	ChimpFACS (Vick et al. 2007)	Facial mobility (Dobson 2008)
20. Lower lip dropped	Pan descriptor	AD	–	AD160	–
21. Upper gum exposed	Pan descriptor – not exclusive	AD	–	–	–
22. Lower gun exposed	Pan descriptor – not exclusive	AD	–	–	–
23. Lower lip folded down and outward	AU B22 (lower lip) lip funnel	Yes	Yes	Yes	Yes
24. Lip corners lifted	AU12 lip corner puller	Yes	Yes	Yes	Yes
25. Lip corners lowered	AU15 lip corner depressor	Yes	No	No	Yes
26. Lips stretched horizontally (withdrawn)	AU20 lip stretch	Yes	No	No	No
27. Lips pressed	AU24 lip press	AD	–	Yes	Yes
28. Lower lip overlapping (the upper lip)	Pan descriptor – not exclusive	AD	–	Yes	Yes
29. Upper lip overlapping (the lower lip)	Pan descriptor – not exclusive	AD	–	Yes	Yes
30. Lip area bulged	Pan descriptor	AD	–	–	–
31. Upper lip (area) bulged	Pan descriptor	AD	–	–	–
32. Mouth closed and relaxed	Baseline lips	AD	–	–	–
33. Cheeks lifted	AU6 cheek raise	Yes	Yes	Yes	No
34. Cheeks asymmetrically lifted (right bias)	Right AU6 (cheek raiser right)	–	–	–	–
35. Cheeks asymmetrically lifted (left bias)	Left AU6 (cheek raiser left)	–	–	–	–
36. Upper lip-below nose area wrinkled	Pan descriptor	AD		–	–
37. Nose wrinkled	AU9 (nose wrinkle)	Yes	Yes	Yes	Yes
38. Nostrils dilated	AU38 (nostril dilate)	Yes	Yes	Yes	–

39. Cheeks relaxed	Baseline cheeks	AD	–	–	–
40. Chin lifted (active)	AU17 chin raise	Yes	Yes	Yes	Yes
41. Chin lifted (skin laterally stretched-synergistically)	AU17 chin raise	AD	–	–	–
42. Chin lifted and wrinkled (active)	AU17 chin raise	AD	–	–	–
43. Chin relaxed	Baseline chin	AD	–	–	–
44. Jaw dropping (chin skin relaxed)	AU 26 (jaw lower)	AD	–	Yes	–
A. Not described	AU13 cheek puffer	Yes	No	No	Yes
B. Not described	AU14 cheek dimpler	Yes	–	No	No
C. AD20 or AD22 or 13?	AU16 lower lip depressor	Yes	Yes	Yes	Yes
D. Not described	AU23 lip tightener	–	–	No	Yes

8.1.4 Problems with Inventories? Variation Within Categories

Most studies of chimpanzee facial expressions, both production and recognition, are based upon categorization of overall configuration; observers are concerned with the overall appearance or expression and may not focus upon individual components or level of intensity. Individual variation may occur in the context or frequency of expressions, i.e., the flexibility of usage. Information from other modalities, such as vocalizations and gestures, may encode individual identity. Pollick and de Waal (2007) compared chimpanzee and bonobo facial configurations using independence from a fixed context as a proxy measure of meaning and function. Some facial expression categories were used similarly across species and between groups while others were species typical. They recorded 18 facial/vocal signals and suggested these were used in similar contexts across species. Unfortunately, only 5 of 18 signals were recorded with sufficient frequency for the contextual analyses across species. Pollick and de Waal noted some cross-species differences including use of the silent pout face. However, without microanalyses of facial configurations, or more fine-grained analysis at the level of the individual, any variation would be masked by pooling of both expressions and individuals. The fact that concordance between facial/vocal expressions and context was less than 100% indicates flexibility in how or when individuals produce expressions in different contexts.

To examine variation in facial expressions, it is necessary to have a measurement tool that differentiates subtle differences in configuration. Gaspar (2001) generated facial expression categories (here *gestalten*), by grouping the facial behavior snapshots according to the quantitative similarity in facial descriptors: Those with identical descriptor composition received facial expression names, such as *semi-pouts*. Whenever *gestalten* matched literature descriptions the first published name was used (e.g., *relaxed face with drooped lip*, Goodall 1986). To incorporate *gestalten* that varied slightly from previous categorizations, two independent judges made classifications; agreement meant the photo was incorporated into one of the *gestalt* collections while disagreement led to adjacent groupings. In six chimpanzee photos and two bonobo photos, there were singular representatives of a gestalt *within* the species. For example, a category labeled *laughing face* had a single chimpanzee observation but was observed 10 times in the bonobos. This procedure resulted in inventories of facial *gestalten* (facial ethograms), with 57 facial *gestalten* for chimpanzees and 46 facial *gestalten* for bonobos.

Gaspar (2001) found considerable facial mobility in chimpanzees and bonobos. For example, *funny faces* – a facial behavior initially described by de Waal (1988) in bonobos was added to the list of *gestalten*, because chimpanzees and bonobos exhibited “face experimenting” sequences in which many expressions were performed with no apparent purpose other than experimenting with facial movement. In fact, one bonobo experimented with facial expressions while observing itself in a mirror. This is similar to testing of contingent movements as mentioned by Nielson et al. (2005), Bard et al. (2006), and Bard (2008). Because Gaspar’s sampling method (described in the previous section) only captures a small proportion of ongoing facial behavior (a video frame is only 1/1,500 of 1 min of focal facial movement) these ethograms may be considered preliminary. The full ethogram of bonobo and chimpanzee facial gestalten, therefore, may be substantially higher than, and considerably surpass the size of, existing inventories of facial displays (Parr et al. 2005; Pollick and de Waal 2007).

Table 8.1 compares the findings about facial movement across studies showing bottom-up facial movement repertoires in chimpanzees including anatomical descriptions of underlying musculature (Burrows et al. 2006) and their movement (Waller et al. 2006) as well as descriptions of observable chimpanzee facial movements (Gaspar 2001; Vick et al. 2007; Dobson 2008). Note that this approach contrasts with studies that categorize expression types in the first instance and use a top-down approach (Preuschoft and van Hooff 1995; Steiner et al. 2001; Ueno et al. 2004). The three observational studies of movement repertoire are not in full agreement. Both Gaspar (2001) and Vick et al. (2007) agree on 12 out of 16 core facial movements presented in Dobson (2008). Disagreement seems largely based on how FACS labels are applied in relation to the precise anatomical basis of a movement (e.g., specific muscle action) rather than whether a particular type of movement is seen (Vick et al. 2007; Dobson 2008).

Even when using a top-down categorical approach, it is clear that there is some configural variation within categories. For example, van Hooff (1967) distinguished a horizontal grin from vertical bared teeth. Goodall (1986) distinguished full and low grins, as well as closed and open grins. Parr et al. (2007) used ChimpFACS to code static images of chimpanzee facial expressions, which were already categorized into expression type. AUs combinations were sufficient for differentiating expressions from one another, indicating its validity as a facial measurement tool. The ChimpFACS approach was also able to identify within category variation for already classified facial expressions by specifying the facial action components of each case, and determining goodness of fit (Parr et al. 2007). For example, there were 34 cases of Bared Teeth display, with analyses indicating two main variants: one variant consisted of raised upper lip (AU10), lip corners pulled toward the ears (AU12) and lips parted (AU25), and the second variant added a lower lip depressed (AU16), resulting in the variant (AU10, AU12, AU16, AU25). In Gaspar's (2001) ethograms, there are considerably more, namely, 12 types of facial *gestalten* with bared teeth (7 types of closed grins and 5 types of open grins). One of the latter *gestalten* "full open grin," for example, upper teeth exposed (AU10) and mouth widely open (AU27) occurred in 100% of the cases, but additional AUs varied: lower teeth exposed, lips horizontally stretched (AU20) occurred in 94% of the cases, and eyebrow lowered (AU4) occurred in only 60% of the cases. At this stage in our studies, the emotional and/or communicative meaning of these different types of bared teeth facial expressions remains unclear. We therefore need to examine more fully systematic differences in the production of expressions by individuals and consider what impact this variation may have on perceivers, who may differentiate, ignore, or not perceive subtle differences in expressions (Fridlund 1994). Additionally, it is important to note that Gaspar (2001) found that although *gestalten* were used in quite different proportions by chimpanzees and bonobos, by far the most frequently sampled *gestalt* was the baseline relaxed face.

8.1.5 Comparing Action Descriptors: ChimpFACS and FACS

Gaspar produced a list of facial ADs shown with ChimpFACS AUs in Table 8.1. Overall there is significant comparability with human facial musculature, appearance changes, and FACS AUs. The majority of AUs were independently found in both of our projects. Similar conclusions followed from the use of these independently created systems. Chimpanzee facial AUs and descriptors (Gaspar 2001, 2006; Gaspar et al. 2004; Vick et al. 2007) overlap extensively with those seen in humans (Ekman and Friesen 1978; Ekman et al. 2002). The differences noted in Table 8.1, relate to the reliance on physical appearance changes by Gaspar (2001) and the reliance on anatomy by the ChimpFACS team (Waller et al. 2006). Two types of species differences were found: absence of some human actions in the chimpanzee face and absence of some chimpanzee actions in the human face.

Clear differences emerged in some appearance changes in the *Pan* species compared to humans, especially in AUs involving the orbital muscle of the lips, *Orbicularis oris* (i.e., AU18 Lip Pucker, AU22 Lip Funneler). Both teams also agreed that chimpanzees exhibit greater independence of movement in the upper and lower lips than is usually seen in humans. Human mouth movements are generally orbital in action, perhaps related to articulatory needs and precision of speech, but in chimpanzees and bonobos it is common to have movement occur in just one lip (Vick et al. 2007). Differences in facial morphology might also account for some differences in lip actions: for example, chimpanzees lack a chin boss so that their lower lip can droop down towards the jaw in a way not seen in humans. Therefore, both the ChimpFACS system and Gaspar distinguished between upper and lower lip for mouth actions (for AU22 Lip Funneler and for AU18 Lip Pucker) which are possible but rarely found in humans.

In the upper face, there are minor differences in the frequency of some actions and in the degree of flexibility of brow movement. For example, humans display more independence in Inner Brow Raise action (AU1), and more complexity in brow actions (e.g., combinations with AU4 Brow Lowerer) than found in chimpanzees. Overall, both groups concluded that the facial AUs of chimpanzees and humans extensively overlap (Gaspar 2001, 2004, 2006; Vick et al. 2007).

8.2 Chimpanzee and Bonobo Facial Behavior Compared

In the following section we compare the facial behavior of chimpanzees and bonobos, and consider the following ways in which individual differences might be apparent: (1) individuals could vary in the size or frequency of *gestalten*; (2) individuals could vary in how closely their facial behavior is tied with specific contexts; or (3) individuals could vary in how often certain gestalten occurred in particular contexts. The results are based on bonobo data from two colonies (Milwaukee County Zoo, Columbus Zoo and Aquarium) and chimpanzee data from three colonies (Cleveland Metroparks Zoo The Detroit Zoo; Burger's Zoo). We excluded AUs and ADs with low total frequencies (<5).

The facial expression repertoires of chimpanzees and bonobos are similar: 41 of the 60 gestalten were found in both species (Table 8.2; Gaspar 2001), but chimpanzees

Table 8.2 Joint chimpanzee and bonobo facial expression (*gestalten*) ethogram. Expression contexts are based on chi-square associations with p values <0.01. This is a summarized version of Gaspar (2001) bonobo and chimpanzee facial expressions' ethograms (pages 69–88 and pages 152–184 respectively). The table omits brow only facial expressions, variations of baseline with tense mouth, variations of Low closed grin, of Full open grin, Full closed grin and of Subducted Lips face, Bulging lips face and Protruded lips face. Bonobo exclusive expressions were also excluded

Joint chimpanzee and bonobo facial expression ethogram			
Photo	Name	Descripton:FACS Action Units equivalents & other descriptors	Signif. Contexts
	**Bulging Lips Face**	AU10+AU17+ AU23 + Upper lip area bulged + lower lip area bulged	Idle; attention shift (chimpanzee); ns associations (bonobo)
	**Relaxed Open Mouth Face**	AU25+26	Group excitment, Watch other's in affiliative interaction, Beg (chimpanzees); Affiliative contact other than grooming, Solitary play, Startle (bonobo)
	**Closed Mouth Smile**	AU12 + AU20	Affiliative contact other than grooming (chimpanzee)
	**Open Mouth Smile**	AU12+ AU25	Beg, Teasing & Quasi agonistic behavior (bonobo); give Invitation to Social play, Solitary play, Excitement (chimpanzee)
	**Play-Face**	AU12 + AU25+26	Invitation for Social play, Calm social play, rough social play (chimpanzee and bonobo); Solitary play, Effort/ physical challenge (chimpanzee)

(continued)

Table 8.2 (continued)

Joint chimpanzee and bonobo facial expression ethogram			
Photo	Name	Descripton:FACS Action Units equivalents & other descriptors	Signif. Contexts
	**Full Closed Grin**	AU10+ AU20+ AU25+ Gums exposed	Fear related (chimpanzee); courtship (bonobo)
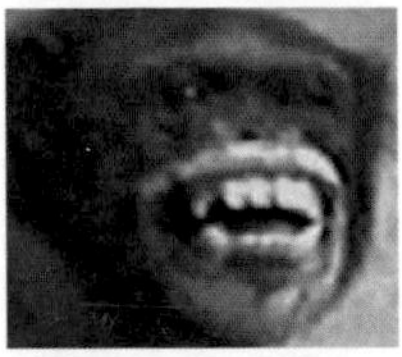	**Full Open Grin**	AU10+ AU20+ AU25+AU27 +Gums exposed	Fear related, (chimpanzee); Rough social play, Anger & threat displays (chimpanzee and bonobo)
	**Low Closed Grin**	AU16+ AU20+ AU25	Fear related (chimpanzee and bonobo), Frustration (chimpanzee)
	**Low Open Grin**	AU16+ AU20+ AU25+26	Anger & threat displays; ns associations in chimpanzee
	**Subducted Lips Face**	AU23 + AUT28 + D28	Effort/physical challenge (chimpanzee); ns associations in bonobo
	**Open Mouth Subducted Lips Face**	AU25+26 + AUT28+AUB28	Effort/Physical (bonobo); ns associations in chimpanzee

(continued)

Table 8.2 (continued)

Joint chimpanzee and bonobo facial expression ethogram			
Photo	Name	Descripton:FACS Action Units equivalents & other descriptors	Signif. Contexts
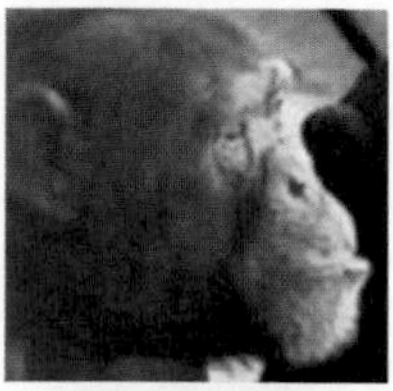	**Overlapping Lower Lip Face (Lip-Flip)**	Lower lip overlapping upper lip (no corresponding AU)	Effort/Physical challenge (bonobo); give Grooming, Mutual grooming (chimpanzee)
	**Pout Face**	AUT18+D18 AU24 AUT22+D22	Give grooming, Courtship, Mild-annoyance (bonobos); ns associations in chimpanzees
	**Protruded Lips Face**	AUT18 + D18	Non-agonistic display, give grooming (bonobos); mutual grooming (chimpanzees)
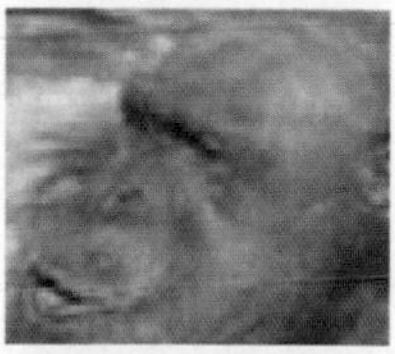	**Hoot-Face**	AU(T18+D18)+ AU(T22+D22)+ AU25+AU26	Anger & threat displays, group excitement, fear related (chimpanzee and bonobo)
	**Very Compressed Lips & Frown Face**	AU4 + AU23	Attention shift (bonobo); ns associations in chimpanzee
	**Stretched Lips Face**	AU20 + AU23	Watch over infant (bonobo); ns associations in chimpanzee

(continued)

Table 8.2 (continued)

Joint chimpanzee and bonobo facial expression ethogram			
Photo	Name	Descripton:FACS Action Units equivalents & other descriptors	Signif. Contexts
	Tongue-Show Face	AU25+ tongue protruded	ns associations in chimpanzee or bonobo, but many different contexts in chimpanzee
FUNNY FACES		many faces, not a gestalt	self-entertainment

and bonobos differed significantly in the frequency of use for 3 of the 9 facial descriptors. Bonobos, as a group, had significantly higher frequencies of AU4 (Brow Lowerer – see Figure 8.1a–c), AU10 (Upper Lip Raiser), and AU12 (Lip Corner Puller) than did chimpanzees, as a group (Gaspar and Bard, unpublished manuscript). There is not a one-to-one correspondence between facial expression *gestalten* (e.g., Bared Teeth facial display) and context (Gaspar 2001; Gaspar and Bard, unpublished manuscript; Parr et al. 2005; Pollick and de Waal 2007). For example, for the "full open grin" *gestalt*, 40% of occurrences were in the fear context, 40% in rough and tumble play context, and 20% were in contexts of anger or aggressive display. It is possible that across all contexts, there is a common highly excited emotional state tied to the facial *gestalt* of full open grin (e.g., Goodall 1986), or this *gestalt* might convey a precise "meaning" message not linked with contexts (perhaps of the sort "Stop what you are doing," e.g., Bard et al., unpublished manuscript).

Individual differences are also apparent in the total diversity of *gestalten* within each context. In this comparison, chimpanzees used a greater diversity of expressions than did bonobos in the contexts of *groom*, and *close-up inspection*. Bonobos used a greater diversity of expressions in *play*. The percentage overlap of expressions between the two species varied across contexts: at least 50% *gestalten* overlap in the contexts of *groom, concentration in activity,* and *anger and threat displays*, and 30–40% of bonobo and chimpanzee *gestalten* overlap in *affiliative contact, receive grooming*, and *fear* contexts. A curious observation is the fact that there are no *gestalten* found in both species in the *startle* context. This evolutionarily based basic reaction should register similarly in the face, so clearly this context needs further exploration in the future (Gaspar 2001).

In conclusion, we find that individual chimpanzees and bonobos varied in their use of different facial movements, in the frequency of different *gestalten*, and in the diversity of *gestalten* across different contexts. The use of the bottom-up approach is critical in this documentation of individual differences.

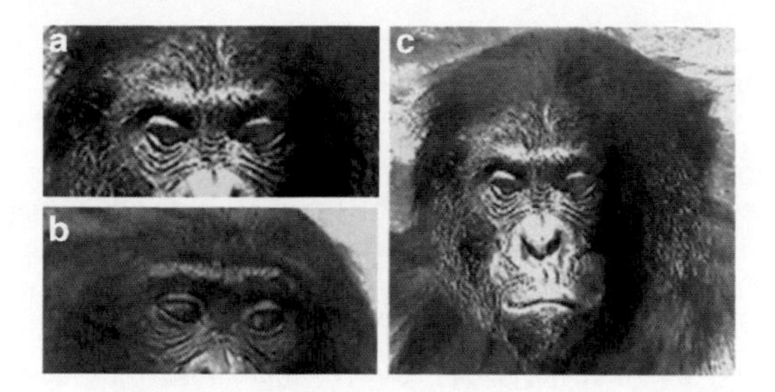

Fig. 8.1 Action Unit (AU) 4 in a bonobo (**a**, **c**). The same individual (Jimmy, at the Columbus Zoo and Aquarium. OH, USA) shown also with his typical relaxed brow shape in (**b**). ADs are detected during video playback and physiognomic differences between individuals require that we have a baseline referent of their facial features while relaxed (for instance, whereas Jimmy´s brows look arched in a baseline condition, but other individual's baseline looks horizontal or even slightly frowned)

8.3 Facial Behavior Is Important in the Study of Individual Differences: Relationships with Personality, Temperament, and Emotionality

In this section, we present our rationale for the study of individual differences in primate facial displays. We note that there is good reason to expect variation among primate individuals. Some primate facial movements are volitional, and therefore, we should not be surprised to find meaningfully large individual differences. We explore how individual differences in personality, temperament, and emotionality might be related to individual differences in facial behavior through higher order constructs such as "dominance," "attractiveness," "expressivity," and "extraversion."

8.3.1 Why Do We Think There Is Variation in Facial Behavior?

As evolutionary adaptations, we would expect phenotypic variation in the production and perception of expressions, with related fitness consequences (Schmidt and Cohn 2001). Given the challenges of studying facial expressions across species, and perhaps the pervasive influence of the universal emotions perspective, the focus has been on understanding the form and function of prototypical facial displays, with little consideration of idiosyncrasies in facial behavior. Important differences in facial behavior relating to sex, dominance, or age have not been fully addressed in nonhuman primates. Life experiences, in addition, may change communicative repertoires; studies of gestural communication reveal individual variation and idiosyncratic gestures (Liebal et al. 2004). Vocalization research has identified individual markers within some types of call (e.g., Owren and Rendall 2003). Like humans, great apes also have long periods of dependency during which time social skills are acquired, resulting in great plasticity in their social behaviors as evidenced by cultures (Whiten et al. 1999; van Schaik et al. 2003). Moreover, in humans, facial expression and individual identity are interlinked. Therefore, we should expect variation in signal production and perception based on the individual characteristics of nonhuman primates.

At the individual level, several variables can shape variation in facial displays (Schmidt and Cohn 2001). Anatomically, human facial musculature reveals some interindividual variation in muscle presence and precise structure (Pessa et al. 1998; Waller et al. 2008), but there has been little systematic investigation of variability for nonhuman primates. There are some inconsistencies in the anatomical literature, but it remains unclear whether these are due to differences or omissions in describing the muscle plan, to individual variation, or small sample sizes (Burrows et al. 2006; Waller et al. 2006). Interestingly, variation in muscle structure may lead to observable differences in expression appearance. For example, in humans, a bifid form of the zygomatic major muscle causes dimples when smiling (Pessa et al. 1998).

Neural control of facial movement may also affect expression production with involuntary and voluntary control over facial actions well described in humans (Rinn 1984) and rhesus macaques (Morecraft et al. 2001). Left–right asymmetry of expressions (Fernandes-Carriba et al. 2002) may be caused by asymmetrical facial structure or muscle innervation, with spontaneous production being more symmetrical than volitional movements. Cortical innervation may also underlie distinctions between the upper and lower face (Ross et al. 2007).

There is clearly variation in expression production. Although traditional ethograms present stereotypical displays, some variations are based on the intensity of expression and temporal patterns. In humans, the onset, offset, and peak of expression impacts expression perception and interpretation (Ekman and Rosenberg 2005). It would be interesting to examine nonhuman primate sensitivities to such subtle variations. Moreover, a combination of facial expressions, head position, and gaze direction may interact with facial movement to allow even greater variation in expression production (Redican 1975). Chimpanzees have distinct forms of gaze (Bethell et al. 2007). Moreover, patterns of mutual gaze are modifiable in early interaction with caregivers (Bard et al. 2005) suggesting that eye movement is another source of variation in dynamic facial displays in nonhuman primates.

In cognitive and behavioral studies, there are variations at the level of individual performance, but the basis for these individual differences is hard to determine. Personality factors are likely to determine an individual's attention, behavioral flexibility, accuracy, and response to reward schedules when performing tasks, and thus personality measures should be able to predict some of this variation (Uher and Asendorpf 2008). Similarly, chimpanzees may differ in overall expressivity which may be correlated with their personality (King and Figueredo 1997). Temperament may interact with previous experiences, so that the emotional valence of a given context can vary across individuals (Bard and Gardner 1996; Bard et al., unpublished manuscript). In addition, individuals may vary in volitional control of facial behaviors, and vary in responsiveness to the expressions of others as a function of attention, emotional sensitivity, or other individual factors.

Emotional contagion/induction or an individual's susceptibility to the emotional experiences of others is also likely to correlate with personality measures, such as extraversion. Even in early infancy, primates demonstrate variability in relation to their ability to imitate facial expressions (Myowa-Yamakoshi et al. 2004; Ferrari et al. 2006; Bard 2007) but we do not know whether this initial variation in tendency to imitate is correlated with personality attributes, or whether it has any impact upon future socioemotional experiences and learning. Contagious yawning is a robust phenomenon in humans and correlates positively with measures of empathy and theory of mind (Platek et al. 2003). Chimpanzees also demonstrate contagious yawning but there is individual variation; only 2 out of 6 female chimpanzees responded to video sequences of conspecific yawns with increased levels of yawning (Anderson et al. 2004). A more recent study in orangutans examined the rapid mimicry of play faces during dyadic play bouts; 16 out of 25 individuals showed rapid (<1 s) facial mimicry to the play face of their playmate but only 9 out of 25 did in a control condition (Davila-Ross et al. 2008). The quality of interactants' social relationships may affect both voluntary and involuntary mimicry of others as reported in humans (McIntosh 2006) but it would be valuable to examine these measures of emotional contagion across time and contexts in the same individuals in relation to personality traits.

The emotional state of receivers is likely to mediate the perception of emotion in nonhuman primates, as it does in humans, for example, by being more responsive to negative emotional expressions when already aroused by negative events. Finally, in humans, personality may bias perception of social signals. For example, trait anxiety leads to heightened perception of hostility when rating happy, neutral, and angry expressions (Knayazev et al. 2008). We do not yet know how chimpanzees' personalities affect their perceptual biases; it may be that individual variation in performance on perception tasks may reflect biases in expression processing that likely affects everyday social interactions with conspecifics. Attending to socially salient events is important for group living primates and is both cognitively costly and imperative for individual fitness (Barrett et al. 2002). We know that primate attention structures may reveal disproportionate interest in dominant individuals, mediated by an individual's rank (Shepherd et al. 2006), or the quality of dyadic relationships (Lane et al. unpublished manuscript). Attention to the expressive behaviors of dominant individuals is probably different than that directed towards other group mates. Some displays may automatically capture attention. For example, angry expressions may be processed rapidly and effectively (Gosselin et al. 1997). However, visual signals allow perceivers some degree of control, by disengaging gaze, to "cut-off" aversive stimuli, such as a facial threat (Kummer 1967; Altmann 1967).

8.3.2 Volition in Facial Expression Production

"A survey of the literature reveals that many who have commented on the signaling of animals ascribe to a view that all of their communicative signals are manifestations of emotion or affect" (Marler and Evans 1997, p. 133). That is, in this view, facial behaviors are spontaneous and uncontrollable displays of internal affective states. New research indicates that some behaviors, such as the raspberry or the extended grunt vocalization found in captive chimpanzees but long dismissed as "merely" emotional displays, are individually invented (Hopkins et al. 2007). The issue of whether nonhuman primates have voluntary control over expressions is a difficult topic to study; if chimpanzees can control their facial displays how could we tell? Awareness or control of facial behavior could result in the (a) deliberate production of a display or (b) intentional showing or (c) concealment of a spontaneous display. Here we briefly consider the evidence in support of each of these possibilities.

8.3.2.1 Deliberate Production of Displays

Chevalier-Skolnikoff (1982) considered apes to be capable of deceptive use of facial signals since they can "manifest facial expressions that are inconsistent with emotional state" (p. 360). Of course, it is difficult to exclude the possibility that the inferred emotional state is erroneous rather than the expression being incongruent. In humans, there is some evidence that spontaneous and voluntary expressions can be identified when analyzed in sufficient detail; timing and symmetry may differentiate between them (Ekman and Rosenberg 2005). The deliberate production of a display would be difficult to ascertain in chimpanzees because even similarly detailed studies of expression production would be limited by the need to make an inference regarding whether the expression was voluntarily produced (the problem of circularity).

The "funny faces" as described by de Waal (1988) in bonobos, and both in chimpanzees and bonobos by Gaspar (2001) and more recently in gorillas (Gaspar et al., unpublished manuscript), also indicate volition over facial movement, although the degree of control over specific facial actions has yet to be described in detail. In addition, some voluntary control over some facial movements can be inferred from imitation studies which show that chimpanzees can copy arbitrary actions (Custance et al. 1995). Two chimpanzees, trained to copy 15 arbitrary actions using a "do this" command, were then presented with 48 novel actions including 6 facial actions. Overall the chimpanzees showed clear evidence of imitation, but performance on the facial actions was weaker with only "lip smacking" being reliably identified by both observers. However, humans also have considerable difficulty copying facial movements. FACS training requires coders to perform each AU and success requires considerable hours of practice (Ekman et al. 2002). Some facial movements are more readily performed than others and questions about voluntary control may need to take the specific action into account. Those expressions that are regularly used with particular combinations of AUs may be more readily imitated than isolated facial movements, or vice versa if selection pressure on detecting cheats means that honest signals (genuine expressions) are costly to produce.

8.3.2.2 Intentional Showing of Displays

Intentional showing of displays could be explored in terms of social context (presence of an audience) or the target of displays, that is, whether signals are intentionally directed towards specific target receivers. Volition can be addressed by examining audience effects because if calls are automatic, the presence of conspecifics should not influence call production (e.g., Marler and Evans 1997). In birds, call production is influenced by audience effects (that are functionally relevant to each call). In chimpanzees, audience effects occur in gesture production (e.g., Leavens et al. 2004) and in vocal production (e.g., Hopkins et al. 2007). However, variation in call patterns may be influenced by arousal levels of an individual which are related to characteristics of the audience. Marler and Evans suggest that the former is unlikely for birds as only calls, and not other concomitant behaviors or physiological responses, differ according to social context.

Audience effects on facial behavior have rarely been directly examined in nonhuman primates, perhaps because of the difficulty in distinguishing among conspecifics as intended recipients of displays. Most primate communication is related to social interaction; emotions are essentially an integral component of any social context so that the social vs. emotional debate is based on an erroneous dichotomy (Parkinson 2005). However, van Hooff and Preuschoft (2003, p. 284) suggest that "the element of emotional expression comes to the fore when the display is performed in the absence of an audience." They suggest that solitary play accompanied by play faces in young chimpanzees is indicative of joyfulness. However, the play face can also be a socially mediated behavior as juvenile chimpanzees modify their signaling in the presence of young playmates' mothers (Jeanotte and de Waal 1996, cited in van Hooff and Peuschoft 2003).

The interaction between social context and facial behaviors and their function is obviously an important direction for future study. We return to this topic later (in Sect. 8.3.2), when discussing possible paths to study facial behavior predictors of personality. In addition to audience effects, facial behavior, unlike vocalizations, can also be directed towards particular recipients (Altmann 1967), who could differ in responsive characteristics. For example, expressions of vulnerability such as pain should be differentially directed at those who might provide support (Krebs and Dawkins 1984). There may also be subtle within-expression variations according to familiarity or relatedness of receivers. Animal signals may have low input specificity (Marler and Evans 1997) with a wide range of stimuli converging into one expressive display, which corresponds to both the emotion and communication view. Each signal can operate in concert with a range of others (Forrester 2008) and these combinations can lead to high specificity. For example, gaze and posture may indicate direction of an expression, age and sex class of sender (or receiver) may alter the meaning of a display, and even individual idiosyncrasies among familiar conspecifics may make the signal more precise. Similarly, Fridlund (1994) suggests that displays would be likely to vary according to other contextual features such as identity of interactants, their previous history, and the situation in which display is given (e.g., competition over food, defense of young).

8.3.2.3 Concealment and Suppression of Displays

Apes are aware of the communicative function of their expressions and may, on occasion, try to conceal the signal (Goodall 1986; de Waal 1992; Mitchell 1999). For example, Tanner and Byrne (1993) reported that a gorilla covered its play face on multiple occasions. Moreover, examples of concealment suggest that the production of the expression itself may not be suppressed. Involuntary facial movements may be more clearly concealed during interactions by simply orienting or moving away from potential recipients, or engaging in displacement activities. However, facial expressions can interfere with other behaviors. For example, at the Chester Zoo we observed a female chimpanzee's unsuccessful attempts to take a drink because her mouth was fixed in an intense Bared Teeth display (Waller and Vick, personal observation). This suggests that, in at least some circumstances, chimpanzees are unable to control their facial behavior. In contrast, humans may use voluntary facial movements to conceal or suppress spontaneous expressions (Ekman et al. 1988); for example, suppressing a smile (AU12 Lip Corner Puller) by the antagonistic action of pulling the mouth corners downwards (AU15 Lip Corner Depressor). As yet, the FACS approach to chimpanzee facial behavior has not revealed any evidence of such masking behaviors. Moreover, the incomplete suppression of an expression may be better conceptualized as conflicting intentions rather than leakage (Fridlund 1994) so that making inferences about the meaning of "deceptive" facial displays would be challenging.

The issue of variation has been related to whether nonhuman primates are capable of voluntary control over their expressions. Steklis and Raleigh (1979, p. 257) dispute the view "that the fundamental contrast between human and nonhuman primates is that the latter's lack significant voluntary control over their vocalizations and facial expressions." According to their view, if responses are involuntary they should be invariable and thus individual variation indicates some degree of volition over expressions. As noted previously, individual variations may be predominantly caused by differences in underlying anatomical structure (Schmidt and Cohn 2001). However, within the variation there may be stereotypical components. Schmidt et al. (2003) found that in human smiling the onset (lip corner movement) did not differ in terms of timing or duration, while peak and offset showed variation. This suggests that within a variable display there may be conspicuous and stereotyped signals. Identifying consistent features would clarify the communicative function of chimpanzee expressions.

8.3.3 Interindividual Differences in Facial Behavior

It has long been proposed that humans and other animals are genetically programmed to decode and respond adequately to the facial behavior of conspecifics (Buck 1984; Dimberg 1988). Ekman (1984) pointed out that humans pay more attention to the face than to other parts of the body when processing other people's nonverbal behavior, suggesting that this is why we are susceptible to being deceived more easily by the face than the body, and why we train face control more than body control. Our perceptions and attributions of personality may rely largely on facial behavior. Examples from studies using prototypical static facial expressions of emotion include perceived "friendly" social dominance and friendliness based on "happy" facial expressions, "unfriendly" dominance based on the display of anger facial expressions, and submissiveness based on the display of fear and surprise expressions (Knutson 1996; Hess et al. 2000; Zebrowitz et al. 2007). In addition, expressive people are seen as more attractive and likable than unexpressive people (Larrance and Zuckerman 1981; Halberstadt 1984; Friedman et al. 1988).

Although relevant information about oneself is often multichannel and what the face conveys is coordinated with what the body conveys (Grammer et al. 1997), relying on the face is fruitful both for sender and receiver, inasmuch as it is a source of accurate predictions of social outcomes. Rejected children display angry facial expressions more often than other children (Hubbard 2001). Popular children display an eyebrow frown less often than their less popular peers (Murphy and Faulkner 2000). A reliable cue to an altruistic nature and a willingness to share seems to be a *Duchenne* Smile (Mehu et al. 2007), a smile *gestalt* comprised of FACS's action units AU6 (Cheeks Raiser) and tightening of lids (with wrinkles and bulges below the lower eyelid) in addition to the AU12 (Lip Corner Puller) that distinguish "felt" spontaneous smiles from other types of smiles, e.g., posed smiles with only AU12 (Ekman and Friesen 1982).

It is plausible that in other primates, regulation of interactions based on expectations follows a similar pattern. This expectation is highest in the case of chimpanzees and bonobos, whom we now know have a very diverse facial behavior (Gaspar 2001; Gaspar and Bard, unpublished manuscript) and a facial musculature nearly identical to that of humans (Burrows et al. 2006). This ought to make possible a range of expressions in chimpanzees that is comparable to the diversity that humans exhibit. Combinations of AUs generated by naturally occurring human facial movements are countless and, most of the time, do not fall neatly into prototypical emotion expressions (Fernández-Dols and Ruiz-Belda 1997; Grammer et al. 1997, 2004). Moreover, within each of the six "basic" emotions (anger, happiness, fear, sadness, surprise, and disgust), there are a large number of different facial configurations interpreted to portray the emotion (Grammer et al. 1997). Finally, within a dynamic exchange, there are striking temporal variations in facial expressions.

A high degree of individuality in behavior has been documented in many nonhuman animals (Maestripieri 1993; Bard 1994; Hammershmidt and Todt 1995; Baker and Aureli 1997; Clarke and Snipes 1998; Bard et al. 2005), including personality dimensions similar to those of humans (Bard and Gardner 1996; Gosling and John 1999). It is clear that nonhuman primate personality at least partly overlaps with the Five-Factor Model of human personality (King and Figueredo 1997; McCrae and John 1992: Weiss et al. 2006). However, individuality in the facial behavior of nonhuman primates has rarely been quantified (with notable exceptions of Gaspar 2001; Gaspar et al. 2004; Jesus 2007; Jesus and Gaspar 2008). This lack of focus on individuality is not related to either the frequency or putative importance of facial individuality in social interaction. Rather, most studies of human facial behavior do not address spontaneous real-life events of facial behavior (more often focusing instead on the categorization of posed facial configurations using a limited list of expression terms). Exceptions, however, show that there is interindividual variation in frequency and specific configuration of emotional facial expressions in human children and adults (Buck 1975; Grammer 1988 cited in Grammer et al. 1997; Fernández-Dols and Ruiz-Belda 1997; Cohn et al. 2002) with notable stability of individual differences in adult facial behavior over time (Cohn et al. 2002).

As with other behaviors that vary individually, a facial action may play a role in the communication of individual qualities to interactors, including aspects of personality, if (a) it has a predictive value in "real world outcomes" or other behaviors; (b) it is consistent over time; and (c) it is consistent in an individual over time (Gaspar and Bard, unpublished manuscript; Gosling et al. 2003; Pervin and John 1997). Individuality in the communication repertoire of chimpanzees and bonobos was expected based on evidence of voluntary control and flexibility in the gestural, and on occasion vocal and facial behavior of gorillas, chimpanzees, bonobos, and orangutans (Hopkins and Savage-Rumbaugh 1991; Bard 1992; Tanner and Byrne 1993; Pika et al. 2003, 2005; Liebal et al. 2006).

An ideal starting point in the research of socially relevant predictions from individualized facial behavior is the assessment of its stability over time. A subsample of several ADs of bonobos was analyzed for consistency. Of the facial descriptors, 8 of the 9 did not differ in frequency of use across the 5 years (Gaspar and Bard, unpublished manuscript). The single exception was cheek raising (AU6), which supports a link of this action with intensity of expressions, rather than with individual expressivity (Messinger 2002). Therefore, we conclude that the frequency of many facial behaviors of individual bonobos remains consistent over time.

Bonobo facial behavior is also intraindividually consistent. There are systematic differences in certain facial action configurations across individuals, with some individuals significantly above the expected frequency for a given movement (Gaspar 2001; Gaspar and Bard, unpublished manuscript). Although male and female chimpanzees appear to use ADs with equivalent frequencies, male and female bonobos differ in some movements; females use Lip Stretch (AU20) more than males, and male bonobos use Upper Lip Raise (AU10) more than females. Female bonobos display more grin faces than males (Kano 1992), but there may be differences in the frequency of the "grin" contexts between males and females. Females display more grin faces in temper tantrums or frustrating contexts, which is fully compatible with Gaspar's (2001) Milwaukee and Columbus bonobo observations. There were no sex differences in the use of common facial *gestalten* in chimpanzees and bonobos. Frequencies of some facial descriptors differ across age categories in both *Pan* species (Goodall 1986; Kano 1992; Gaspar 2001; Gaspar and Bard, unpublished manuscript). For example, brow furrowing (AU4) in bonobos occurs most frequently in adults (54%) and least frequently in infants (10%). This is an interesting result since infant brow movements are easier to perceive. Infant bonobos performed more (67%) lip corner movements (AU12) compared to adolescents (18%) and adults (15%). Perhaps this is not surprising, as AU12 is a component of *play face* and *laughing face*, and social play is much more frequent in infants. Chimpanzees displayed large age differences in (1) brow raising (AU1 + 2): adults account for most of the observations (70%), compared to adolescents and infants (both at 15%) and (2) lip corner raising and stretching movements (AU12 and AU20) with infants accounting for most of the occurrences, but adult chimpanzees showing more than adolescents. We do not know whether these age-dependent facial movement differences relate to emotional, contextual, or other individual characteristics that may be age dependent as well.

8.3.4 Predicting Individual Traits from Facial Behavior

A major question that follows from findings that individuals are not only different from their age-class peers in facial behavior but also are consistent across long time-spans is whether individual differences in facial behavior are related to personality traits, other individual attributes, or specific behaviors.

Top-down approaches to chimpanzee personality based on human questionnaires and bottom-up approaches based on ethological methods differ (see Uher 2011) but, in our view, are complementary rather than mutually exclusive approaches. Those interested in using personality measures to predict specific behaviors in chimpanzees might be best off using a combined approach. To our knowledge, Carvalho (2008), a graduate student at Coimbra University, Portugal, is the first to investigate the relation between personality traits and three well-known chimpanzee facial expressions: play face, hoot face (Goodall 1986), and silent bared-teeth (van Hooff 1972). Carvalho adopted a quasi bottom-up approach. She selected personality trait descriptors from King and Figueredo's (1997) questionnaire and modified the descriptors so that they referred to directly observable behavior units. She and another researcher investigated the relation between traits and these expressions in 15 rehabilitant chimpanzees at the Jane Goodall Institute's Chimpanzee Eden Sanctuary in South Africa. As she predicted, there was large interindividual variation in personality traits and in the use of the facial expressions, and some significant relations. For example, the trait *active* was positively correlated with play face, the trait *dominant* was positively correlated with Hoot-face (Carvalho 2008).

The key point in the application of an ethological approach to assessing personality is that it circumvents a limitation of the questionnaire approach viz. the rater needs to be well acquainted with each subject. When personality traits are linked with specific behavioral measures, the behavioral measures can be used by any researcher, including those having no acquaintance with the target individuals (Carvalho et al. 2008). In addition, facial expressions can be useful in providing external validation for personality questionnaire items. Ideally personality traits could be assessed with both approaches (ethological behavior and personality questionnaires) for a large sample of chimpanzees. The larger the overlap of results, the more opportunities there will be for studying personality in great apes in different kinds of settings. Given validation, the behavior measures may be especially useful for longitudinal and ontogenetic studies that require the participation of different observers over time.

Research on the prediction of other individual attributes by facial behavior is still in its early days. In several nonhuman primate species, several associations were found between facial behavior and formal dominance status, with certain facial expressions being more frequent in dominant individuals (Chevalier-Skolnikoff 1973; van Hooff 1973; Jacobus and Loy 1981; Preuschoft 1992; Reichler et al. 1998) or differing according to context in those individuals (Gaspar 2001).

Human children vary in social attractiveness which ranges from peer rejection to being the most popular. Albeit there is controversy surrounding the factor(s) that contribute towards lowering or enhancing a child or teenager's social attraction (Babad 2001), peer relational status and interactive style are highly stable across many years and contexts (Cillessen et al. 2000; Englund et al. 2000; Dodge et al. 2003). These findings suggest that social attraction must be strongly dependent on individual personality traits.

For these reasons, Gaspar and colleagues investigated Social Attraction in chimpanzees and bonobos. Social Attraction is an individual attribute defined as "the proportion of affinitive interactions in which the target individual was engaged, at the time of sampling, that were *not* initiated by the target individual," i.e., receptivity to the affinitive invitations of others (Gaspar et al. 2004). The question was whether Social Attraction could be predicted from facial behavior traits, such as *expressivity* (a measure of facial behavior diversity) or the frequencies of play faces (the combination of mouth opening movements such as AUs 25 + 26 present in relaxed open-mouth face, play face, and open mouth smile connected to positive affect) and AU12 (Lip Corner Puller), present in open mouth smile and play face. In bonobos, Social Attraction correlated significantly with expressivity, and frequency of AU25 + 26 and AU12. For chimpanzees, Social Attraction was negatively correlated with frowning (AU4 Brow Lowerer). Social attraction in bonobos and chimpanzees is influenced by individual qualities of facial movements. One can argue that individuals use more play faces because they play more, regardless of whom starts the interaction, but the regression analysis on the bonobo data showed two interesting effects: The presence of AU25 + AU26 increases an individual's Social Attraction, but also when an individual is engaged in affinitive interactions (i.e., deemed by others to be socially attractive) there is an increased occurrence of AU25 + AU26 (Gaspar et al. 2004). Those results do not rule out the communicative function of facial behavior, specifically that an individual's facial behavior causes changes in social activities and/or relationships. However, it also shows that there is a two-way influence between individual quality and facial behavior. It provides support for the role of facial behavior in the expression of internal states, specifically that individual differences are more likely to result from variation in the intensity or frequency of emotions. Of course individuals will feel different emotions as a consequence of their different social interactions and different social roles (e.g., Parkinson 2005). Regardless of the directionality of the causal link, it is clear that the facial behaviors of chimpanzees and bonobos express their individuality.

With the facial behavior traits we used (Gaspar et al. 2004), it is clear that chimpanzees are less transparent in facial behavior than bonobos, and chimpanzee's

Social Attraction does not seem to be strongly affected by expressivity or specific facial actions (with the exception of brow furrowing). In bonobos there is a two-way connection between expressivity and certain facial actions on one hand, and Social Attraction on the other. This difference may reflect differences between the two species in social pressures. Bonobos are typically allowed a considerable amount of behavioral freedom. Chimpanzees, on the other hand, are tightly bound to a formalized social hierarchy that may not usually permit them to express their individuality. This contrast has been anticipated by the *Power Asymmetry Hypothesis* (Preuschoft and van Hooff 1995, 1997), which predicts that there should be greater plasticity in the use of signals in species with an egalitarian society or at least a nonrigid formal hierarchy than in species with an accentuated formal hierarchy. This is due to differential costs in being misunderstood by interactors, i.e., low in bonobos and high in chimpanzees.

The relation between these facial behavior traits and Social Attraction in apes was similar to that seen in human preschool children (Gaspar et al. 2004). There is no reason why these analogs should not be homologies in great apes and humans. Predictions from facial behavior to personality in chimpanzees, bonobos, and human children should be expected since chimpanzees and bonobos are humans' closest phylogenetic relatives (sharing 96–99% of DNA and a common ancestor that lived about 5 million years ago: Sarich and Wilson 1967; Sibley and Ahlquist 1984; Gagneux et al. 1999; Kumar et al. 2005). Once we adapt our studies to human adults, we should understand these factors more fully.

At a more basic level of analysis, predicting specific social behaviors from facial actions, we have still to address the following general questions: (1) Are certain types of interaction more common in individuals that display a particular expression more often than expected? and (2) Are there predictable causal relations between the intensity of facial movements and the intensity of emotion? Does the degree of mouth opening, for example, relate to emotional intensity? We are beginning to see other more specific questions addressed in a few promising studies: For example, does a high frequency of brow lowering (AU4) predict increased aggression or dominance or gender (Campbell et al. 1999)? Or can the dropped jaw/open mouth configuration (AU25+26) of the play face predict that the actor will be subsequently involved in a play bout? This seems a plausible prediction, since the invitation for social play stands out among other social invitations received by those bonobos and children who display it (Gaspar et al. 2004, Gaspar 2005; and see ongoing studies of chimpanzees, Thorsteinsson and Bard 2009; Davila-Ross et al. 2011).

Gaspar (2006) made a case for the advantages of reliable personality cueing in facial behavior based on game theory and on evidence that people make inferences about other people's personality traits promptly at zero acquaintance or after brief viewings of photos or videos of facial behavior (Ekman and Friesen 1978; Laser 1982; Gaspar 1994; Borkenau and Liebler 1995; Uleman et al. 1996; Mueller and Mazur 1997; Zebrowitz 1997; Krull and Dill 1998; Yamagishi et al. 2003; Grammer 2004). People spontaneously attribute personality or other traits to individuals even if specifically instructed to provide only emotional terms in an open questionnaire (Gaspar 1994) attributing for example "mean" to "anger" faces and "nice" to "happy"

(Ekman and Friesen 1975) and make accurate predictions for the other's future behavior (Grammer 2004). In the case of smiles (AU12) people at zero acquaintance also make good matches with target self-reports of extraversion (Borkenau and Liebler 1993). Bearing in mind that there are positive and negative social outcomes associated with the use of some facial actions and that the "best facial action readers" are also more popular children (Boyatzis and Satyaprazad 1994; Underwood 1997), we may be looking at a co-adaptation between expressivity (sensu transparency) and decoding capacity.

This co-adaptation of expressivity and decoding of emotion may enhance both actor's and receiver's fitness in a cooperative setting. The cost in vulnerability of honest signaling can be overcome by the returns in the form of trustworthy reputation and preference as social partner in cooperative tasks that clearly benefit the actor and his/her family. Forging good alliances for protection and food gathering is a great asset in a resource-limited changing environment. However, honest signaling is constrained by group size, as individual recognition is required (Dunbar 1988, 1993). Communication of intentions may be crucial in large complex societies, but studies of honest signaling in *Pongo*, *Gorilla,* and *Hylobatidae* facial behavior highlight a role for idiosyncratic variation in emotional responses in terms of generating individualized facial behavior. A study by Mehu et al. (2007) indicated that honest signaling is an asset for those collecting the benefits of cooperative relationships. Mehu et al. found that the Duchenne Smile (AU6 Cheek Raise + AU12 Lip Corner Puller) vary interindividually and was affected by situational factors, such as an hypothetical altruistic act, rather than positive emotion.

Emotion-related factors, interindividual differences in temperament, and appraisal-related personality traits may have a higher impact in the diversity of facial behavior than has been acknowledged. For example, we recently found (Gaspar and Esteves, in press) that "joy/playful" (Panksepp 2005), a prosocial oriented emotion, is the most convergent in terms of the facial actions that are used by toddlers. This prosocial emotional condition was the one where spontaneous emotion-related behavior best matches the universal facial configuration of "happy" (AU6 + AU12 + AU25 and eventually + AU26) as proposed by Ekman and Friesen (1975). Although "happy face" received only 27% of hits, far fewer hits occurred for the "fear face" (11%) or "surprise face" (5%). This leads us to hypothesize that only emotions that are directed at immediately changing an interactor's behavior will be highly stereotyped, indicating such action tendencies as readiness to interact socially or to play, etc.

Some emotions can be more susceptible to facial behavior modulation than others, as illustrated in Peleg and colleagues' (2006) elegant study of the heritability of emotional facial behavior. The authors compared the facial movements in born-blind individuals with those of their sighted relatives and nonrelatives and found that, for at least three emotions (anger, sadness, and "think-concentrate"), facial behavior is highly heritable.

Although some interindividual differences in emotional facial behavior can be attributed to facial anatomy (differences in muscles, fat tissue, etc.), personality traits related to temperament and situation appraisal are at the motivational basis of individual differences in the facial display of emotions. Therefore, the application of personality to the study of emotional facial behavior could be an important new development. It would release facial behavior research from its current stalemate between two underlying views: one view that discrete emotions have corresponding universal facial expressions that are consensually and "correctly" appraised, and another view that emotions have a componential nature (e.g., activation and valence) that results in a large diversity of facial behavior and appraisals.

Extraversion appears to translate well into predictable facial behavior. Compared with introverts, extraverts are more active and excitable (Eysenck 1975), and therefore they are expected to be more emotionally expressive. However, extraversion per se is uninformative about whether individuals are honest signalers. Future studies should focus on behavioral phenotypes that can be characterized differentially by quantitative aspects of the facial AUs and configurations they use, especially by the interaction of facial actions by context by appraisal. Whether individuals have higher or lower rates of spontaneous emotional configurations compared to a reference population should be useful in characterizing phenotypes at the high and low poles of neuroticism and extraversion. This reasoning is based on the assumption that not all that emotions include a "package" of typical facial actions and emotion, and that some emotions may not even involve facial actions. Furthermore, this could vary from individual to individual, since it would not be tied to an unequivocal message destined to elicit a typical reaction from the observer. For example, the relative inexpressiveness of introverts (Riggio and Riggio 2002) highlights the need to relate "invisible" facial actions to personality and emotion, which may be achievable using electromyography. Emotions that are not directed at modifying the behavior of the interactor (e.g., fear) could vary much more interindividually and contextually than those that evolved to modify the behavior of the interactant in specific ways (e.g., anger). Individuals with high neuroticism could use prototypical emotional expressions as an efficient means to recruit more attention and assistance. Individuals high in agreeableness may not have more frequent facial actions, but if agreeableness is linked with altruism and sympathy, these individuals may display more facial mimicry (Mehu et al. 2007; Davila-Ross et al. 2008).

8.4 Future Directions

While variation in presence and differentiation of underlying facial musculature occurs in humans (Schmidt and Cohn 2001), the paucity of data on nonhuman primates precludes any comparison in terms of phenotypes for facial displays at this level. However, a detailed anatomically based approach to recording facial behavior means that variance at the level of the display itself is detectable. For example, we do not yet know whether all individual chimpanzees show the same basic set of prototypical expressions, or whether there are consistent subtypes that could result from either variance in underlying musculature or behavioral idiosyncrasies. It would be interesting to note whether humans and chimpanzees share common variance in

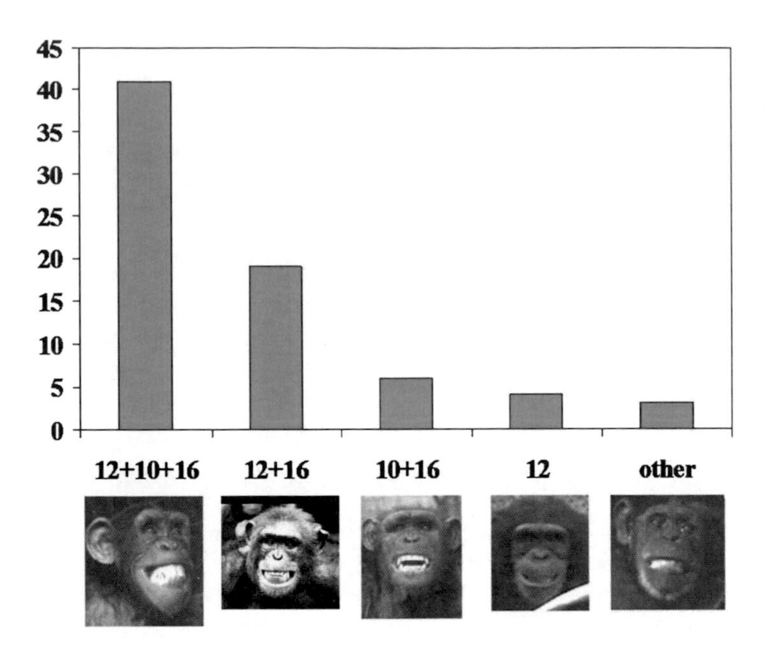

Fig. 8.2 A preliminary screening of approximately 70 photographs of chimpanzee Bared Teeth displays revealed a variety of combinations of AUs

facial myology or whether human variation is the result of more recent adaptations. For example, a relatively common variation in the zygomatic major muscle of humans is a bifid zygomatic major which causes dimpling during smiling (Pessa et al. 1998). This may serve to enhance the signal value of a smile (Schmidt and Cohn 2001). Given the universal emotions view, how can we reconcile variation with common displays? In fact, a recent anatomical study suggests that the muscles involved in the production of the "basic emotion" expressions are those that have the least individual variation (Waller et al. 2008). Facial displays may be fairly robust signals and perceivers may either not detect variation, or not attribute meaningfulness to variation (Fridlund 1994). Alternatively, since intensity, frequency, diversity, and efficacy of facial displays can be predicted by personality (e.g., extraversion or neuroticism: Riggio and Riggio 2002), we expect that quantitative variation will be found in "universal" displays of primate emotion expressions.

The most widely studied set of primate facial actions are those that fall under the nomenclature "grin" or *silent-bared teeth* display. Progress has been slowed because of the lack of comparability in definitions across studies. Here we present distinct types of bared teeth faces, based on a common coding system of facial actions that establishes quantitatively distinct variants (Figure 8.2). Future research can more appropriately consider whether "grins" systematically differ across contexts (Gaspar 2006).

In humans, the onset, offset, and peak of expression influences expression perception and interpretation (Ekman and Rosenberg 2005). It would be interesting to examine nonhuman primate sensitivities to such subtle variations. The perception of variation in facial displays by conspecifics could also be tested empirically to determine whether variations in configuration that can now be identified by micro-analysis of facial appearance can also be reliably categorized by conspecifics. Understanding the causal relation between emotional intensity and variation in displays is also a necessary next step in conceptualizing variation in facial expression in nonhuman primates.

The future holds great promise for the extension of facial coding systems to additional species. Dobson (2008) has suggested that the degree of facial mobility is related to body size, so that the large bodied apes have the largest facial movement repertoires. Orangutans may be particularly expressive in terms of facial mobility (Maple 1980; Call personal communication). Preliminary studies of gorilla facial behavior indicate that gorilla faces are capable of virtually the same facial actions used by chimpanzees and bonobos and that their facial behavior varies interindividually (Jesus and Gaspar 2008; Gaspar et al. unpublished manuscript).

Multiple evolutionary questions remain to be addressed in future research. Why should faces be "transparent" to individual qualities, and why might this vary between Hominoidea and other primate families? What are the biological advantages of honest signaling in the face and do they differ based on whether emotion or intention is being signaled? What inferences can be made about the social scenarios in the evolutionary history of *Homo* and *Pan* and how can we understand the social pressures that may have contributed to shape facial behavioral evolution?

8.5 Conclusions

Excellent ethograms of peak or prototypic facial expressions exist for chimpanzees. However, until recently no common language existed for exact identification of appearance changes. This has made it difficult to compare expressions across groups or individuals. If facial displays are to be considered as adaptations, then such phenotypic variation needs to be understood and its potential fitness consequences assessed (Schmidt and Cohn 2001). This cross-over in methodology, with ethological human studies and new coding techniques in chimpanzee research, should facilitate more meaningful comparisons between species and generate testable hypotheses for future study.

Schmidt and Cohn (2001) place human facial expressions within an evolutionary framework. They view "coordinated facial displays" as behavioral phenotypes but also recognize individual variation at three levels: facial structure, including age and sex differences, movement, and perception. Thus, the study of nonhuman primate facial expressions needs to allow for individual variation in production and also discriminate any factors influencing perception.

The success of the FACS itself means that there are many studies with humans that can be readily adapted for chimpanzee facial expressions. That is, the development of these coding systems allows for directly comparable methods to be employed in cross-species research and contribute to the questions surrounding both chimpanzee communication and the evolution of human facial expressions (Oster and Ekman 1978; Fridlund 1994). To date, questions of homology have primarily been answered on the basis of similarity in appearance of expressions (Preuschoft and van Hooff 1995; Steiner et al. 2001). More detailed, standardized, and anatomically based means of comparison is overdue. It is only by such cross-species comparisons that we can gain a better understanding of what is and is not unique to humans.

Acknowledgments We appreciate the support of Marcia Smith Pasqualini, Lisa Parr, and Bridget Waller in the development of ChimpFACS. Augusta Gaspar was supported by EU grants PRAXIS XXI/BD/9406/96, POCTI/PSI/57547/2002, and SFRH/BPD/26387/2005 through the Portuguese Council for Science and Technology. Kim Bard and Sarah-Jane Vick were supported, in part, by a Research Interchange Grant, F/00 678/E, from The Leverhulme Trust, and by the FEELIX GROWING grant from the European Commission, FP6-IST-045169.

References

Altmann SA (1967) The structure of primate social communication. In: Altmann SA (ed) Social communication among primates. University of Chicago Press, Chicago

Anderson JR, Myowa-Yamakoshi M, Matsuzawa T (2004) Contagious yawning in chimpanzees. Proc Biol Sci 271:468–470

Andrew RJ (1963) The origin and evolution of the calls and facial expressions of the primates. Behaviour 20:1–109

Babad E (2001) On the conception and measurement of popularity: More facts and some straight conclusions. Soc Psychol Educ 5:3–30

Bakeman R, Gottman J (1997) Observing Interaction: An introduction to sequential analysis. Cambridge University Press, Cambridge

Baker KC, Aureli F (1997) Behavioral indicators of anxiety: An empirical test in chimpanzees. Behaviour 134:1031–1050

Bard KA (1992) Intentional behavior and intentional communication in young free-ranging orangutans. Child Dev 63:1186–1197

Bard KA (1994) Evolutionary roots of intuitive parenting: Maternal competence in chimpanzees. Early Dev Parenting 3:19–28

Bard KA (1998) Social-experiential contributions to imitation and emotion in chimpanzees. In: Braten S (ed) Intersubjective communication and emotion in early ontogeny: A source book. Cambridge University Press, Cambridge

Bard KA (2003) Development of emotional expressions in chimpanzees (*Pan troglodytes*). Ann N Y Acad Sci 1000:88–90

Bard KA (2005) Emotions in chimpanzee infants: The value of a comparative developmental approach to understand the evolutionary bases of emotion. In: Nadel J, Muir D (eds) Emotional development. Oxford University Press, Oxford

Bard KA (2007) Neonatal imitation in chimpanzees (*Pan troglodytes*) tested with two paradigms. Anim Cogn 10:233–242

Bard KA (2008) Understanding reflections of self and other objects. In: Lange-Kuttner C, Vinter A (eds) Drawing and the non-verbal mind. A life-span perspective. Cambridge University Press, Cambridge

Bard KA, Gardner KH (1996) Influences on development in infant chimpanzees: Enculturation, temperament, and cognition. In: Russon AE, Bard KA, Parker ST (eds) Reaching into thought: The minds of the great apes. Cambridge University Press, New York

Bard KA, Myowa-Yamakoshi M, Tomonaga M et al. (2005) Group differences in the mutual gaze of chimpanzees (*Pan troglodytes*). Dev Psychol 41:616–624

Bard KA, Todd B, Bernier C et al. (2006) Self-awareness in human and chimpanzee infants: What is measured and what is meant by the mirror-and-mark test? Infancy 9:185–213

Bard KA, Bulbrook S, Maguire V et al. (unpublished manuscript). Developmental milestones in social skills and social 'manners' of young chimpanzees

Barrett L, Dunbar RIM, Lycett J (2002) Human evolutionary psychology. Palgrave, New York

Berdecio S, Nash LT (1981) Chimpanzee visual communication. Facial, gestural and postural expressive movement in young, captive chimpanzees (*Pan troglodytes*). Anthropological Research Papers n.26. Arizona State University, Tempe, AZ

Bethell E, Vick SJ, Bard KA (2007) Measurement of eye gaze in chimpanzees (*Pan troglodytes*). Am J Primatol 69:562–575

Borkenau P, Liebler A (1993) Consensus and self-other agreement for trait inferences from minimal information. J Pers 61:477–496

Borkenau P, Liebler A (1995) Observable attributes as manifestations and cues of personality and intelligence. J Pers 63:1–25

Boyatzis CJ, Satyaprazad C (1994) Children's facial and gestural decoding and encoding: Relations between skills and with popularity. J Nonverbal Behav 18:37–55

Buck R (1975) Nonverbal communication of affect in children. J Pers Soc Psychol 31:644–653

Buck R (1984) The communication of emotion. Guilford Press, New York

Burrows AM, Waller BM, Parr LA et al. (2006) Muscles of facial expression in the chimpanzee (*Pan troglodytes*): Descriptive, comparative and phylogenetic contexts. J Anat 208:153–167

Campbell R, Benson PJ, Wallace SB et al. (1999) More about brows: How poses that change brow position affect perceptions of gender. Perception 28:489–504

Carvalho C (2008) Relação entre traços de personalidade e expressões faciais em chimpanzés: um estudo com medidas comportamentais. [relation between personality traits and facial expressions in chimpanzees: a study with behavioral measures]. Master's Dissertation, Coimbra University, Portugal

Carvalho C, Gaspar A, Ney J et al. (2008) Chimpanzee personality traits and related facial expressions. A study with behavior measures. Paper presented at the International Primatological Society, 2008 Conference (August 3–8, Edinburgh, UK)

Chevalier-Skolnikoff S (1973) Facial expression of emotion in nonhuman primates. In: Ekman P (ed) Darwin and facial expression. Academic Press, New York

Chevalier-Skolnikoff S (1982) A cognitive analysis of facial behavior in Old World monkeys, apes, and human beings. In: Snowdon CT, Brown CH, Petersen MR (eds) Primate communication. Cambridge University Press, Cambridge

Cillessen AH, Bukowski WM, Haselager GJ (2000) Stability of sociometric categories. In: Cillessen AH, Bukowski WM (eds) Recent advances in the measurement of acceptance and rejection in the peer system. Jossey-Bass, San Francisco

Clarke AS, Snipes M (1998) Early behavioral development and temperamental traits in mother- vs peer-reared rhesus monkeys. Primates 39:433–448

Cohn J, Schmidt K, Gross R et al. (2002) Individual differences in facial expression: Stability over time, relation to self-reported emotion, and ability to inform person identification. Paper presented at the 2002 IEEE International Conference on Multimodal Interfaces, Pittsburgh

Custance D, Whiten A, Bard KA (1995) Can young chimpanzees (*Pan troglodytes*) imitate arbitrary actions? Hayes and Hayes (1952) revisited. Behaviour 132:837–859

Darwin C (1872/1965) The expression of the emotions in man and animals. University of Chicago Press, Chicago

Davila-Ross M, Menzler S, Zimmermann E (2008) Rapid facial mimicry in orangutan play. Biol Lett 4:27–30

Davila-Ross, M., Allcock, B., Thomas, C., & Bard, K.A. (2011). Aping expressions? Chimpanzees produce distinct laugh types when responding to laughter of others. Emotion, online first Feb 28, 2011, http://psycnet.apa.org/doi/10.1037/a0022594

Dawkins MS, Guilford T (1991) Receiver psychology and the evolution of animal signals. Anim Behav 42:1–14

de Waal FBM (1988) The communicative repertoire of captive bonobos (*Pan paniscus*) compared to that of chimpanzees. Behaviour 106:183–251

de Waal FBM (1992) Chimpanzee politics: Power and sex among apes. Johns Hopkins University Press, Baltimore
Dimberg U (1988) Facial expressions and emotional reactions: A psychobiological analysis of human social behavior. In: Wagner H (ed) Social psychophysiology: Theory and clinical practice. Wiley, New York
Dobson SD (2008) Allometry of facial mobility in anthropoid primates: Implications for the evolution of facial expression. Am J Phys Anthropol 138:70–81
Dodge KA, Lansford JE, Burks VS et al. (2003) Peer rejection and social information-processing factors in the development of aggressive behavior problems in children. Child Dev 74:374–393
Dunbar RIM. (1988) Primate social systems. Chapman and Hall, London
Dunbar RIM (1993) Coevolution of neocortex size, group size and language in humans. Behav Brain Sci 16:681–735
Ekman P (1984) Expression and the nature of emotion. In: Scherer K, Ekman P (eds) Approaches to emotion. Lawrence Erlbaum, Hillsdale
Ekman P, Friesen WV (1975) Unmasking the face. Consulting Psychologists Press, Palo Alto
Ekman P, Friesen WV (1978) Facial action coding system. Consulting Psychologists Press, Palo Alto
Ekman P, Friesen WV (1982) Felt, false and miserable smiles. J Nonverbal Behav 64:238–252
Ekman P, Friesen WV, Hager JC (2002) Facial action coding system. Research Nexus, Salt Lake City
Ekman P, Friesen WV, O'Sullivan M (1988) Smiles when lying. J Pers Soc Psychol 54:414–420
Ekman P, Friesen WV, O'Sullivan M et al. (1987) Universals and cultural differences in the judgments of facial expressions of emotion. J Pers Soc Psychol 53:712–717
Ekman P, Rosenberg EL (eds) (2005) What the face reveals: Basic and applied studies of spontaneous expression using the Facial Action Coding System (FACS). Oxford University Press, Oxford
Ekman P, Sorenson RE, Friesen WV (1969) Pan-cultural elements in facial displays of emotions. Science 164:86–88
Englund MM, Levy AK, Hyson DM et al. (2000) Adolescent social competence: Effectiveness in a group setting. Child Dev 71:1049–1060
Eysenck HJ (1975) The inequality of man. Edits Publishers, San Diego, CA
Fernandez-Carriba, S, Loeches A, Morcillo A et al. (2002) Asymmetry in facial expression of emotions by chimpanzees. Neuropsychologia 40:1523–1533
Fernández-Dols JM, Ruiz-Belda MA (1997) Spontaneous facial behavior during intense emotional episodes: Artistic truth and optical truth. In: Russell JA, Fernández-Dols JM (eds) The psychology of facial expression. Cambridge University Press, New York
Ferrari PF, Visalberghi E, Paukner A et al. (2006) Neonatal imitation in rhesus macaques. PLoS Biol 4:302
Friedman HS, Riggio RE, Casella DF (1988) Nonverbal skill, personal charisma, and initial attraction. Pers Soc Psychol Bull 14:203–211
Forrester GS (2008) A multidimensional approach to investigations of behavior: Revealing structure in animal communication signals. Anim Behav 76:1749–1760
Fossey D (1983) Gorillas in the mist. Houghton Mifflin, Boston
Fridlund AJ (1994) Human facial expression: An evolutionary view. Academic Press, San Diego
Gagneux P, Wills C, Gerloff U et al. (1999) Mitochondrial sequences show diverse evolutionary histories of African hominoids. Proc Natl Acad Sci USA 96:5077–5082
Gaspar A (1994) Interpretação de Combinações de Unidades de Acção Facial -estudo preliminar numa população de estudantes universitários (Interpretation of action unit combinations – a preliminary study with college students). In: Almada V, Oliveira R (eds) Biologia e comportamento actas do I Congresso Nacional de Etologia. ISPA: Lisboa
Gaspar AD (1996). [Behavior and personality attribution – a study with common chimpanzees] Comportamento e atribuição de personalidade - estudo em chimpanzés comuns (*Pan troglodytes*). Lisboa, Instituto Superior de Psicologia Aplicada. MSc Thesis
Gaspar AD (2001) Comportamento facial em *Pan* e *Homo*. Contribuição para o estudo evolutivo das expressões faciais (Facial behavior in *Pan* and *Homo*. Contribution to the evolutionary study of facial expressions). Doctoral dissertation, Universidade Nova de Lisboa, Lisboa

Gaspar AD (2005). [Facial Behavior and social attraction - Project final report POCTI/PSI/47547]. Lisboa: Fundação para a Ciência e Tecnologia
Gaspar AD (2006) Universals and individuality in facial behavior – past and future of an evolutionary perspective. Acta Ethol 9:1–14
Gaspar A, Bard KA (unpublished manuscript). Chimpanzee and bonobo facial behavior compared: Analyzing a multi-functional hypothesis
Gaspar A, Carvalho C, Melo S (2004) Facial action and social attraction in chimpanzees, bonobos and humans. Paper presented at the International Society for Human Ethology, Ghent, Belgium, July 2004
Gaspar AD, Esteves F (in press). Preschoolers' faces in spontaneous emotional contexts: How well do they match adult facial prototypes? In J Behav Dev
Gaspar A, Jesus G, Esteves F (unpublished manuscript). Gorilla (*Gorilla gorilla gorilla*) facial behavior – main actions, their contexts and an emotional structure?
Goodall J (1986) The chimpanzees of Gombe: Patterns of behavior. Harvard University Press, Cambridge MA
Gosling SD, John OP (1999) Personality dimensions in nonhuman animals: A cross species review. Curr Dir Psychol Sci 8:69–75
Gosling SD, Lilienfeld S, Marino L (2003) Personality. In: Maestripieri D (ed) Primate psychology. Harvard University Press, Cambridge
Gosselin P, Kirouac G, Dore FY (1997) Components and recognition of facial expression in the communication of emotion by actors. In: Ekman P, Rosenberg E (eds) What the face reveals. Oxford University Press, New York
Grammer K (2004) Body motion and personality. Paper presented at the congress the future of mind and behavior sciences, Torre do Tombo, Lisbon, March 2004
Grammer K, Filova V, Fieder M. (1997) The communication paradox and possible solutions. In: Schmitt A, Atzwanger K, Grammer K, Schaefer K (eds) New aspects of human ethology. Plenum Press, London
Halberstadt AG (1984) Family expression of emotion. In: Malatesta CZ, Izard CE (eds) Emotion in adult development. Sage, Beverly Hills
Hammershmidt K, Todt D (1995) Individual differences in vocalizations of young Barbary macaques (*Macaca sylvanus*): A multi-parametric analysis to identify critical cues in acoustic signaling. Behaviour 132:381–399
Hess U, Blairy S, Kleck RE (2000) The influence of facial emotion displays, gender and ethnicity on judgments of dominance and affiliation. J Nonverbal Behav 24:265–283
Hinde RA (1985) Was 'The expression of the emotions' a misleading phrase? Anim Behav 33: 985–992
Hubbard, JA (2001) Emotion expression processes in children's peer interaction: The role of peer rejection, aggression and gender. Child Dev 72:1426–1438
Hopkins WD, Savage-Rumbaugh ES (1991) Vocal communication as a function of differential rearing experiences in *Pan paniscus*: A preliminary report. Int J Primatol 12:559–583
Hopkins WD, Taglialatela J, Leavens DA (2007) Chimpanzees differentially produce novel vocalizations to capture the attention of a human. Anim Behav 73:281–286
Jacobus S, Loy J (1981) The grimace and gecker: A submissive display among patas monkeys. Primates 22:393–398
Jesus G (2007). Estudo tológico do comportamento facial de um grupo de gorilas da África Ocidental (Gorilla gorilla) em semi-cativeiro [An ethological study of facial behavior in a group of semi-captive gorillas]. Graduation Thesis in Biology. Lisboa. Universidade Lusofona
Jesus G, Gaspar A (2008) A little idiosyncrasy during a preliminary study of the facial behavior of a semi-captive group of Western Lowland Gorillas Poster presented at the International conference "Behavior and Individuality in Primates and other Mammals" (Lusofona University, Lisboa, Portugal, March 17–18)
Kano T (1992). The last ape: Pygmy chimpanzee behavior and ecology. Stanford University Press, Stanford

King JE, Figueredo AJ (1997) The Five-Factor Model plus Dominance in chimpanzee personality. J Res Pers 31:257–271
Knayazev GC, Bocharov AV, Slobodskaya HR et al. (2008) Personality-linked biases in perception of emotional facial expressions. Pers Indiv Diff 44:1093–1104
Knutson B (1996) Facial expressions of emotion influence interpersonal trait inferences. J Nonverbal Behav 20:165–182
Krebs JR and Dawkins R (1984) Animal signals: Mind reading and manipulation. In: Krebs JR, Davies NB (eds) Behavioral ecology: An integrated approach. Blackwell, Oxford
Krull DS, Dill JC (1998) Do smiles elicit more inferences than do frowns? The effect of emotional valence on the production of spontaneous inferences. Pers Soc Psychol Bull 24:289–300
Kumar S, Filipski A, Swarna V et al. (2005) Placing confidence limits on the molecular age of the human-chimpanzee divergence. Proc Natl Acad Sci USA 102:18842–18847
Kummer H (1967) Tripartite relationships in hamadryas baboons. In: Altmann SA (ed) Social communication among primates, University of Chicago Press, Chicago
Lane, L, Bard KA, Reddy V et al. (Unpublished manuscript) An affiliative basis for visual social monitoring in free ranging ring tailed lemurs (*Lemur catta*) at Berenty reserve, Madagascar
Larrance DT, Zuckerman M (1981) Facial attractiveness and vocal likability as determinants of nonverbal sending skills. J Pers 49:349–362
Laser PS (1982) Face facts: An unbidden role for features in communication. J Nonverbal Behav 7:3–19
Leavens DA, Hostetter AB, Wesley MJ et al. (2004) Tactical use of multimodal and bimodal communication by chimpanzees. Anim Behav 67:467–476
Liebal K, Call J, Tomasello M (2004) Use of gestural sequences in chimpanzees. Am J Primatol 64:377–396
Liebal K, Pika S, Tomasello M (2006) Gestural communication of orangutans (*Pongo pygmaeus*). Gesture 6:1–38
Maestripieri D (1993) Maternal anxiety in rhesus macaques (*Macaca mulatta*). II. Emotional bases of individual differences in mothering style. Ethology 95:32–42.
Maple T (1980) Orang-utan Behavior. Van Nostrand Reinhold Company, New York
Marler P, Evans CS (1997) Communicative signals of animals: Contributions of emotion and reference. In: Segerstrale U, Molnar P (eds) Nonverbal communication: Where nature meets culture. Lawrence Erlbaum Associates, Mahwah
Martin P, Bateson P (1994) Measuring behavior. An introductory guide (Second edn). Cambridge University Press, Cambridge
McCrae RR, John OP (1992) An introduction to the Five-Factor Model and its applications. J Pers 60:175–215
McIntosh DN (2006) Spontaneous facial mimicry, liking and emotional contagion. Pol Psychol Bull 37:31–42
Mehu M, Grammer K, Dunbar RIM (2007) Smiles when sharing. Evol Human Behav 28:415–422
Messinger D (2002) Positive and negative: Infant facial expressions and emotions. Curr Dir Psychol Sci 11:1–6
Mitchell RW (1999) Deception and concealment as strategic script violation in great apes and humans. In: Parker ST, Mitchell RW, Miles HL (eds) The mentalities of gorillas and orangutans. Cambridge University Press, New York
Morecraft RJ, Louie JL, Herrick JL et al. (2001) Cortical innervation of the facial nucleus in the nonhuman primate. Brain 124:176–208
Mueller U, Mazur A (1997) Facial dominance in *Homo sapiens* as honest signaling of male quality. Behav Ecol 8:569–579
Murphy S, Faulkner D (2000) Learning to collaborate: Can young children develop better communication strategies through collaboration with a more popular peer. Eur J Psychol Educ 15:389–404
Myowa-Yamakoshi M, Tomonaga M, Tanaka M et al. (2004) Imitation in neonatal chimpanzees (*Pan troglodytes*). Dev Sci 7:437–442

Nielson M, Baker EC, Davis JM et al. (2005) Imitation recognition in a captive chimpanzee (*Pan troglodytes*). Anim Cogn 8:31–36
Oster H (2005) The repertoire of infant facial expressions: An ontogenetic perspective. In: Nadel J, Muir D (eds) Emotional development. Oxford University Press, Oxford
Oster H, Ekman P (1978) Facial behavior in child development. In: Collins A (ed) Minnesota Symposia on Child Psychology, Vol. 11. Lawrence Erlbaum Associates, Hillsdale
Owren MJ, Rendall D (2003) Salience of caller identity in rhesus monkey (*Macaca mulatta*) coo and screams: Perceptual experiments with human listeners. J Comp Psychol 117: 380–390
Panksepp J (2005) Affective consciousness: Core emotional feelings in animals and humans. Conscious Cogn 14:30–80
Parkinson B (2005) Do facial movements express emotions or communicative motives? Pers Soc Psychol Rev 9:278–311
Parr LA, Cohen M, de Waal FBM (2005) Influence of social context on the use of blended and graded facial displays in chimpanzees. Int J Primatol 26:73–103
Parr LA, Waller BM, Vick SJ et al. (2007) Classifying chimpanzee facial expressions using muscle action. Emotion 7:172–181
Peleg G, Katzir G, Peleg O et al. (2006) Hereditary family signature of facial expression. Proc Natl Acad Sci USA 103:15921–15926
Pervin L, John OP (1997) Personality: Theory and research (7th edn) Wiley, New York
Pessa JE, Zadoo VP, Garza PA et al. (1998) Double or bifid zygomaticus major muscle: Anatomy, incidence and clinical correlation. Clin Anat 11:310–313
Pika S, Liebal K, Tomasello M (2003) Gestural communication in young gorillas (*Gorilla gorilla*): Gestural repertoire, learning, and use. Am J Primatol 60:95–111
Pika S, Liebal K,Tomasello M (2005) Gestural communication in subadult bonobos (*Pan paniscus*): Repertoire and use. Am J Primatol 65:39–61
Platek SM, Critton SR, Myers TE et al. (2003) Contagious yawning: The role of self-awareness and mental state attribution. Cogn Brain Res 17:223–227
Pollick AS, de Waal FBM (2007) Ape gestures and language evolution. Proc Natl Acad USA 104:8184–8189
Preuschoft S (1992) "Laughter" and "smile" in Barbary macaques (*Macaca sylvanus*). Ethology 91:220–236
Preuschoft S, van Hooff JARAM (1995) Homologizing primate facial displays: A critical review of methods. Folia Primatol 65:121–137
Preuschoft S, van Hooff JARAM (1997) The social function of "smile" and "laughter": Variations across primate species and societies. In: Segestrale U, Molnár P (eds) Nonverbal communication: Where nature meets culture. Lawrence Erlbaum Associates, Mahwah
Redican WK (1975) Facial expression in nonhuman primates. In: Rosenblum LA (ed) Primate behavior: Developments in field and laboratory research (vol 4). Academic Press, New York
Reichler S, Ducker S, Preuschoft S (1998). Agonistic behavior and dominance style in gelada baboons (*Theropithecus gelada*). Folia Primatol 69:200
Riggio H, Riggio RE (2002) Emotional expressiveness, extraversion, and neuroticism: A meta-analysis. J Nonverbal Behav 26:195–218
Rinn WE (1984). The neuropsychology of facial expression: A review of the neurological and psychological mechanisms for producing facial expressions. Psychol Bull 95:52–77.
Ross ED, Prodan CI, Monnot M (2007) Human facial expressions are organized functionally across the upper-lower facial axis. The Neuroscientist, 13:433–446
Sarich VM, Wilson AC (1967) Immunological time scale for hominid evolution. Science 158:1200–1203
Schaller GB (1964) The year of the gorilla. University of Chicago Press, Chicago
Schmidt KL, Cohn JF (2001) Human facial expressions as adaptations: Evolutionary questions in facial expression research. Yearb Phys Anthropol 44:3–24
Schmidt KL, Cohn JF, Tian Y (2003) Signal characteristics of spontaneous facial expressions: Automatic movement in solitary and social smiles. Biol Psychol 65:49–66.

Shepherd S, Deaner R, Platt M (2006) Social status gates social attention in monkeys. Curr Biol 16:119–120
Sibley CG, Ahlquist JE (1984) The phylogeny of hominid primates as indicated by DNA-DNA hybridization. J Mol Evol 20:2–15
Steiner JE, Glaser DHME, Berridge KC (2001) Comparative expression of hedonic impact: Affective reactions to taste by human infants and other primates. Neurosci Biobehav Rev 25:53–74
Steklis, HD, Raleigh MJ (1979) Behavioral and neurobiological aspects of primate vocalization and facial expression. In: Steklis HD, Raleigh MJ (eds) Communication and behavior: An interdisciplinary series. Academic Press, New York
Tanner JE, Byrne WB (1993) Concealing facial evidence of mood: perspective taking in a captive gorilla. Primates 34:451–457
Thorsteinsson K, Bard KA (2009) Coding infant chimpanzee facial expressions of joy. In: Banninger-Huber E, Peham D (eds) Current and future perspectives in facial expression research: Topics and methodological questions. Proceedings of the International Meeting at the Institute of Psychology, University of Innsbruck/Austria, Innsbruck University Press, Innsbruck Austria
Ueno A, Ueno Y, Tomonaga M (2004) Responses to four basic tastes in newborn rhesus macaques (*Macaca mulatta*) and chimpanzees (*Pan troglodytes*). Behav Brain Res 154:261–271
Uher J, Asendorpf, JB (2008) Personality assessment in the Great Apes: Comparing ecologically valid behavior measures, behavior ratings, and adjective ratings. J Res Pers 42:821–838
Uher J (2011) Personality in nonhuman primates: What can we learn from personality psychology. In: Weiss A, King JE, Murray L (eds) Personality and temperament in nonhuman primates. Springer, New York
Uleman JS, Hon A, Roman RJ et al. (1996) On-line evidence for spontaneous trait inferences at encoding. Pers Soc Psychol Bull 22:377–394
Underwood MK (1997) Peer social status and children's understanding of the expression and control of positive and negative emotions. Merrill-Palmer Q 43:610–634
van Hooff JARAM (1962) Facial expressions in higher primates. Symp Zool Soc Lond 8:97–125
van Hooff JARAM (1967) The facial displays of the catarrhine monkeys and apes. In: Morris D (ed) Primate ethology. Aldine, Chicago
van Hooff JARAM (1972) A comparative approach to the phylogeny of laughter and smile. In: Hinde RA (ed) Non-verbal communication. Cambridge Univ. Press, Cambridge
van Hooff JARAM (1973) A structural analysis of the social behavior of a semi-captive group of chimpanzees. In: von Cranach M, Vine I (eds) Social communication and movement. European Monographs in Social Psychology 4. Academic Press, London
van Hooff JARAM, Peuschoft S (2003) Laughter and smiling: The intertwining of nature and culture. In: de Waal FBM, Tyack PL (eds) Animal social complexity: Intelligence, culture, and individualized societies. Harvard University Press, Cambridge
van IJzendoorn MH, Bard KA, Bakermans-Kranenburg MJ et al. (2009) Enhancement of attachment and cognitive development of young nursery-reared chimpanzees in responsive versus standard care. Dev Psychobiol 51:173–185
van Lawick-Goodall J (1968) A preliminary report on expressive movements and communication in the Gombe Stream Chimpanzees. In: Jay PC (ed) Primates: Studies in adaptation and variability. Holt, Rinehart & Winston, New York
van Schaik CP, Ancrenaz M, Borgen G et al. (2003) Orangutan cultures and the evolution of material culture. Science 299:102–105
Vick SJ, Anderson JR (2003) The use of visual orientation cues in a competitive task by olive baboons (*Papio anubis*). J Comp Psychol 117:209–216
Vick SJ, Paukner A (2010) Variation and context of yawns in captive chimpanzees (*Pan troglodytes*). Am J Primatol 72:262–269
Vick SJ, Toxopeus I, Anderson JR (2006) Pictorial gaze cues do not enhance long tailed macaques' performance on a computerised object-location task. Behav Process 73:308–314

Vick SJ, Waller BM, Parr LA et al. (2007) A cross-species comparison of facial morphology and movement in humans and chimpanzees using the Facial Action Coding System (FACS). J Nonverbal Behav 31:1–20
Waller B, Bard KA, Vick SJ et al. (2007) Physical comparisons between chimpanzee and human facial expressions are related to emotional interpretation. J Comp Psychol 121:398–404
Waller BM, Cray JJ, Burrows AM (2008) Selection for universal facial emotion. Emotion 8: 435–439
Waller BM, Vick SJ, Parr LA et al. (2006) Intramuscular electrical stimulation of facial muscles in humans and chimpanzees: Duchenne revisited and extended. Emotion 6:367–382
Weiss A, King JE, Perkins L (2006) Personality and subjective well-being in orangutans (*Pongo pygmaeus* and *Pongo abelii*). J Pers Soc Psychol 90:501–511
Whiten A, Goodall J, McGrew WC et al. (1999) Cultures in chimpanzees. Nature 399:682–685
Yamagishi T, Tanida S, Mashima R et al. (2003) You can judge a book by its cover. Evidence that cheaters may look different from cooperators. Evol Hum Behav 24:290–301
Zebrowitz L (1997) Reading faces. Westview, Boulder
Zebrowitz L, Kikuchi M, Fellous JM (2007) Are effects of emotion expression on trait impressions mediated by babyfaceness? Evidence from connectionist modeling. Pers Soc Psychol Bull 33:648–662

Chapter 9
Nonhuman Primate Personality and Immunity: Mechanisms of Health and Disease

John P. Capitanio

Abstract There is growing recognition that consistent individual differences in behavioral dispositions, often referred to as personality, play an important role in primates' adaptations to their environments. Studies are reviewed that suggest that the dispositions Sociability/Affiliation, Emotionality/Reactivity, and Behavioral Inhibition are associated consistently with variation in immune function. It is likely that personality influences immunity by affecting how individuals appraise and cope with situations, and by the physiological consequences of differences in appraisal and coping. The important physiological mechanisms involve the two major stress-response systems: the hypothalamic-pituitary-adrenal system and the sympathetic-adrenal-medullary system. While many statistically significant effects have been found, the biological significance of personality–immune relationships, as responses to pathogen challenge, is less clear.

9.1 Introduction and Goals

One of the earliest theoretical schemes to understand health and disease was put forth by the Greek physician Hippocrates of Cos, who lived around 400 BC. The body was composed of four paired and opposing fluids (or humors) – blood and black bile, and phlegm and yellow bile – and each of these humors was also associated with the four elements, namely air and earth, and water and fire, respectively. Good health resulted from a proper balance of the four humors, and ill health resulted from one of the humors predominating. This imbalance had to be corrected to return to a state of health. A common cold, for example, resulted from an overabundance

J.P. Capitanio (✉)
California National Primate Research Center, University of California,
One Shields Avenue, Davis, CA 95616, USA
e-mail: jpcapitanio@ucdavis.edu

A. Weiss et al. (eds.), *Personality and Temperament in Nonhuman Primates*, Developments in Primatology: Progress and Prospects, DOI 10.1007/978-1-4614-0176-6_9,

of phlegm. Treatment of such a condition involved using the opposite element. Since phlegm was associated with water, and its opposite, yellow bile, was associated with fire, treatment might include staying in bed to generate warmth (Boylan 2006).

More than 500 years later, Galen, a physician working in Rome, elaborated upon Hippocrates's ideas and postulated that individual differences in character could also be explained by the balance of the four humors. The four humors are often associated with adjectives suggestive of temperament – blood with sanguine, black bile with melancholic, phlegm with phlegmatic, and yellow bile with choleric – although in Galen's time, the term "temperament" had a slightly different meaning (Siegel 1973). Galen himself apparently wrote very little on character, although he did describe the character associated with a melancholic temperament (resulting from an excess of black bile) as expressing fear, depression, and discontent with life. It was others, such as Immanuel Kant, who later elaborated the psychological characteristics associated with the four humors (Stelmack and Stalikas 1991).

The significance of the writings of Hippocrates and Galen was that the same underlying causes were proposed as contributing to both one's character and one's health. More modern treatments of personality and physical health, particularly in humans, have mostly focused on identifying associations between personality characteristics and poor health outcomes, and have focused less on specific pathophysiologic mechanisms (Capitanio 2008). The classic work in this area reported on a relationship between the so-called "Type A" behavior pattern (comprising competitiveness, hostility, impatience) and coronary heart disease (Friedman and Rosenman 1959). Later research showed that it was primarily the "hostility" component of the behavior pattern that was the most toxic risk factor for heart disease (Smith et al. 2004). Other personality-related risk factors have been associated with health outcomes: a "socially dominant" style, reflected in loud, emphatic speech and a tendency to "talk over" others during conversation is associated with coronary heart disease; negative affect/neuroticism is associated with shorter life span and increased incidence of illness in general; and positive characteristics such as optimism and conscientiousness have been related to a variety of positive health outcomes (see review by Smith 2006).

Unlike many other areas of biology, animal research has made few direct contributions to our understanding of personality and disease, primarily because of the rather sporadic way that personality or temperament (from this point on, these terms will be used interchangeably: see Capitanio 2004) has been studied in nonhuman species (Gosling 2001). Recently, this has changed. Interest in studying personality in animals has grown substantially in the past decade (Weinstein et al. 2008), and seminal studies in nonhuman primates by Stevenson-Hinde and colleagues (e.g., Stevenson-Hinde and Zunz 1978) have shown that such studies can be done in a rigorous and quantitative manner. At about the same time, a strong interest developed in understanding how psychological processes, especially those associated with "stress," affect immune functioning (Ader 1981). As might be expected, however, the overlap between these two relatively young areas of investigation – animal personality and psychoneuroimmunology – has been small.

The goal of the present chapter is to review studies from nonhuman primates that bear on the issue of personality and immunity. Two sets of studies are relevant. The first set has an explicit focus on demonstrating how immune function is related to naturally occurring variation in personality characteristics in normally reared animals. The second set of studies focuses on differences in immune responses between groups of animals that have had different early experiences, which have been shown to result in differences in personality. We note at the outset, however, that the number of nonhuman primate species for which there are data on either personality or immune function is quite limited. The data reviewed below come from a handful of Old World species, primarily macaques. Studies of other species (e.g., New World monkeys) would be helpful in testing the generality of the conclusions discussed below.

9.2 A Brief Overview of the Immune System

The immune system comprises a diverse set of tissues whose principal function is to protect the organism from attack by potentially harmful microbes and their toxins. A detailed treatment of the immune system is obviously beyond the scope of the present review, but can be found in any recent immunology textbook. Below, however, I will provide some general background that will help make sense of the specific types of measures that have been used in research on personality and immunity.

There are two broad types of immune responses, typically referred to as "innate" vs. "specific." Innate immune responses are usually the first line of defense, allowing the organism to combat the microbe while more targeted, specific, immune responses develop. For example, within hours of becoming infected with a virus, a cell will produce high levels of proteins called interferons, which inhibit replication of the virus and generally induce an antiviral state. Interferons also activate natural killer (NK) cells, which kill virally infected cells. These processes are "innate" in the sense that they occur relatively automatically, and with consistent "strength" to virtually every viral infection. This is in contrast to specific immunity, whose key cell type is the lymphocyte. When a pathogen is first encountered, only a few lymphocytes are present in the organism that can detect that pathogen, and these cells become activated and clonally expand. Over the course of the next couple of weeks, a highly specific response develops to that specific pathogen (or more accurately, to a protein portion of the pathogen, referred to as an antigen). Once the pathogen has been eliminated, specific immune responses linger – lymphocytes demonstrate "memory." Should that same pathogen be encountered a second time months or years later, memory lymphocytes will be able to quickly up-regulate their responses, and so this very intense and targeted immune response will occur much more rapidly with succeeding exposures. This is the rationale behind vaccination – present the organism with an attenuated or killed version of a pathogen to develop a specific immune response; later, should the organism encounter the wild-type pathogen, all of the resources of specific immunity can be brought to bear virtually immediately.

Specific immunity comes in two varieties, humoral and cellular. Humoral immunity involves the production of antibodies (also referred to as immunoglobulins, or Ig) by B lymphocytes. Antibodies are proteins that are secreted by B cells and are specific to a single antigen. Antibodies attach themselves to the antigen for which they are specific, and consequently neutralize the ability of that pathogen to infect or cause disease in the host. There are five broad types of antibodies, but the type that provides the bulk of humoral immunity (at least in the blood compartment) is the "G" subtype, referred to as IgG. Antibodies do not enter infected cells; consequently, they are most effective against extracellular microbes. Cellular immunity, on the other hand, does not rely on a secreted product to neutralize a microbe, but rather on the development of a specific immune response designed to kill the infected cell. The main effector cell is the CD8+ T cell, often referred to as a cytotoxic T lymphocyte (CTL). ("CD" refers to a set of proteins on the surface of the cell that can be used to identify the type of immune cell that it is.) This function is similar to the NK cell's function, but an NK cell will kill any virally infected cell, whereas a CTL will only kill a cell expressing the specific antigen that it is able to detect.

It is important to note that the two distinctions made so far, namely between innate and specific, and between cellular and humoral, are in reality much less distinct. Both the innate and specific arms of the immune system typically work together – a specific immune response can amplify, focus, and direct the innate immune response, for example. Similarly, a particular pathogen may activate both cellular and humoral responses. Viruses, for example, are obligatory intracellular parasites, requiring a cell's machinery to replicate. While they are inside a cell, they cannot be reached by Ig; rather cellular immune responses focus on killing the infected cell. Once the virus has replicated within a cell, however, it will spread itself by leaving the cell and circulating to target other, as-yet uninfected cells. During this extracellular stage, it is susceptible to neutralization by antibody. And it is also important to recognize that one very important T cell, the CD4+ T lymphocyte, referred to as a "helper" T cell, helps facilitate a variety of immune responses, both cellular and humoral, innate and specific.

How are immune responses measured? There are a number of measures that I will refer to below. One of the easiest measures is simply to draw a blood sample and determine the numbers of specific types of cells that are in the circulation. Do animals that vary along a personality dimension such as "Sociability" also co-vary in the number of CD4+, CD8+, or NK cells? While such a relationship may be found, its meaning is unclear because enumerative measures such as these bear little relationship to immune *function*, the ability of the cells of the immune system to respond to an antigen. Only a small percentage of immune cells are present in the circulation – most of the "action" in an immune response takes place in lymphoid tissue such as lymph nodes or spleen, and there is generally no relationship between cell numbers and the ability of the organism to mount an effective immune response to a pathogen (Westermann and Pabst 1990).

Most contemporary studies focus on functional immune measures. One simple functional measure of humoral immunity is to immunize (vaccinate) an animal and measure the amount of IgG that is produced several weeks later. This in vivo measure reflects the integrated response of an individual's immune system, and can assess a primary response (if this is the first time the animal has been immunized) or a secondary response (how well the immune system responds to an antigen that it had already seen some time in the past). A second commonly used measure is to draw a blood sample, isolate the peripheral blood mononuclear cells (PBMCs, most of which are lymphocytes), and stimulate them with either mitogens or antigens. When a lymphocyte is stimulated, one of its first responses is to undergo mitosis. An antigen stimulates a small, specific set of cells, whereas a mitogen stimulates numerous sets of cells. A variety of mitogens are used, including Concanavalin A (Con A) and Phytohemagglutinin (PHA), both of which are T cell mitogens and Pokeweed mitogen (PWM), which stimulates both T and B cells. Once stimulated, a marker is incorporated into the culture that allows the amount of proliferation to be measured. A third functional measure assesses the ability of NK cells to lyse particular target cells, such as K562 or Raji cells, which are human tumor cell lines.

While there are a number of ways to assess the activity of the immune system, it is important to realize that it is not often clear how any of these measures relate to actual protection from infection or illness. It is possible, for example, that above some threshold, additional IgG or additional cytotoxic activity may not be useful. Still, there is no question that the immune system is responsible for maintaining health and neutralizing microbes, and while the relationship between variation in such measures of immunity and actual health outcomes is not always known at this time, this relationship is in principle knowable.

9.3 How Can Personality Affect Immunity?

What is personality? Others in this volume have discussed their views of what personality is; the definition that I prefer is from Allport (1937, p. 48): "the dynamic organization within the individual of those psychophysical systems that determine his unique adjustments to his environment." There are many important aspects to this definition, but a critical one is that personality is not the same as behavior ("adjustments to his environment"), although we infer it from behavior. "Personality *is* something and *does* something. It is not synonymous with behavior or activity… It is what lies *behind* specific acts and *within* the individual" (Allport 1937, p. 48, emphasis in the original). This suggests that personality is a higher level phenomenon, and historically in psychology, it has been conceptualized as comprising motives (e.g., the three "social motives" in psychology are achievement, affiliation, and power) and/or traits (e.g., the Big Five traits of Extraversion, Neuroticism, Conscientiousness, Agreeableness, Openness). While "trait" and "motive" approaches to personality have often been considered separately, Winter et al. (1998, p. 231) suggest that the two views are complementary: "traits channel or direct the ways in which motives are expressed in particular actions" (for more discussion, see Capitanio 2004).

How then can some higher level construct, reflecting motives and traits, influence health and immune function? The key lies in the obvious fact that organisms exist within environments. We can, of course, choose our environments. If we are high in sensation-seeking, for example, we might select environments that include roller coasters. Sometimes, though, our environments present us with situations that may be challenging. Health psychologists often talk about how personality affects an individual's appraisal of a situation, and how one copes with the situation. If one appraises a situation as threatening, and/or if one is unable to cope effectively with the situation, one may experience stress. For example, a college environment can present a student with challenges such as getting along with a new roommate; balancing one's time between class, work, and recreational activities; and dealing with exams. If one has a low affiliation motive, or is low in Agreeableness, roommate problems might follow. If one is low in Conscientiousness or in achievement motivation, insufficient study time may be planned with a consequence of poor performance on exams. The result of the poor fit between an individual's dispositions and the needs of the situation may be the psychological experience of stress.

Stress involves activation of multiple physiological systems, including neural and endocrine systems that can interact with the immune system. One stress response system is the sympathetic-adrenal-medullary (SAM) system, and includes release of the neurotransmitter norepinephrine onto target organs, one of which is the adrenal medulla, which then secretes the hormone epinephrine into the circulation. Cells of the immune system possess adrenoreceptors that can bind epinephrine and norepinephrine, resulting in altered functioning of the cells. For example, in a simple experiment, Cole et al. (1998) found that adding norepinephrine to a cell culture of HIV-infected PBMCs led to significantly elevated replication of the virus by the cells. Pharmacologically blocking the adrenoreceptors eliminated this effect. Recently, Sloan et al. (2006) identified the in vivo analog of this experiment. Sympathetic nerve fibers innervate lymphoid tissue such as lymph nodes and the spleen; thus, the immune system "knows" when the sympathetic nervous system is activated. Sloan et al. (2006) reported that, in the lymph nodes of adult male rhesus (*Macaca mulatta*) monkeys, the likelihood of simian immunodeficiency virus (SIV) replication was nearly fourfold higher in the vicinity of norepinephrine release sites, compared to lymph node areas that were farther removed from nerve fibers. Clearly, activation of the SAM response can have significant consequences for immune function and health.

The second major stress-response system is the hypothalamic-pituitary-adrenal (HPA) system. The hypothalamus receives input from a variety of structures, such as the amygdala, that are involved in processing information about emotion. When activated, the hypothalamus secretes corticotrophin releasing hormone, which stimulates the pituitary to release adrenocorticotrophic hormone into the circulation, resulting in the release of the glucocorticoid cortisol from the adrenal cortex. Cortisol is a steroid hormone whose main mechanism of action involves entering a cell and binding to an intracellular receptor. The hormone–receptor complex migrates to the nucleus, where it can then affect transcription of a cell's genes and production of proteins. Virtually all the cells of the body possess glucocorticoid receptors, including cells of the immune system. One class of proteins that immune cells produce is cytokines (which includes the interferons described above), which serve a variety of functions – they are responsible for inter-cellular communication, can attract other immune cells to the area, can promote inflammation, etc. Cortisol typically has immunosuppressive and anti-inflammatory actions for the organism as a result of its ability to suppress transcription of genes that code for cytokines.

One answer, then, to the question of how personality can affect immunity is that personality affects an individual's appraisals of its environment (e.g., *what* in the environment is considered a challenge), as well as the ability of the individual to cope with what has been appraised as challenging in its environment. Negative appraisals and/or poor coping can result in the psychological experience of stress and the activation of stress response systems. Both the major stress response systems can impact immune function through reasonably well worked out molecular and genetic mechanisms, and in the presence of a pathogen, a disease condition may be more likely to develop or an existing disease may be exacerbated.

9.4 Nonhuman Primate Personality and Immunity

9.4.1 Introduction and Caveats

As described earlier, I will review two sets of studies pertaining to personality and immunity – those studies that have explicitly focused on natural variation in personality in normally reared monkeys and studies of variation in personality that was induced by different early rearing experiences. It is important to note at the outset that the immune system is finely attuned to protect the organism from attack by pathogens. Data showing that the immune system affects, and is affected by, many other organ systems in the body indicates that the immune system is a sensory organ that "forms part of an integrated homeostatic network" (Husband 1995, p. 397). While this is a good thing from the perspective of protection of the organism, it can be problematic methodologically – the immune system is very sensitive to small differences in procedures. For example, we demonstrated that simply being in a room and drawing blood samples on monkeys can result in significantly different counts of cells between monkeys sampled earlier in the session compared to monkeys sampled later (Capitanio et al. 1996). Moreover, different results can be obtained using different reagents in an assay. One measure of NK cell function is to quantify in vitro their ability to lyse specific target cells. Several different target cell lines are available for use, however, and Lubach et al. (1996) found different results when two different cell lines (K562 and Raji) were used. The difference appeared to be related to CD8+ T cells (in addition to NK cells) having a role in killing the K562 target cells, but not the Raji targets. These caveats are raised only to indicate that cross-laboratory comparisons are difficult, owing to differences in blood sampling procedures, assay procedures, reagents used, etc. Thus, the principal focus in the review below will be on within-laboratory contrasts.

9.4.2 *Studies of Personality and Immunity*

Several different aspects of personality (or individual differences in patterns of response suggestive of personality) have been examined – Sociability/affiliation, emotionality/reactivity, and "dominance."

9.4.2.1 Sociability

Sociability reflects a tendency to affiliate and, in our laboratory at the California National Primate Research Center, it has been assessed using a rating methodology (Gosling 2001): trained behavioral observers conduct focal animal observations for a specified period of time, in order to become familiar with the animals, after which they rate each animal, using a one to seven scale, on a set of traits (e.g., aggressive, bold, nervous, timid). Traits that demonstrate adequate inter-rater agreement and reliability are then selected for factor analysis. We have conducted three such studies of adult male rhesus monkeys, one of which employed a confirmatory factor analysis to confirm a specific four-factor structure found in an earlier study (Capitanio et al. 1999; Maninger et al. 2003; Capitanio and Widaman 2005). One factor, Sociability, reflects ratings by the observers on the traits "warm" and "affiliative," as well as "solitary" (which is reverse coded when constructing the factor score). In our laboratory, personality is always assessed while the animals are living in their familiar, outdoor social groups, after which they are relocated and immune studies are done weeks to years later while they live and interact in different settings.

Under the hypothesis that removal of an animal from its familiar, natal group of 100+ animals to individual caging indoors might have a greater impact for animals that are higher in Sociability compared to those that are less Sociable, we immunized adult male rhesus monkeys with tetanus toxoid a few weeks following relocation. This vaccine is routinely given to all animals in our colony at an early age, so our study assessed the secondary antibody (specifically IgG) response – the ability of the immune system to increase antibody production to an antigen that it has "seen" before. We found that high Sociable (HS) animals had a significantly higher IgG response than did low Sociable (LS) animals (Maninger et al. 2003). This result was not explainable by differences in preexisting antibody levels, or by past immunization history. Although it is tempting to conclude that the HS animals had a better IgG response to social separation and relocation, other studies have found that higher antibody responses can be associated with chronic stress and alterations in HPA regulation, such as might be seen following relocation (Cunnick et al. 1991; Capitanio et al. 1998). Therefore this result is consistent with the idea that HS animals experienced greater stress following separation from familiar companions and relocation to individual, indoor housing.

How might Sociability affect immune function when animals are given social experience? Evidence indicates that being more Sociable can be protective, particularly under conditions of social stress (Capitanio et al. 1999, 2008). We formed adult male rhesus monkeys (some of which had been inoculated with SIV) into social groups that were either Stable or Unstable in composition. In a Stable group, the same three animals met daily for 100 min. In Unstable groups, the number and identity of animals changed daily – one animal might be in a group with one, two, or three others on any given day, with the companions drawn randomly from a pool of nine animals. The lack of predictability in number and identity of partners is stressful for animals in Unstable conditions, as indicated by increased frequencies of conflictual behavior and altered HPA regulation (Capitanio et al. 1998). In an early, post hoc analysis of the role of Sociability in simian AIDS (Capitanio et al. 1999), we were interested in determining whether Sociability could be related to two measures of antibody response. For this analysis, we statistically controlled for social condition, in order to maximize the use of our relatively small sample of SIV-infected monkeys ($n=18$, with half in Stable and half in Unstable conditions). One immune measure was the antibody response to cytomegalovirus (CMV), and the second measure was the antibody response to SIV itself. CMV is a herpes viruses with which most rhesus monkeys (like most humans) are naturally infected from an early age. As with other herpes viruses, the immune system (if intact) successfully manages the latent infection, and a low level of antibody is present in plasma. Sometimes, however, if an individual's immune system is compromised (such as occurs with SIV infection in rhesus monkeys and HIV infection in humans) herpes viruses such as CMV can re-express themselves. We assessed antibody levels 4 and 8 weeks following inoculation with SIV, and compared them to antibody levels obtained during a baseline period prior to SIV inoculation. We found that Sociability was strongly correlated with anti-CMV IgG levels – specifically, animals higher in Sociability had an increase in IgG compared to pre-SIV levels, and animals lower in Sociability had a decrease from baseline in IgG at these time points (Capitanio et al. 1999). An inability to mount a strong antibody response to CMV reactivation following SIV infection is associated with shorter survival (Baroncelli et al. 1997), suggesting that Sociability was a protective factor in this disease. We also measured the primary IgG response to SIV, but found no relationship with Sociability.

Our post hoc analysis of the role of Sociability in SIV infection led to a second study, in which animals were selected as being relatively high or relatively low in Sociability, with a goal of determining specifically whether Sociability would be protective in stressful circumstances. Animals were formed into Stable or Unstable groups as described above. Within a few weeks of group formations, we found that Sociability was related to a variety of outcomes, but only for animals in Unstable conditions. One result was that Sociability was negatively correlated with submissive behavior – Low Sociable (LS) animals showed more present-sex posture than did HS animals. A second result was that Sociability was also positively correlated with basal plasma cortisol concentrations. Our earlier study (Capitanio et al. 1998; see also the literature on posttraumatic stress disorder: Yehuda et al. 1995) demonstrated that *basal* concentrations of cortisol (measured when the animals are not in their daily social groups) are reliably lower in animals in Unstable social conditions, which is presumably a reflection of the greater activation of the HPA system during the actual group formations; thus, in this model, higher concentrations are associated with better outcomes, and in this personality study, it was the HS animals in the Unstable conditions that had the higher concentrations. A third result was that Sociability was negatively correlated with the expression of interferon-stimulated genes (ISG) in PBMCs. As described earlier, interferons are part of the innate immune system, and are secreted by cells upon infection with a virus. One of the main functions of interferon-alpha is to stimulate transcription of genes that code for proteins that are effector molecules. We assessed transcription of three of these "ISG," and found higher transcription among LS animals in Unstable social conditions. Since these animals were infected with a virus, and since the interferon-alpha system is one of the first lines of defense against a viral pathogen, this might suggest that LS animals were better protected. However, studies have found that, for HIV and SIV, higher ISG levels reflect a poor outcome, not better protection (Abel et al. 2002). We found the same result (Capitanio et al. 2008). Thus, this study revealed that, in stressful social conditions, Low-Sociability was a risk factor for faster disease progression – it was associated with increased submissive behavior, changes in cortisol secretion suggestive of chronic stress, and increased ISG transcription, which in turn was associated with higher viral load and shorter survival. Importantly, no relationships between Sociability and these measures were found in the Stable social condition. Finally, as with the previous study, no relationship was found between Sociability and the primary IgG response to SIV.

The results of the above study are worth emphasizing because they show that the importance of a personality factor may be evident only in certain environments. In Stable social conditions, Sociability was uncorrelated with cortisol, ISG expression, and submissive behavior. What this means is that, to the extent that "Low Sociability" reflects some social deficit (see Capitanio 2002), its biobehavioral impact is negligible when animals are in stable, familiar, social conditions. But when situations are stressful, Sociability becomes an important contributor to biobehavioral outcomes – correlations that were nonsignificant among animals in Stable conditions were all significant for animals in Unstable conditions. Being low in Sociability affects coping in socially stressful situations in that LS animals displayed more present-sex posture, which was usually given upon the approach of a higher ranked animal. HS animals, in contrast, seemed better able to anticipate and avoid those situations, and consequently showed better, more protective, physiological responses (Capitanio et al. 2008).

A protective role of greater affiliation under conditions of social stress has been found by others as well. Kaplan et al. (1991) studied 30 adult male cynomolgus monkeys (*Macaca fascicularis*) that were formed into five-member social groups. Every 5 weeks, however, the groups were reorganized so that an animal's new cage-mates would be three or four previously unfamiliar animals. Across the entire 26-month period of the experiment, behavioral data were recorded, indices of affiliation and aggression were calculated, and median splits were used to identify four groups of animals: those high in affiliation and aggression (i.e., above the medians on each index), those low in affiliation and aggression (i.e., below the medians on each index), as well as those that were low affiliation/high aggression, and those that were high affiliation/low aggression. Blood was sampled in the fourth week following the twentieth and final reorganization of the groups, and two in vitro measures of cellular immune function were made. Animals that were high in affiliation showed a healthier immune response, particularly if they were also low in aggressiveness – such animals displayed greater proliferation of lymphocytes in response to stimulation with the mitogens Con A and PHA, and greater NK cytolytic activity. In a follow-up study, Cohen et al. (1992) contrasted two sets of monkeys, those that were in the stressful, reorganization condition, and those that remained housed in their stable social groups. As expected, the proliferation responses to Con A were highest in the stable social condition, and were higher among animals (regardless of social condition) that were above the median in affiliation. The lowest proliferation responses were seen among low-affiliation animals in the reorganization condition.

To summarize, these data indicate that individual differences in the tendency to affiliate are related to immune function, particularly under challenging social circumstances. This relationship does not appear to be specific to any particular type of immune response, having been seen in measures of humoral immunity (the secondary IgG response to tetanus toxoid and the CMV-specific IgG response, though not for the primary IgG response to SIV in two studies), measures of cellular immune function (in vitro proliferation responses to mitogen stimulation, and NK cytotoxic responses), and gene expression for mechanisms of innate immunity (ISG).

How might Sociability influence immune responses in stressful social situations? As described earlier, it is likely through modulation of stress-response systems. Recently, we obtained data suggesting that Sociability is related to both SAM and HPA responses within an immune context. In a study done prior to SIV inoculation and prior to Stable/Unstable social group formations (Maninger et al. 2005), we stimulated a fixed number of PBMCs from adult male rhesus monkeys with the mitogen-like substance staphylococcal enterotoxin B (SEB). Some of the cultures also contained varying concentrations of dexamethasone, which is a synthetic glucocorticoid, and like cortisol, binds to the intracellular glucocorticoid receptor. When PBMCs are stimulated with SEB, they will produce and secrete interferon-gamma (IFN-g), a cytokine that is involved in both innate and specific immune responses, and that can promote inflammation. Because glucocorticoids generally inhibit cytokine production we would expect that, in the presence of dexamethasone, IFN-g production would decrease, which it did for all animals. In response to the lowest quantity of dexamethasone, however, HS animals produced less IFN-g than did Low Sociable animals, suggesting that the immune cells for HS animals are more sensitive to even small changes in glucocorticoid concentrations. Thus, it is possible that greater Sociability is protective because there is some fundamental difference in HPA-immune regulation in these animals, compared to LS animals.

Sociability also appears to be related to the SAM system. We biopsied lymph nodes from adult male rhesus monkeys in the Unstable social conditions 37 weeks following SIV (or saline) inoculation, and, in collaboration with Steve Cole and Erica Sloan at UCLA, we examined patterns of innervation by the sympathetic nervous system. When we examined lymph nodes from the saline-inoculated controls, we found that nodes from LS animals had a significantly greater density of catecholaminergic varicosities (the part of the peripheral nerve that secretes norepinephrine) than did nodes from HS animals (Sloan et al. 2008). Together, the lymph node data and the HPA data (described in the previous paragraph) suggest that the protective effects of affiliation on immune function in challenging situations may be mediated by more basic individual differences in how the individuals' nervous and hormonal systems are constructed, and how these systems interact with, and regulate, the immune system. The origins of these individual differences, and the degree of plasticity in these processes, remain to be fully explored

9.4.2.2 Emotional Reactivity

A handful of studies have explicitly examined how individual differences in emotionality relate to immune function. "Emotionality" typically refers to the degree to which individuals display visible signs of arousal – distress vocalizations, activity, etc. This is the same construct that will be discussed in detail below when considering the role of early experience in personality and immune function. The studies discussed in the present section typically focused on variation in behavioral responses to some event, such as infant–mother separation, and correlated those responses with some measure of immunity that is obtained in response to the same event. The assumption is that the degree of behavioral responsiveness to the event is a reflection of a stable, underlying trait of emotionality or reactivity.

In a study that reported responses to maternal separation for pigtailed (*Macaca nemestrina*) and bonnet (*Macaca radiata*) macaques, Laudenslager et al. (1990) found both species differences and individual differences in measures of immune function. The measures used were proliferative responses of PBMCs to the mitogens Con A and PHA. For both the species, the change in the amount of proliferation from baseline to week two postseparation was inversely related to the frequency of vocalization on the first day of maternal separation – that is, animals that vocalized more on the first day after their mothers were removed, had lower proliferative responses 2 weeks later, compared to those animals that vocalized less on Day 1. To the extent that elevations in distress vocalization reflect increased emotional reactivity that is trait-like, these data suggest that greater reactivity in response to a stressor might result in compromised cellular immune function.

The association between greater emotionality and suppressed immunity was also seen in a study of rhesus monkeys on Cayo Santiago. When rhesus monkeys reach about 6 months of age, their mothers typically resume estrous, resulting in a change in the mother–infant relationship and increased distress behavior by the infants (Berman et al. 1994). Laudenslager et al. (1993) studied ten infants that were immunized with tetanus toxoid when they were about 1 year old, and assessed tetanus-specific IgG 10–14 days after immunization. As indexed by rates of distress vocalizations (screaming and geckering) that were displayed during their mothers' estrous, animals that were more reactive (higher levels of distress vocalizations during this time) had reduced tetanus-specific IgG. Interestingly, the animals that had the higher distress vocalizations also were rejected the most by mothers and spent more time at a greater distance from mother during her estrous. Thus, it is difficult to determine whether this measure reflects only infant temperament, or rather some preexisting tension in the mother–infant relationship, although, of course, these explanations are not mutually exclusive.

Not all studies have found a relationship between measures of reactivity and immunity. For example, Westergaard et al. (2002) studied 27 2-year-old rhesus females that had been captured from their island colony. Measures of reactivity included assessments of fleeing or aggression in response to a human standing in front of the animals' cages. No relationship was found between such behavioral measures and the numbers of CD4+ and CD8+ T cells.

In summary, despite the negative finding of Westergaard et al. (2002), which examined enumerative measures of immunity, the two studies described that focused on immune *function* suggest that animals that display greater behavioral responsiveness to challenging circumstances (which presumably reflects the temperament trait of emotional reactivity) have impaired immune responses.

9.4.2.3 Dominance

A number of studies have examined whether animals of different social ranks differ in terms of immune function. Often, such differences are expressed in terms of "dominance," which is a concept that has multiple meanings in animal behavior. Most commonly, it refers to a difference in power between two or more individuals. If "A" is dominant to "B," then "A" can presumably gain access to more resources than "B," even taking a resource away from "B"; by the inherently asymmetric nature of dominance, however, "B" cannot do the same to "A." Thus, "dominance" describes something about a relationship between individuals – one individual has priority of access over other individuals. Does "dominance" exist within an individual? Consider the following. If you have three groups of three rhesus monkeys, each group will possess a dominance hierarchy that will almost certainly be linearly ordered – in each group there will be a first-ranked, a second-ranked, and a third-ranked animal. If you take the third-ranked animal from each group, and put them together in a new group, you will have an identical dominance hierarchy, with a first-ranked, a second-ranked, and a third-ranked animal. One animal that was third-ranked in its original group is now first-ranked in a new group. Has its "dominance" changed? Has its personality changed? Certainly its rank has changed, as has the context in which the rank is determined.

Consider again the three top-ranked males from the initial groups. All are at the top of their respective dominance hierarchies. But other than that, do they share any other characteristics? Not necessarily; studies by Sapolsky and Ray (1989), for example, have described different "styles" of dominance, relating mostly to social skillfulness. Dominant animals with good social skills tended to have lower basal cortisol concentrations, while dominant animals without those skills had basal cortisol concentrations that were comparable to those found among subordinate animals.

Thus, dominance is a concept that has great descriptive (and some predictive) value in primatology, and one's rank can have significant consequences for one's access to resources, and the extent to which one has stressful experiences. But in addition to whatever personal characteristics lead to an individual's rank, it is clear that there are elements of the context (identities and characteristics of other animals, availability of resources) that are equally important contributors to an animal's rank. The contextual contributions (and limitations) of a specific hierarchy are often ignored when one talks about dominance as though it were a trait of the individual. (We do note that some investigators have identified a personality trait in chimpanzees that they have labeled "Dominance," which describes a chimpanzee that is "independent, confident, and fearless" (King and Figueredo 1997, p. 261), a usage that more truly reflects an individual-level trait.) Because "dominance" has such a strong contextual component to it, and because the underlying personality characteristics of "dominant" individuals can be variable, one might expect that results relating "dominance" to immune (or other physiological) measures might be inconsistent. This appears to be the case.

While Kaplan et al. (1991) reported that affiliation was related to immune function in their studies of group reorganization of cynomolgus monkeys, no effects of social status were found for either NK cell activity or proliferative responses of PBMCs to the mitogens PHA or Con A. Similarly, Clarke et al. (1996) found no significant correlations between rank and proliferative responses to Con A and PWM in rhesus monkeys following group reorganization. Prior to the reorganization, however, while the animals were living in stable social groups, a correlation was found – high-ranked monkeys had higher proliferative responses. In contrast to these results, Gust et al. (1991) did find rank-related differences in immune measures following formation of a new social group of adult female rhesus monkeys. By 8 days following reorganization, the numbers of lymphocytes in peripheral blood for the four high-ranked animals had risen to baseline (pregroup-formation) values, while the numbers of cells for the four low-ranked animals remained below baseline for several weeks. No effect was seen for the other immune measure, total concentrations of antibody in plasma. The authors speculate that, since the only statistically significant behavioral difference between low- and high-ranked monkeys was that high-ranked animals had higher durations of grooming, the more rapid return to baseline values for lymphocyte numbers may have been due to this more affiliation-related aspect of rank. Unfortunately, there was some confounding in this study between rank and prior housing, and the authors suggest that it is possible that prior housing condition may have affected not only rank acquisition, but also the values for lymphocyte subset numbers themselves.

Boccia et al. (1992) identified dominance hierarchies among females in stable social groups of bonnet and pigtailed macaques by restricting access to water for 24 h, then observing the pattern of drinking once water was again made available. Immune measures included proliferative responses to mitogens and NK cell function. Unlike Clarke et al.'s (1996) results with rhesus monkeys, no relationship was found between rank and immune function assessed during baseline conditions (i.e., when immune function was measured from blood drawn prior to the water restriction). Changes in immune function were found 48 h after the water restriction for the pigtailed macaques only, but these changes were related to frequencies of agonistic behavior, not to dominance rank per se (Laudenslager and Boccia 1996).

In contrast, rank was related to NK function during the baseline phase of a study that focused on spatial restriction of food in groups of bonnet macaques. Given that the baseline conditions of both the water restriction and the food restriction studies should be comparable (i.e., they reflect the usual pattern of resource availability with which the animals are quite familiar), these results appear to be in conflict (Laudenslager and Boccia 1996).

Cunnick et al. (1991), working with Kaplan, explored the relationship between dominance and specific antibody production in cynomolgus monkeys that experienced social reorganization as described above. Prior to the initial reorganization, all animals had been housed in stable social groups for several months. Initial rank was determined in these groups. Around the time of the first reorganization, when all animals' groups were reformed, the animals were given a primary immunization with tetanus toxoid, and IgG was measured 4 weeks later. Dominant animals (those with first and second ranks) had significantly lower IgG compared to subordinate animals (ranks 3–5) in the five-member groups. Beginning 5 weeks after the first reorganization, most of the animals remained in their five-member groups, while the remaining animals underwent additional reorganizations at 4-week intervals. One week after the tenth reorganization, the animals were given a booster immunization with tetanus toxoid. At the same time, the animals in the stable groups also were boosted. The secondary IgG response was measured from blood samples taken 1 and 4 weeks later. No rank effects were found in this study.

Finally, Masataka et al. (1990) found that dominance rank did correlate negatively with serum antibody levels in five stable, captive groups of chimpanzees – higher ranked animals had lower immunoglobulin levels. Animals from four of the groups were later housed individually, and blood samples were drawn approximately 3–5 weeks after the housing change. Once individually housed, there was no correlation between Ig levels and earlier social rank.

In conclusion, the relationship between social rank and measures of immunity were inconsistent. Following social reorganization, Gust et al. (1991) and Cunnick et al. (1991) found effects of rank, but Kaplan et al. (1991), Clarke et al. (1996), and Masataka et al. (1990) did not. Under conditions of social stability, prior to either water- or food-restriction challenges or social reorganization, Clarke et al. (1996), Laudenslager and Boccia (1996), and Masataka et al. (1990) found relationships, but Boccia et al. (1992) did not. Of course, there were many differences between the various studies – species, immune measures, details of procedures – but the lack of consistency, unlike that seen for the other two traits (sociability and emotionality), suggests that position in a hierarchy may be less informative about immune functioning than other psychological characteristics that contribute to attaining and maintaining one's rank.

9.4.3 Studies of Early Experience and Immunity

Data collected over several decades have shown that variation in early experience is associated with persisting differences in behavioral and physiological measures. Some of the behavioral changes in particular are remarkably long lasting and are best described as reflecting experience-induced changes in temperament or personality. A variety of experimental paradigms have been utilized, including nursery-rearing and infant–mother separation, and there are several studies that have employed immunologic endpoints. These data will be reviewed in this section, which will be organized by experimental procedures rather than by personality construct. The emphasis in this section will be on long-term outcomes of such experiences rather than acute responses to, for example, maternal separation.

9.4.3.1 Prenatal Effects

The behavioral and immunologic effects of prenatal stress have been studied experimentally by Schneider, Coe, and colleagues since the early 1990s. The procedure that has been used most often involves moving pregnant females from their living cages to another room for a 10-min period, during which three brief bursts of noise (at about 110 db) are randomly played. The females experienced this simple event 5 days per week for about 6 weeks, typically in mid-gestation. Offspring of mothers that received such experience have been shown to have impaired neuromotor development and attentional deficits at birth (Schneider 1992). These animals have been followed through late adolescence and have been described as behaviorally inhibited: "Like inhibited children, PNS (prenatally stressed) monkeys showed more inactivity, engaged in less exploratory behavior, less play and other social behavior, and showed more evidence of disturbance under conditions of novelty than did controls" (Clarke and Schneider 1997, p. 491).

In addition to behavioral changes, prenatal stress results in impaired immune function. Within days of birth, PBMCs from monkeys that were prenatally stressed during mid-gestation showed a decreased proliferative response in a mixed lymphocyte culture compared to nonstressed controls (Coe et al. 1999). This difference in immunity persisted through 2 years of age, when cytokine responses were examined at baseline and under conditions of acute stress (Coe et al. 2002). PBMCs from prenatally stressed and from control monkeys were incubated with lipopolysaccharide (LPS), and the concentrations of the pro-inflammatory cytokines tumor necrosis factor-alpha (TNF-a) and interleukin-6 (IL-6) were measured. During both the baseline and the acute stress conditions, monkeys that had experienced prenatal stress had significantly lower cytokine production in response to LPS. PBMCs from the monkeys were also stimulated with LPS and with varying concentrations of dexamethasone added to the wells. Dexamethasone inhibited TNF-a and IL-6 production in a dose-dependent fashion, as expected. However, addition of dexamethasone to the wells for the control animals resulted in cytokine concentrations that were similar to those found in the prenatally stressed animals' wells that did not contain dexamethasone. This result suggests that the alterations in immune response found for prenatally stressed animals may be related to the effects that prenatal stress has on the HPA axis. Together, these data suggest that experiencing stress prenatally can lead to an inhibited personality and reduced immunologic responsiveness.

9.4.3.2 Nursery Rearing

There are probably as many different protocols for nursery-rearing as there are laboratories that engage in the practice. In general, nursery-rearing typically involves: (1) removal of an animal from its mother at birth or soon after, (2) rearing for a brief period of time in an isolator in order that the animal can receive the frequent and individualized care necessary to ensure that it will feed properly, and (3) eventual exposure to one or more companions to help ensure later social competence. Several variations on this basic procedure are possible, relating to the timing of these events, whether the subsequent peer experience is intermittent (i.e., for a limited time per day) or continuous, and the size and composition of the peer group. In general, however, rearing without access to adults generally leads to animals developing poor emotion regulation, and a temperamental profile that reflects heightened emotionality (Suomi 1991).

Compared to animals that were mother-reared (MR) in social groups, monkeys reared in a nursery do appear to show differences in immunity. The first report came from Coe's laboratory (Coe et al. 1989). Eight animals were removed from their mothers at birth and lived in individual cages. Beginning at 2–3 weeks of age, the animals were placed in peer-groups 3 times per week, but were individually housed otherwise. These nursery-reared (NR) animals, at about 1 year of age, showed significantly higher lymphocyte proliferative responses to the mitogens PHA, Con A, and PWM compared to MR monkeys. Animals were again contrasted 1.5 years later, after the NR animals had been socially housed in peer groups, and the proliferative response to Con A remained significantly elevated. No significant effects were seen for NK cytotoxicity.

In a second study, Coe et al. (1992) found similar results, despite the fact that the rearing strategy was slightly different – specifically, from 30 days of age on, the animals were housed continuously in peer groups. NR monkeys, regardless of whether they had experienced prenatal stress, had elevated proliferative responses to Con A and PWM at 5–8 months of age. There was a nonsignificant reduction in NK activity among NR animals at 9–12 months of age, and no rearing differences were found in IgG levels to a primary immunization with tetanus toxoid. A third study in this series confirmed and extended this result (Lubach et al. 1995) – NR animals were individually housed until about 3 months of age, after which they received direct socialization twice weekly. At 14–15 months of age, the eight animals were socially housed together with an unfamiliar adult peer to attempt to resocialize the animals. Blood was sampled at 6-month intervals over the first 2 years of life. Compared to MR animals, NR monkeys had significantly higher CD4/CD8 ratios across the 24 months of this study, related primarily to a lower percentage of CD8+ cells in peripheral blood. NR animals also had consistently higher proliferative responses to PWM, and responses to PHA were higher at two time points. Unlike in the previous studies, NR animals had significantly lower NK cytotoxicity, compared to MR monkeys. Antibody responses to immunization for influenza, tetanus toxoid, or diphtheria were not significantly different between NR and MR animals.

In summary, nursery-rearing results in heightened emotionality in monkeys, and an enhancement of in vitro cellular immune responses, as well as some evidence of differences in immune cell numbers in peripheral blood, and possibly decreased NK activity. Data are quite consistent, however, in showing no effects on humoral responses to immunization.

9.4.3.3 Isolation Rearing

Many readers may recognize that modern nursery-rearing procedures are considerably different from isolation-rearing, which had been studied in the 1960s, and which revealed the critical importance of early social experience on later behavioral functioning (see review by Capitanio 1986). Elsewhere, we have suggested that adult rhesus monkeys that were reared with limited or no social opportunities (and in a restricted laboratory environment) for the first year of life could be described (using terms from the Five-Factor Model of human personality) as showing little social interest and competence (very low Extraversion, Agreeableness, and Conscientiousness) as well as low adaptability and high volatility (high Neuroticism, low Openness) (Weinstein et al. 2008). Might there be immune correlates of such a personality style? Lewis et al. (2000) studied several immune measures in isolate-reared (IR) monkeys that were 18–24 years old at the time of study. Originally part of a cohort of 48 monkeys, the monkeys that remained comprised 12 isolates and 10 MR aged animals. IR animals were removed from mothers shortly after birth, and lived in individual cages for 9 months. Some animals could see and hear other monkeys, but some could not. Human handling for all IR animals was minimal. MR animals lived with mothers and peers, but from 9 months of age, housing for MR and IR animals was identical – namely, in individual wire cages in colony rooms. Except for their participation in learning tests, occasional tests of social behavior, and a period of social housing lasting about 1 year, the animals remained in individual cages where they could see, hear, and smell other animals. Three blood samples were taken at approximately 20 years of age. In terms of enumerative measures, IR animals had a lower CD4/CD8 ratio (attributed to increased CD8+ numbers), and higher NK cell numbers. Functionally, IR animals had greater NK cell activity, which the authors suggest could have been due to the higher number of cells; no rearing differences were found in the proliferative response to Con A, PWM, or tetanus toxoid. These unique data are remarkable because they show persisting differences arising from early experiences that occurred 20 years earlier. As the authors note, however, all-cause mortality was significantly greater among IR animals compared to MR animals; consequently these results reflect data from animals that survived into their third decade, and may not truly reflect possible impairments in immunity resulting from early experience.

9.4.3.4 Social Separation

It has been known for some time that separation from an attachment object (a mother or a peer, in the case of peer-reared monkeys) results in substantial physiological disruption, including changes in measures of immunity, such as decreased lymphocyte proliferative responses to mitogens (Worlein and Laudenslager 2001). There is some evidence that early maternal separation can result in an altered temperament. Pigtailed macaques that had been subjects in studies of maternal or peer separation were observed from about 2 to 5 years of age. Some animals had been separated for a 10–14-day period at 5–7 months of age, whereas the remainder had served as nonseparated controls. At the time of the follow-up assessments, the animals had been living together in social groups. Compared to the control animals, previously separated animals were rated as less Sociable (Caine et al. 1983), showed less social play and had more restricted social networks (Capitanio and Reite 1984), displayed more disturbance behavior when confronted with novel situations, and had a longer latency to reach for a preferred food item (Capitanio et al. 1986). Spencer-Booth and Hinde (1971) also found that previously separated rhesus monkeys showed greater disturbance and less willingness to interact with novel objects. Together, these behavioral data suggest greater behavioral inhibition and lower sociability as a consequence of early social separation.

Studies of immune function in previously separated animals reveal persisting changes. Among 5–7-year-old pigtailed monkeys, prior separation was associated with reduced lymphocyte proliferation to the mitogens PHA, Con A, and Pokeweed, but no differences in cell counts or plasma immunoglobulin levels (Laudenslager et al. 1985). Coe et al. (1989) also found that 18-month-old rhesus monkeys that had experienced multiple maternal separations between 3 and 7 months of age had lower proliferative responses to Con A and to PWM. In assessing NK cytoxicity, however, Laudenslager et al. (1996) found that previously separated animals showed enhanced lysis of target cells compared to nonseparated animals through 4 years of age. In conclusion, the few studies focused on long-term effects of early maternal separation reveal an inhibition in cellular immune function, no effects on immunoglobulin levels, and some enhancement of NK function.

9.5 Conclusions

Together, the data reviewed here suggest that some aspects of personality are consistently associated with immunity. The data are strong, for example, that individuals that are more Sociable/Affiliative show better immune function, especially under conditions of social challenge and when companions are available. Evidence also suggests that greater emotional reactivity during stressful circumstances, such as maternal separation, is associated with lower immune function. Long-term effects of differences in early experience are apparent for behavior and immune function. Animals that experience early maternal separation or prenatal stress are characterized as behaviorally inhibited and show reduced sociality, while animals that are NR show heightened emotionality. Interestingly, these two contrasting personality patterns show opposite immune effects: animals subjected to maternal separation or prenatal stress show lower cellular immune function and increased NK function, while NR animals show enhanced cellular immune function and reduced NK function. The most inconsistent evidence relating personality to immunity concerns the construct of "dominance." I have argued that this is due to a mis-specification of dominance (or rank) as an enduring, personality characteristic of an individual.

There is some evidence suggesting that the effects of personality on immune function may be mediated by stress-response systems, the operating characteristics of which may be "initialized" during development, and which later can reflect the "fit" between the individual and its circumstances. Clearly, more work is needed to understand the mechanisms that are involved in the personality–immune relationship. We reiterate, however, that the data reported here reflect, for the most part, *statistical* relationships. It is unclear whether the differences are of sufficient magnitude to impact an actual disease process that might affect an individual's fitness. To date, very few studies have examined disease-related end-points (though see Capitanio et al. 1999, 2008).

Finally, to what extent might the relationships reviewed here apply to cross-species comparisons? Personality is usually considered to be a construct that describes variation within a species. But the same terms are often used to describe modal tendencies across species (Capitanio 2004): bonnet macaques are affiliative, rhesus macaques are aggressive, squirrel monkeys are emotionally reactive, etc. Several published studies have described species differences in immune function. For example, a recent study reported that NK function among hamadryas baboons, chimpanzees, and vervets was significantly lower than among rhesus and cynomolgus macaques, mandrills, and humans (Poaty-Mavoungou et al. 2001). Unfortunately, there are almost no comparative studies describing both behavioral and immune differences within the Primate Order. One important exception is a research program by Laudenslager and Boccia (1996), contrasting bonnet and pigtailed macaques. Under nonstressed conditions, pigtails tend to show greater NK activity compared to bonnets, whereas bonnets tend to show greater proliferative responses of PBMCs to mitogen stimulation. Pigtails are also typically described as less affiliative than bonnets. This between-species contrast is reminiscent of the contrast between monkeys that experienced early maternal separation compared to those that did not – within species, early separated animals show reduced proliferative responses and greater NK activity, and are described as lower in Sociability compared to non-separated controls – a pattern of behavior and immune function similar to that shown by pigtailed macaques. Standardized comparisons of personality and immune function in multiple species would be valuable for determining whether intraspecific variation in biobehavioral organization described above (reflecting personality–immune relationships, and including relevant endocrine and neural mechanisms) is paralleled by similar relationships inter-specifically.

Acknowledgments JPC was supported by NIH Grants RR000169 and MH049033. I thank G. Lubach and J. Worlein for helpful comments on an earlier version of this manuscript.

References

Abel K, Alegria-Hartman MJ, Rothaeusler K et al. (2002) The relationship between simian immunodeficiency virus RNA levels and the mRNA levels of alpha/beta interferons (IFN-alpha/beta) and IFN-alpha/beta-inducible Mx in lymphoid tissues of rhesus macaques during acute and chronic infection. J Virol 76:8433–8445

Ader R (1981) Psychoneuroimmunology. Academic, New York

Allport GW (1937) Personality: A psychological interpretation. Henry Holt, New York

Baroncelli S, Barry PA, Capitanio JP et al. (1997) Cytomegalovirus and simian immunodeficiency virus coinfection: Longitudinal study of antibody responses and disease progression. J Acquir Immune Defic Syndr Hum Retrovirol 15:5–15

Berman CM, Rasmussen KLR, Suomi SJ (1994) Responses of free-ranging rhesus monkeys to a natural form of social separation. I. Parallels with mother-infant separation in captivity. Child Dev 65:1028–1041

Boccia ML, Laudenslager ML, Broussard CL et al. (1992) Immune responses following competitive water tests in two species of macaques. Brain Behav Immun 6:201–213

Boylan M (2006) Hippocrates. The internet encyclopedia of philosophy, http://www.iep.utm.edu/h/hippocra.htm, retrieved 10 October 2008.

Caine NG, Earle H, Reite ML (1983) Personality traits of adolescent pig-tailed monkeys (*Macaca nemestrina*): An analysis of social rank and early separation experience. Am J Primatol 4:253–260

Capitanio JP (1986) Behavioral pathology. In: Mitchell G, Erwin J (eds) Comparative primate biology, Vol. 2A. Alan R. Liss, New York

Capitanio JP (2002) Sociability and responses to video playbacks in adult male rhesus monkeys (*Macaca mulatta*). Primates 43:169–177

Capitanio JP (2004) Personality factors between and within species. In: B. Thierry, M. Singh, W. Kaumanns (eds) Macaque societies: A model for the study of social organizations. Cambridge University Press, Cambridge

Capitanio JP (2008) Personality and disease. Brain Behav Immun 22:647–650

Capitanio JP, Reite ML (1984) The roles of early separation experience and prior familiarity in the social relations of pigtail macaques: A descriptive multivariate study. Primates 25:475–484

Capitanio JP, Widaman KF (2005) Confirmatory factor analysis of personality structure in adult male rhesus monkeys (*Macaca mulatta*). Am J Primatol 65:289–294

Capitanio JP, Rasmussen KLR, Snyder DS et al. (1986) Long-term follow-up of previously separated pigtail macaques: Group and individual differences in response to novel situations. J Child Psychol Psychiatry 27:531–538

Capitanio JP, Mendoza SP, McChesney M (1996) Influences of blood sampling procedures on basal hypothalamic-pituitary-adrenal hormone levels and leukocyte values in rhesus macaques (*Macaca mulatta*). J Med Primatol 25:26–33

Capitanio JP, Mendoza SP, Lerche NW et al. (1998) Social stress results in altered glucocorticoid regulation and shorter survival in simian acquired immune deficiency syndrome. Proc Natl Acad Sci USA 95:4714–4719

Capitanio JP, Mendoza SP, Baroncelli S (1999) The relationship of personality dimensions in adult male rhesus macaques to progression of simian immunodeficiency virus disease. Brain Behav Immun 13:138–54

Capitanio JP, Abel K, Mendoza SP et al. (2008) Personality and serotonin transporter genotype interact with social context to affect immunity and viral set-point in simian immunodeficiency virus disease. Brain Behav Immun 22:676–689

Clarke AS, Schneider ML (1997) Effects of prenatal stress on behavior in adolescent rhesus monkeys. Ann N Y Acad Sci 807:490–491

Clarke MR, Harrison RM, Didier ES (1996) Behavioral, imunological, and hormonal responses associated with social change in rhesus monkeys (*Macaca mulatta*). Am J Primatol 39:223–233

Coe CL, Lubach GR, Ershler WB et al. (1989) Influence of early rearing on lymphocyte proliferation responses in juvenile rhesus monkeys. Brain Behav Immun 3:47–60

Coe CL, Lubach GR, Schneider ML et al. (1992) Early rearing conditions alter immune responses in the developing infant primate. Pediatrics 90 (Supp):505–509

Coe CL, Lubach GR, Karaszewski JW (1999) Prenatal stress and immune recognition of self and nonself in the primate neonate. Biol Neonate 76:301–310

Coe CL, Kramer M, Kirschbaum C et al. (2002) Prenatal stress diminishes the cytokines response of leukocytes to endotoxin stimulation in juvenile rhesus monkeys. J Clin Endocrinol Metab 87:675–681

Cohen S, Kaplan JR, Cunnick JE et al. (1992) Chronic social stress, affiliation, and cellular immune response in nonhuman primates. Psychol Sci 3:301–304

Cole SW, Korin YD, Fahey JL et al. (1998) Norepinephrine accelerates HIV replcations via protein kinase A-dependent effects on cytokine production. J Immunol 161:610–616

Cunnick JE, Cohen S, Rabin BS et al. (1991) Alterations in specific antibody production due to rank and social instability. Brain Behav Immun 5:357–369

Friedman HS, Rosenman RH (1959) Association of specific overt behavior pattern with increases in blood cholesterol, blood clotting time, incidence of arcus senilis and clinical coronary artery disease. J Am Med Assoc 169:1286–1296

Gosling SD (2001) From mice to men: What can we learn about personality from animal research? Psychol Bull 127:45–86

Gust DA, Gordon TP, Wilson ME et al. (1991) Formation of a new social group of unfamiliar female rhesus monkeys affects the immune and pituitary adrenocortical systems. Brain Behav Immun 5:296–307

Husband AJ (1995) The immune system and integrated homeostasis. Immunol Cell Biol 73:377–382

Kaplan JR, Heise ER, Manuck SB et al. (1991) The relationship of agonistic and affiliative behavior patterns to cellular immune function among cynomolgus monkeys (*Macaca fascicularis*) living in unstable social groups. Am J Primatol 25:157–173

King JE, Figueredo AJ (1997) The Five-Factor Model plus Dominance in chimpanzee personality. J Res Pers 31:257–271

Laudenslager ML, Boccia ML (1996) Some observations on psychosocial stressors, immunity, and individual differences in nonhuman primates. Am J Primatol 39:205–221

Laudenslager M, Capitanio JP, Reite M (1985) Possible effects of early separation experiences on subsequent immune function in adult macaque monkeys. Am J Psychiatry 142:862–864

Laudenslager ML, Held PE, Boccia ML et al. (1990) Behavioral and immunological consequences of brief mother-infant separation: A species comparison. Dev Psychobiol 23:247–264

Laudenslager ML, Rasmussen KLR, Berman CM et al. (1993) Specific antibody levels in free-ranging rhesus monkeys: Relationships to plasma hormones, cardiac parameters, and early behavior. Dev Psychobiol 26:407–420

Laudenslager ML, Berger CL, Boccia ML et al. (1996) Natural cytotoxicity toward K562 cells by macaque lymphocytes from infancy through puberty: Effects of early social challenge. Brain Behav Immun 10:275–287

Lewis MH, Gluck JP, Petitto JM et al. (2000) Early social deprivation in nonhuman primates: Long-term effects on survival and cell-mediated immunity. Biol Psychiatry 47:119–126

Lubach GR, Coe CL, Ershler WB (1995) Effects of early rearing environment on immune responses of infant rhesus monkeys. Brain Behav Immun 9:31–46

Lubach GR, Coe CL, Karaszewski JW et al. (1996) Effector and target cells in the assessment of natural cytotoxic activity of rhesus monkeys. Am J Primatol 39:275–287

Maninger N, Capitanio JP, Mendoza S et al. (2003) Personality influences tetanus-specific antibody response in adult male rhesus macaques after removal from natal group and housing relocation. Am J Primatol 61:73–83

Maninger N, Capitanio JP, Brennan CM (2005) Sociability influences interferon-gamma secretion and glucocorticoid sensitivity in adult male rhesus macaques. Am J Primatol 66 (supp 1) : 190–191

Masataka N, Ishida T, Suzuki J et al. (1990) Dominance and immunity in chimpanzees (*Pan troglodytes*). Ethology 85:147–155

Poaty-Mavoungou V, Onanga R, Yaba P et al. (2001) Comparative analysis of natural killer cell activity, lymphoproliferation and lymphocyte surface antigen expression in nonhuman primates housed at the CIRMF Primate Center, Gabon. J Med Primatol 30:26–35

Sapolsky RM, Ray JC (1989) Styles of dominance and their endocrine correlates among wild baboons (*Papio anubis*). Am J Primatol 18:1–13

Schneider ML (1992) The effect of mild stress during pregnancy on birthweight and neuromotor maturation in rhesus monkey infants (*Macaca mulatta*). Infant Behav Dev 15:389–403

Siegel RE (1973) Galen on psychology, psychopathology, and function and diseases of the nervous system. Karger, Basel
Sloan EK, Tarara RP, Capitanio JP et al. (2006) Enhanced SIV replication adjacent to catecholaminergic varicosities in primate lymph nodes. J Virol 80:4326–4335
Sloan EK, Capitanio JP, Tarara RP et al. (2008) Social temperament and lymph node innervation. Brain Behav Immun 22:717–726
Smith TW (2006) Personality as risk and resilience in physical health. Curr Dir Psychol Sci 15:227–231
Smith T, Glazer KRJ, Gallo L (2004) Hostility, anger, aggressiveness and coronary heart disease: An interpersonal perspective on personality, emotion and health. J Pers 72:1217–1270
Spencer-Booth Y, Hinde R (1971) Effects of brief separations from mothers during infancy on behaviour of rhesus monkeys 6-24 months later. J Child Psychol Psychiatry 12:157–172
Stelmack RM, Stalikas A (1991) Galen and the humour theory of temperament. Pers Indiv Diff 12:255–263
Stevenson-Hinde J, Zunz M (1978) Subjective assessment of individual rhesus monkeys. Primates 19: 473–482
Suomi SJ (1991) Early stress and adult emotional reactivity in rhesus monkeys. Ciba Foundation Symposium 156. Wiley, Chichester
Weinstein, TAR, Capitanio, JP, and Gosling, SD. (2008) Personality in animals. In: John, OP, Robins, RW, and Pervin, LA (Eds). Handbook of personality: Theory and research. Guilford Press, New York
Westergaard GC, Suomi SJ, Higley JD (2002) Handedness is associated with immune functioning and behavioural reactivity in rhesus macaques. Laterality 7:359–369
Westermann J, Pabst R (1990) Lymphocyte subsets in the blood: A diagnostic window on the lymphoid system? Immunol Today 11:406–410
Winter DG, John OP, Stewart AJ et al. (1998) Traits and motives: Toward an integration of two traditions in personality research. Psychol Rev 105:230–250
Worlein JM, Laudenslager ML (2001) Effects of early rearing experiences and social interactions on immune function in nonhuman primates. In: Ader R, Felten DL, Cohen N (eds) Psychoneuroimmunology. Academic, San Diego
Yehuda R, Kahana B, Binder-Brynes K et al. (1995) Low urinary cortisol excretion in Holocaust survivors with posttraumatic stress disorder. Am J Psychiatry 152: 982–986

Chapter 10
Impulsivity and Aggression as Personality Traits in Nonhuman Primates

J. Dee Higley, Stephen J. Suomi, and Andrew C. Chaffin

Abstract Studies of macaques show that aggressiveness, along with its related cousin impulsivity, is trait-like, showing stable interindividual differences across time and situations. Two variations of aggressive temperament have been described: The first, aggressive temperament or overall aggressiveness, is characterized as competitive, marked by competition and a goal to win. While competitive and aggressive, such individuals seldom engage in violence. In competitive interchanges they often emerge as winners, and are typically high in social dominance. A second type of aggressive temperament leads to impulsive and unrestrained violence. This form of aggression has a strong relationship with impulse-control deficits. Evidence suggests that the two different forms of aggressiveness are mediated by differing systems, with competitive aggression mediated by testosterone. Impulsive aggression is mediated, at least in part by deficits in the serotonin system, with clear genetic and environmental underpinnings. These serotonin-impaired macaques show a variety of antisocial-like personality differences, exhibiting social alienation, sociosexual impairments, as well as impulse-control deficits, violence, and premature death, typically due to violent means. A variety of new molecular genetic studies show that the second form of aggressiveness is modulated, at least in part, by genetic × environmental interactions.

10.1 Introduction

In this paper we will describe a general trait, aggressiveness, along with its cousin impulsivity. We will describe two forms of aggression. Aggressive temperament or aggressiveness is characterized as being competitive, marked by competition and a goal to win. Such aggression is constrained and typically lacks severe physical

J.D. Higley (✉)
Department of Psychology, Brigham Young University,
1042 SWKT, Provo, UT 84602, USA
e-mail: james_higley@byu.edu

A. Weiss et al. (eds.), *Personality and Temperament in Nonhuman Primates*,
Developments in Primatology: Progress and Prospects, DOI 10.1007/978-1-4614-0176-6_10,

trauma. Individuals high in this trait are competitive, fearless, and are often high in dominance rank. Alternatively, aggressiveness can be violent, impulsive and unrestrained, a trait we have described using the term impulsive aggression or violent aggression. Violent or impulsive aggression is often described as out of proportion to the setting. Impulsive aggression is often dangerous and, because of its intensity and risk for injury, appears irrational or unplanned. Frequently it begins at a low level and escalates to dangerous levels. This latter type of aggressiveness describes an individual who is quick to exhibit anger and prone to violence. Such individuals are often described as hostile, unfriendly, harsh, antagonistic, violent, "hot-heads," or as having a short fuse.

Studies from our laboratory have a long tradition of focusing on aggression and its relationship to impulse-control deficits. These studies showed that impulsivity and aggression (particularly violent aggression) are at least in part modulated by serotonin, with impaired central nervous system serotonin (CNS 5-HT) being the hallmark of impulsive and aggressive subjects (Soubrié 1986; Higley 2003). As with fearfulness, rodent studies showed that impulsivity and aggressiveness could be selectively bred (Ferrari et al. 2005; Popova 2006; Caramaschi et al. 2007). There is an indication that these selective breeding studies may be breeding for differences in CNS serotonin functioning. For example, as wild silver foxes were selectively bred for tameness each generation that increased in domestic placidity (or put another way, reduced aggressiveness) also increased in CNS concentrations of 5-HT and its metabolite, 5-hydroxyindoleacetic acid (hereafter 5-HIAA) (Namboodiri et al. 1985; Popova et al. 1991a, b).

10.2 Observation of Aggression

Aggression has been measured in a wide variety of settings. For example, behavior has been directly measured, subjectively rated (Raleigh et al. 1989; Raleigh and McGuire 1990; Bolig et al. 1992; Higley et al. 1992b; Capitanio and Widaman 2005) and, wounds and scars have been counted, with their bodily locations indicating flight or attack (Crockett and Pope 1988; Higley et al. 1992b, 1996b, c). The most frequent context in which an aggressive temperament is expressed is in defense of status. While such aggression is typically exhibited in challenge, with a winner and loser, surprisingly, unlike rodents, the best fighters and largest animals do not necessarily become the highest-ranking animals (Kaufman 1967; Chapais 1983, 1986; Higley and Suomi 1989, 1996; Higley et al. 1994; Raleigh and McGuire 1994; Bastian et al. 2003). Nevertheless, aggression is critical to defend and maintain status, but such aggression seldom results in trauma or serious wounds (Bernstein 1981; Higley et al. 1996a, c). Individual differences in competitive aggression, for example, when measured in the context of social dominance show stability across months and years (Higley et al. 1996a; Westergaard et al. 1999b; Bastian et al. 2003).

While often similar in form, violent or impulsive aggression is more severe and is founded on different underlying mechanisms. While competitive aggression seldom results in trauma and injury, impulsive or violent aggression often does.

For example, monkeys who have a high number of wounds (a measurement of past aggressive encounters) are more likely to engage in violent behavior in the future and to die by violent means (Higley et al. 1996b; Westergaard et al. 1999a, b; Howell et al. 2007). Ethically, it is difficult to experimentally study violent aggression because it often leads to trauma and such subjects are often isolated socially by their conspecifics. Moreover, violent aggression is rare, making its study practically difficult as well. For those reasons, impulsive aggression is more often studied in natural settings where large numbers of subjects can be observed to detect rarely occurring events and where intervention is not practically possible. This form of aggressiveness is also stable across time and situations (Higley 2003; Howell et al. 2007). Somewhat surprisingly, males and females do not differ in absolute levels of aggressiveness, although when males fight, they are more likely to inflict serious wounds and trauma than females, a function of males' more elongated and dangerous canines (Higley 2003). Other data indicate that males are more likely than females to engage in contact aggression (Barr et al. 2004b).

10.3 Stability of Aggression

Individual monkeys show wide variation in competitive and impulsive aggression (Higley et al. 1994, 1996a, b; Westergaard et al. 1999a; Howell et al. 2007). Consistent with Mischel's (1968) definition of a trait, aggression displays long-term stability across time and situation (Mehlman et al. 1994; Howell et al. 2007). Similarly, violent aggressiveness, first measured early in life when macaques are still dependent on their mothers, is predictive of future violence in monkeys as much as a decade later (Howell et al. 2007). In this study, 7–10 years after the male monkeys had migrated to new troops and were middle-aged, if they were not killed by violent means, they were still the most violent among the large population of research subjects (Howell et al. 2007).

10.4 Rearing and Aggressive Temperament

Rearing plays an important role in the acquisition of temperamental aggressiveness. Peer-reared monkeys, for example, are more likely to be removed from their social group for violent aggression (Higley et al. 1994). Surrogate-mother-reared monkeys are less likely to show competitive aggression than mother-reared and peer-reared monkeys (Bastian et al. 2003). Moreover, monkeys deprived of social experience show high levels of impulsive aggression when confronted by other monkeys, even if the stimulus monkey is an adult male three to four times the size of the socially deprived juvenile (Suomi 1982). While mother-reared male and female subjects are likely to pick up an infant under conditions where they are required to "babysit," peer-reared monkeys not only ignore the infant but they are likely to act aggressively toward it when the infant approaches (Higley et al. 1994). Under the stressor of a new baby, they are also more likely to reject or neglect their first offspring

(Suomi and Ripp 1983). The mechanism for these differences in aggressive temperament may lie in part in the serotonin system.

10.5 Biological Correlates of Temperament

10.5.1 Trait-Like Nature of Physiological Underpinnings

One of the most replicated findings in biological psychology and psychiatry is that low or impaired CNS serotonin is correlated with impulse-control deficits and impulsive aggression (Higley and Barr 2008). To the extent that central serotonin functioning underlies stable personality traits such as impulsivity and aggression, across time and between situations, interindividual stability should be observed not only in behavior, but in the central serotonin system as well.

There is evidence for this. The brain is surrounded by fluid called cerebrospinal fluid (CSF) that bathes the brain. It removes used biological "trash" and protects the brain from trauma. In the same sense that we can look in someone's trash to know what they have eaten or what they have spent their money on, we can take small samples of CSF from the base of the brain in a region known as the cisterna magna, and look for the terminal products of neurotransmitters to assess what systems are activated under the conditions measured. One of those products is 5-HIAA, the final metabolite of serotonin. High concentrations of 5-HIAA reflect high activity in the serotonin system. As with aggressiveness, and consistent with Mischel's definition of a trait, the most replicated finding from our laboratory is that CSF 5-HIAA concentrations show strong interindividual stability both across time and situation (Kraemer et al. 1989; Higley et al. 1991, 1992c, 1993, 1994, 1996a, d; Mehlman et al. 1994; Cleveland et al. 2004; Shannon et al. 2005). To illustrate, CSF 5-HIAA was first measured when the subjects lived alone in single cages, and again after they were placed into novel social groups. Even in these different environments, CSF 5-HIAA concentrations were positively correlated across repeated samples.

Interindividual stability in CSF 5-HIAA is not limited to the controlled setting of the laboratory. Free-ranging monkeys were tested as infants, adolescents, and adults. During the initial 2 or 3 years of the study, most of the subjects, now adolescents, migrated from their natal troops to join new social groups (Higley et al. 1994). This is a period of high social stress, as the young males must form new relationships and face social challenges including trauma and premature mortality (Higley et al. 1996b; Howell et al. 2007). Remarkably, although the monkeys went through a wide variety of uncontrolled events, becoming independent of their mothers, moving from their family troop, often to several other troops, experiencing numerous fights and other social challenges, and even a hurricane, interindividual differences in CSF 5-HIAA levels remained stable from infancy into adolescence showing modest positive correlations, typically around $r=0.50$. (Mehlman et al. 1994; Higley et al. 1996a). Moreover, samples taken in infancy predicted interindividual differences in middle-aged adult males nearly a decade later (Howell et al. 2007).

As with temperament, interindividual differences in CNS 5-HT appear shortly after birth and stabilize early in life. When CSF 5-HIAA was obtained from neonatal monkeys on postnatal days 14, 30, 60, 90, 120, and 150, interindividual differences were stable across time with an average correlation of $r = 0.50$ across samples (Shannon et al. 2005). Our unpublished data show that samples taken from the subjects when they were only 7 days old also exhibit similar interindividual stability. These early differences remain stable, with mean concentrations of CSF 5-HIAA taken in late infancy (6 months of age) predicting concentrations a year later in middle childhood (Higley et al. 1992c), and into adulthood as well (Higley et al. 1996a, e).

While most early research on CNS serotonin differences was initially performed using rhesus macaques, stability in this measure of CNS 5-HT functioning occurs in other nonhuman primate species as well. When CSF was obtained from female pigtail and rhesus macaques, within each species, interindividual stability was virtually identical across repeated samples (Westergaard et al. 1999b). Raleigh and colleagues found a high degree of interindividual stability in CSF 5-HIAA concentrations in male vervet monkeys (Raleigh et al. 1992; Raleigh and McGuire 1994). Similar data for humans, with repeated samples, show stable interindividual differences early in life that are maintained across time (Bertilsson et al. 1982; Träskman-Bendz et al. 1984; Riddle et al. 1986).

The presence and early developmental emergence of trait-like, stable interindividual CSF 5-HIAA concentrations have important implications for those studying personality and temperament. Primary among those is an explanation for two individuals in the same setting responding differently. To illustrate, two monkeys, one high and one low in CNS serotonin functioning both find themselves confronted by a dilemma: an adult female that has been shadowed by the alpha male shows interest in mating. Despite the same risk of a fight and trauma from the alpha male, the first male possessing low CNS serotonin functioning and as a consequence under-regulated impulses, immediately responds to her solicitations by sitting close and grooming her. He is quickly chased away by the alpha male or, worse, is caught and receives serious trauma and wounds. The male with relatively high CNS serotonin functioning postpones his response for several hours until the female is eating by herself in a tree and as a consequence receives no retaliatory response from the alpha male. On the other hand, that same male with high CNS serotonin activity may wait so long that he loses his opportunity to mate entirely. The merits of waiting could be debated from an evolutionary perspective, but the point is that the situation is identical. The sexual motivation is strong in both cases; the difference is the activity of the serotonin system and its ultimate regulatory influence on these two individual males.

10.5.2 Serotonin and Aggression

Soubrié (1986) in his classic, comprehensive review of serotonin and impulse control concludes that aggression is higher in animals with an impaired or compromised central serotonin system. For example, rodent research shows that pharmacological agents that increase serotonin activity decrease aggression, while agents that decrease serotonin activity or block its action increase aggression (Miczek and Donat 1990; Olivier and Mos 1990; Olivier et al. 1990; Nikulina et al. 1992).

Naturally occurring differences between individuals in the functioning of the serotonin system have also been linked to differences in aggression and violent behavior. Among the most replicated findings in human psychiatry is that men with low CSF 5-HIAA concentrations exhibit increased unplanned aggression and impulsive violence (Brown et al. 1979; Linnoila et al. 1983; Lidberg et al. 1985; Roy et al. 1988; Limson et al. 1991). Similarly, nonhuman primates with low CSF 5-HIAA concentrations display increased rates of wounding, unprovoked and unrestrained aggression, and violent deaths (Higley et al. 1992b, 1996a–d; Mehlman et al. 1994, 1997; Westergaard et al. 2003a, b; Howell et al. 2007). For example, vervet monkeys that consumed experimental diets high in the serotonin precursor tryptophan exhibited decreased aggression, whereas individuals placed on diets low in tryptophan exhibited increased aggression, with the effects at times stronger for males than for females (Raleigh et al. 1985, 1986, 1991; Chamberlain et al. 1987). Similar changes have been demonstrated with other serotonin treatments, including decreases in aggression with short-term administration of serotonin reuptake inhibitors (which augment serotonin levels in the synapse) (Raleigh et al. 1980, 1985, 1986, 1991; Chamberlain et al. 1987; Higley et al. 1998) and increases in aggression after administration of the serotonin synthesis inhibitor *p*-chlorophenylalanine which acts to diminish serotonin production (Raleigh et al. 1980; Raleigh and McGuire 1986). Such findings suggest a causal, and not simply a correlational relationship between low or impaired CNS serotonin and aggression, especially violent or impulsive aggression.

Other studies have assessed the serotonin system using another method originally called the prolactin challenge. This procedure, widely used in humans, involves the administration of a serotonin-enhancing drug, such as fenfluramine hydrochloride or another reuptake inhibitor and then levels of prolactin in the blood are measured. High levels of prolactin are interpreted as evidence of an active serotonin system and low levels are seen as evidence of an impaired central serotonin system. Numerous studies using this method have shown impaired central serotonin functioning in aggressive men and violent patients (Newman et al. 1998). In monkeys, adult male cynomolgus monkeys administered fenfluramine who respond with low prolactin concentrations were more likely to exhibit aggressive responses to pictures of humans making threatening gestures (Kyes et al. 1995). Frontal cortex damage results in impaired impulse controls and results in violent aggression (Linnoila et al. 1993; Critchley et al. 2000; Woermann et al. 2000). We postulate that aggressiveness and violent temperaments seen in monkeys (and probably humans as well) with low CNS serotonin functioning are related in part to impairments with frontal cortex regulation of impulses that control emotions such as aggression.

While the relationship between serotonin and impulsive aggression has not been systematically studied in human females, nonhuman primate studies suggest that this relationship may extend to females as well as males. In female cynomolgus monkeys, low prolactin after fenfluramine administration was associated with high rates of aggression (Botchin et al. 1993; Shively et al. 1995). Adolescent and female macaques removed from their group for excessive aggression or those that exhibit high rates of impulsive aggression and unrestrained violence show lower CSF 5-HIAA than females who are not involved in aggression (Higley et al. 1996a, d; Westergaard et al. 2003a, b). The general pattern of violent aggression being more likely in subjects with low CSF 5-HIAA concentrations extends to female as well as to male macaques (Higley et al. 1996a, d; Westergaard et al. 2003a, b). Paralleling findings in males, high levels of impulsive behavior were correlated with high rates of violent aggression in females (Westergaard et al. 2003b). On the other hand, in the natural setting, where female aggression is more often between kin, females with low CSF 5-HIAA concentrations engage in high levels of aggression toward other females, but the aggression seldom escalates to violent levels that could result in trauma (Westergaard et al. 2003b).

10.5.3 Impaired Serotonin and Impulsive Temperament

Soubrié (1986) suggests that animals with impaired CNS serotonin are more aggressive because their impulse-control capabilities are defective, thus leading to uncontrolled emotional responses when provoked. Supporting evidence is that low CSF 5-HIAA is not correlated with overall levels of aggression; instead, it is only spontaneous, impulsive aggression that often escalates out of control and is negatively correlated with CSF 5-HIAA concentrations (Mehlman et al. 1994; Higley et al. 1992b, 1996b, c; Westergaard et al. 2003a; Howell et al. 2007). While many studies have investigated the relationship between diminished serotonin functioning and increased aggression and violence, monkeys with impaired CNS serotonin functioning show a cluster of closely related behaviors and traits centering around impaired impulse control (Apter et al. 1990). Cloninger (1986) reviews evidence suggesting personality disorders related to dysfunctional social relationships, aggression, and diminished social bonding are neurobiologically based on diminished serotonin functioning and possibly reduced dopamine functioning. Studies of both human children (Kruesi et al. 1990) and rhesus macaques (Higley et al. 1994, 1996a, d) show that individuals high in social deviancy or who exhibit less competent social behavior have relatively low CSF 5-HIAA. Cloninger's biosocial theory has become influential in understanding the underlying biological basis of personality. Many of his predictions hold up well with nonhuman primates, particularly predictions about low or impaired CNS serotonin and temperamental traits of impulsivity and violence (Higley and Barr 2008).

Soubrié's (1986) comprehensive review of serotonin and impulsivity in animals and subsequent studies (Higley and Barr 2008) show that serotonin plays a role in controlling a wide variety of impulses including aggression. For example, serotonin-enhancing pharmacological treatments decrease alcohol consumption in both rodents and nonhuman primates (Gill and Amit 1989; McBride et al. 1989;

Higley et al. 1992a). Conversely, in rodents, pharmacologically reducing CNS serotonin activity increases the frequency of performing a response despite the threat of punishment for responding (Soubrié 1986; Gleeson et al. 1989; Miczek et al. 1989). Additionally, men with low CSF 5-HIAA concentrations exhibit evidence of impaired impulse control such as increased unplanned fire setting (Virkkunen et al. 1987) and increased violent criminal recidivism (Virkkunen et al. 1989).

Like humans, nonhuman primates with low CSF 5-HIAA concentrations are more likely to exhibit behaviors characteristic of impaired impulse control including spontaneous, unprovoked long leaps at dangerous heights and repeated jumping into baited traps (Mehlman et al. 1994; Higley et al. 1996c; Fairbanks et al. 1999, 2001; Manuck et al. 2003). In the laboratory setting, when a novel black box is placed in a room, monkeys with low CSF 5-HIAA concentrations take less time to approach the box and to touch it than do monkeys with high CSF 5-HIAA concentrations (Bennett et al. 1998). Rhesus macaques with low CSF 5-HIAA concentrations are also more likely to over consume alcohol once drinking begins (Higley et al. 1996e).

In one recent study, Fairbanks and colleagues used a standardized test of impulse control to measure the relationship between impulsivity and CNS serotonin functioning. This test, known as the intruder paradigm, measures the latency to approach an unfamiliar same-aged, same sex unfamiliar monkey (see Fairbanks and Jorgensen 2011). In this study, they found that monkeys with low CSF 5-HIAA concentrations or with blunted response to a serotonin agonist approached a potentially dangerous stranger more quickly (Fairbanks et al. 2001; Manuck et al. 2003). If excessive aggression is a result of an impulse-control deficit, impulsivity would be positively correlated with rates of aggression and violence. In three separate studies (two with males and one with females), we found that rates of spontaneous, unprovoked, dangerous long leaps were positively correlated with rates of violent behavior (Mehlman et al. 1994; Higley et al. 1996c; Westergaard et al. 2003b). Similarly, monkeys who quickly approached a potentially dangerous male were more likely to act aggressively (Fairbanks et al. 2001), thus suggesting that high levels of aggression and violence in subjects with impaired central serotonin functioning is a result of deficient internal impulse control.

10.5.4 Low CSF 5-HIAA Concentrations, High Aggression and Social Alienation

Violent, impulsively aggressive males are not preferred companions and seldom affiliate with adult females (Chapais 1986; Mehlman et al. 1995, 1997). Because monkeys with low CSF 5-HIAA concentrations are more aggressive and exhibit impulse-control deficits that are likely to interfere with basic social processes such as turn-taking and reciprocity, it seems reasonable to hypothesize that males with low CSF 5-HIAA concentrations would also show impairments in Sociality.

Chamove et al. (1972) identified Sociality as one of three core personality traits in rhesus monkeys. Sociality in monkeys is similar to the Extraversion factor that Eysenck identified in humans (Eysenck and Eysenck 1971).

Several studies among nonhuman primates have demonstrated that Sociality is positively correlated with CNS serotonin function. For example, in a sample of free-ranging adolescent male monkeys, aggressive subjects with low CSF 5-HIAA concentrations exhibited reduced levels of four measures of Sociality: time spent grooming other monkeys; time spent in close proximity to other group members; time spent in general affiliative social behaviors; and mean number of companions within a 5-m radius (<5 m) (Mehlman et al. 1995, 1997). In the laboratory, among juveniles, low rates of positive social interactions are also correlated with low CSF 5-HIAA concentrations in both sexes (Higley et al. 1994, 1996a). The relationship between CNS serotonin and aggression may also be a causal one. For example, across repeated studies of captive vervet monkeys, enhancing serotonin functioning by administering the serotonin precursor tryptophan, the reuptake inhibitor fluoxetine, or the serotonin agonist quipazine, decreased aggression while increasing positive social behaviors such as approaching and grooming other monkeys (Raleigh et al. 1980, 1983, 1985). When serotonin functioning was reduced by administering the tryptophan hydroxylase enzyme inhibitor PCPA to monkeys, the inhibitor produced opposite effects. That is the monkeys withdrew, avoiding social proximity and affiliative social interactions (Raleigh et al. 1980, 1985; Raleigh and McGuire 1990).

Studies of both violent humans (Brown et al. 1982a) and nonhuman primates (Raleigh et al. 1989; McGuire et al. 1994) have shown that individuals high in social deviancy, or low in competent social behaviors (Kruesi et al. 1990) have relatively low CSF 5-HIAA concentrations. Social dominance is acquired and maintained through the formation of affiliative bonds with other troop members who then support the dominant male during hostile and challenging social encounters (Packer 1979; Smuts 1987; Walters and Seyfarth 1987; Raleigh and McGuire 1991; Raleigh et al. 1991). As noted previously, building coalitions and maintaining social support is crucial to acquire and maintain a high social dominance rank (Chapais 1986, 1988; Raleigh and McGuire 1986; Raleigh et al. 1991; Higley and Suomi 1996). Thus it is reasonable to predict that monkeys low in CSF 5-HIAA concentrations would be more likely to be low in social dominance. Indeed, in a series of studies within our laboratory and others, a frequently replicated finding has been that naturally occurring low CNS serotonin functioning, as measured by low CSF 5-HIAA (Raleigh et al. 1983; Higley et al. 1994, 1996a; Raleigh and McGuire 1994; Westergaard et al. 1999a), or pharmacologically reducing CNS serotonin functioning (Raleigh et al. 1983, 1986) is linked to low social dominance ranking. This finding may not be true for all species however, because in cynomolgus monkeys, studies have not always found relationship between low rank and impaired CNS serotonin (Botchin et al. 1993; Shively et al. 1995). This may also be context-dependent. In natural settings, aggressive monkeys with low CSF 5-HIAA concentrations may at times become high ranking through violent means, however, their tenure is limited and they are more likely to be killed or removed by violent means (Howell et al. 2007). One might say of such methods,

> "*… for all those who take the sword will perish by the sword.*" (Matthew, 26:52).

10.5.5 Serotonin and Other Deficits Possibly Related to Aggressive Personality

Aggressive monkeys with low CSF 5-HIAA concentrations also show other deficits. They exhibit high levels of stereotypy, perhaps reflecting their irritable natures (Erickson et al. 2001). They are also dysregulated in daily activity and circadian cycles. Juvenile monkeys with low CSF 5-HIAA concentrations take longer to fall asleep at night and exhibit more motor activity during the day. In the field, the aggressive macaques with low CSF 5-HIAA concentrations woke-up more often and spent more time in motor activity during the night than did their high CSF 5-HIAA conspecifics. They also napped more often during the daytime (Zajicek et al. 1997; Mehlman et al. 2000).

10.5.6 Testosterone

Historically testosterone has received the most research as a hormone modulating violent aggression. This is largely because medically or chemically castrated animals are unlikely to engage in aggression. However, when it is replaced, agonistic behavior reappears (Archer 1991). It is now clear that testosterone is not the major cause of violent aggression in human and nonhuman primates. Instead testosterone probably underlies competitiveness or an overall agonistic motivation in humans (Christiansen and Knussmann 1987b; Olweus et al. 1988; Archer 1991; Buchanan et al. 1992). Studies showing a positive correlation between aggression and testosterone often fail to replicate (Olweus et al. 1988; Archer 1991), but significant correlations between testosterone and aggression usually occur when both are measured during competition or during challenges to social status (Scaramella and Brown 1978; Mazur 1983) or in response to provocation or threat (Olweus et al. 1988). The other exception to this rule is that among violent criminals, a positive correlation between aggression and testosterone is sometimes reported (Olweus et al. 1988; Archer 1991). This may, however, be due to other personality anomalies relating to impulse control. Indeed, most individuals with high testosterone are not violent but are restrained in their use of aggression, expressing it in socially acceptable ways. This may be a function for which testosterone-mediated aggression has been selected (i.e., maintenance of social status and defense of competitive challenges). Moreover, correlations between testosterone and behavior are not limited to aggression. Testosterone is also correlated with positive traits such as toughness (Dabbs et al. 1987), social dominance (Ehrenkranz et al. 1974; Christiansen and Knussmann 1987a; Lindman et al. 1987; Booth et al. 1989), social assertiveness (Lindman et al. 1987), and competitiveness and physical vigor (Mattsson et al. 1980; Booth et al. 1989).

Among nonhuman primates, both aggression and testosterone peak during the breeding season (Kaufman 1967; Gordon et al. 1976, 1978; Bernstein et al. 1977;

Rose et al. 1978; Paul 1989; Kuester and Paul 1992; Wickings and Dixson 1992). The parallel increase in both testosterone and aggression may, however, be related more to the sexual motivation and consequent competition over females than a direct effect of testosterone on aggression. As in studies of humans, many animal studies report no relationship between aggression and testosterone (Rose et al. 1972, 1975; Eaton and Resko 1974; Gordon et al. 1976). Augmenting testosterone has no effect on either aggression or acquisition of social dominance rank in talapoin monkeys living in an all male group (Keverne et al. 1983). On the other hand, when cynomolgus monkeys were injected with testosterone propionate, aggression increased, but aggression varied across individual subjects, with higher rates in dominant than in subordinate monkeys (Rejeski et al. 1988). Consistent with our belief that testosterone mediates competition rather than aggression per se in primates, the motivating influence of testosterone is mediated by social dominance. Aggression increased in dominant monkeys, but decreased in the subordinate monkeys during testosterone treatment (Rejeski et al. 1990).

Perhaps the strongest finding concerning testosterone and aggression is that during competition winning augments testosterone and losing reduces testosterone. Steklis et al. (1985) failed to find a relationship between overall aggression and testosterone, or social dominance and testosterone in vervet monkeys. On the other hand, and consistent with an interpretation that winning augments testosterone output, in that same study, on days that males fought, testosterone was higher in the dominant male than in the subordinate male. In talapoin monkeys, levels of preexisting testosterone failed to predict which monkeys became dominant; however, the monkeys who became dominant showed the highest levels of testosterone as the social dominance ranking became established (Eberhart et al. 1985). Similarly, when rhesus males were placed into a social group, the male who became dominant showed progressive increases in testosterone, while the males who became subordinate showed decreases in testosterone levels (Rose et al. 1975). In talapoin monkeys, males who increased their rank showed an increase in testosterone levels (Yodyingyuad et al. 1982) and males who were aggressed against and defeated exhibited a drop in testosterone (Martensz et al. 1987). Other studies have also shown that in aggressive rank competitions, losers show a fall in testosterone (Rose et al. 1972; Bernstein et al. 1979).

Testosterone and aggression are more likely to show a relationship when continued competition is present. Rose found that during the first 9 months following group formation, aggression was correlated with testosterone (Rose et al. 1971). In baboons, during a period of social instability, after the dominant male had been overthrown, dominant males initiated the most fights and had the highest testosterone (Sapolsky 1983). In other long-term studies, testosterone was less likely to show a relationship with social dominance rank and aggression (Eaton and Resko 1974; Gordon et al. 1976). Perhaps this is because after the dominance hierarchy is established, rank no longer needs to be won in each aggressive encounter. Sapolsky concluded that in groups that have been together for over a year (who presumably have had sufficient time to establish a long-term stable dominance hierarchy), the relationship between dominance and testosterone disappears (Sapolsky 1992).

10.5.7 Evidence for Separate Roles of Serotonin and Testosterone in Aggression

The above findings and others (e.g., Olweus et al. 1980, 1988; Soubrié 1986; Archer 1991), suggest that testosterone may be correlated with aggressive motives and competitiveness, rather than aggression per se. Serotonin, on the other hand, may function to inhibit its initiation, limiting aggression to proper time, setting, and intensity. As a direct test of this hypothesis, we measured both CSF testosterone and 5-HIAA in a group of free-ranging monkeys and measured rates of impulsivity, and restrained competitive aggression as well as violent aggression (Higley et al. 1996c). We found that (1) central testosterone was positively correlated with overall aggressiveness, and with competition for status, but not with measures of impulsivity such as spontaneous long leaps or rates of capture. (2) Central serotonin was negatively correlated with impulsive behaviors (spontaneous long leaps and rates of capture), and severe, violent aggression, but not with overall rates of aggression. Moreover, high rates of impulsive behavior were positively correlated with severe, unrestrained aggression, but not overall rates of aggression. (3) Central serotonin was negatively correlated with aggression, and high CSF testosterone further augmented rates and intensity of aggression in subjects with low serotonin, except for the most severe forms of aggression which are unaffected by high testosterone.

One possible interpretation of these data is that testosterone and serotonin may contribute differentially to the expression of aggressive behaviors, with testosterone contributing to aggressive drive and motivation and serotonin regulating the threshold, intensity, and resulting frequency of the behavioral expression. Thus, individuals with low testosterone would be unlikely to engage in aggression, regardless of their serotonin levels, but if they had impaired central serotonin functioning, they might act impulsively in other behaviors. On the other hand, individuals with above average testosterone but normal serotonin may express aggression in a variety of settings, but generally, would not express violence or unrestrained aggression. They would also exhibit more assertive behaviors that characterize socially dominant males, such as threats, displacements, or mounting of other male monkeys. Individuals with lower than average serotonin functioning would be expected to exhibit impaired impulse control resulting in a low threshold to display aggression and ultimately engage in more frequent violence. However, high testosterone, because it increases aggressive motivation, would further augment the propensity to engage in aggression in subjects with low central serotonin. Furthermore, once an aggressive act had begun, subjects with low central serotonin would exhibit deficits in stopping the aggression before it escalated into violence that has a high probability of injury. Put simply, testosterone provides the push to act competitively and serotonin provides the brakes that determine the timing and intensity.

10.6 Molecular Genetics

10.6.1 rh5-HTT, Other Genes and Impulsive–Aggressive Temperament

The best characterized and most studied gene is that of the serotonin transporter. It has been characterized as having two repeat polymorphisms that are orthologous to those found in humans, one with a short and the other a long allele (hereafter rh5-HTT). The short variant is less efficient, exhibiting reduced transporter transcriptional efficiency and decreased rh5-HTT expression. As a consequence, monkeys with the short allele are likely to exhibit low CSF 5-HIAA concentrations (Bennett et al. 2002) and a variety of serotonin mediated temperamental and stress-axis hormonal differences (Champoux et al. 2002; Barr et al. 2004a, c, d).

While the effects of the serotonin transporter genotype are most often investigated in relation to reactivity/anxiety temperaments, more recently evidence has emerged that the genotype may not be specific for anxiety/reactivity. As might be expected given our earlier discussions of aggression and impaired CNS serotonin, the effects of the genotype may extend to impulse control and other related traits. For example, Jarrell et al. (2008) found that rhesus females with one or two copies of the rh5-HTT short allele variant were more likely to exhibit high rates of both aggression and submissive behaviors than those females homozygous for the long allele. Subjects with the rh5-HTT short allele are also more likely to exhibit aggression and bully their way to high social dominance (Miller-Butterworth et al. 2007). McCormack et al. (2009) showed that abusive mothers were more likely to possess the rh5-HTT short allele than nonabusive mothers and were less likely to hold and care for their infants. In a battery of tests measuring temperament in these abusive mothers' infants, levels of anxiety and irritability were higher in infants from both abusive and control mothers if they possessed a copy of the rh5-HTT short allele. This growing body of research indicates that the effects of the rh5-HTT gene are not limited to anxious temperament, but extend to aggressive tendencies, suggesting a more general role for the transporter in negative emotionality.

The relationship between high impulsivity and aggression suggests that rh5-HTT is related to impulsive cognitions. Subjects with the rh5-HTT short allele exhibit errors indicative of dose-dependent impulse-control deficits, such as perseveration in reversal tasks, and a failure to extinguish previously reinforced behavior, as well as increased measures of aggression. Subjects with one copy exhibit significantly more of these errors while those with two copies of the short allele exhibit even more than those with one copy (Izquierdo et al. 2007; Jedema et al. 2010). In a study of monkeys in a natural setting, males homozygous for the short allele exhibited increased impulsive behaviors. These males were also more likely to leave the safety of their natal troop before they were fully mature increasing the risk for premature death (Trefilov et al. 2000).

The common mechanism that underlies both the violent and the anxious behaviors mediated by the serotonin transporter short allele may be a result of amygdala deficits. The relationships between aggression and the short allele of the serotonin transporter have led Lesch to conclude in a recent review that the serotonin transporter genotype may be more involved in negative emotionality than in anxiety per se (Lesch 2007). Further studies to determine the common and different contributions of the transporter genotype to aggression and anxiety are necessary before we will understand how it functions to modulate such divergent temperaments.

Other serotonin transporter genotypes may also be involved in nonhuman primate personality and cognition. Some species, such as cynomolgus macaques, do not show the rh5-HTT short allele variants seen in other species of macaques, but even in this species, there are numerous genetic variants in the serotonin transporter gene. In fact, they and the rhesus show about double the number of genetic variants seen in humans (Miller-Butterworth et al. 2007). Diplotype groupings of these serotonin transporter variants in cynomolgus macaques showed an association with the acquisition of social dominance rank (Miller-Butterworth et al. 2007). Genotype effects on aggression are not limited to the serotonin transporter gene, but extend to other genes that regulate the serotonin system, such as MAOa genes and the tryptophan hydroxylase 2 (TPH2) gene. Newman et al. (2005), for example, found that a low-activity allele variant found in the promoter region leads to higher aggression, although as discussed below the effect of the genotype on aggression was limited to animals with adverse rearing conditions. Similarly, there was an association of the TPH2 gene with aggression, although more so for animals reared without their mothers in peer groups.

10.6.2 Gene × Environment Interactions

Traditional models that outline genetic effects have used quantitative genetics to understand how genes and the environment interact to produce phenotypic outcomes. These additive models posit average effects of genotypes considered as uniform across a wide variety of environments. It is clear, however, that the effects of genes are often not monolithic across all environments. For example, while the average heritability of human alcoholism is about 0.50 or $h^2=0.50$ (Lorenz et al. 2006), that value varies across environments. Among Muslims and Mormons, who abstain from alcohol, the phenotypic expression of alcoholic genes is negligible (Linsky et al. 1986). For bartenders and brewers who have a high rate of alcoholism (Mandell et al. 1992), on the other hand, the phenotypic expression of alcoholic genes are likely to be higher than the rest of the population. With the capacity to characterize people according to genotype, researchers could directly assess

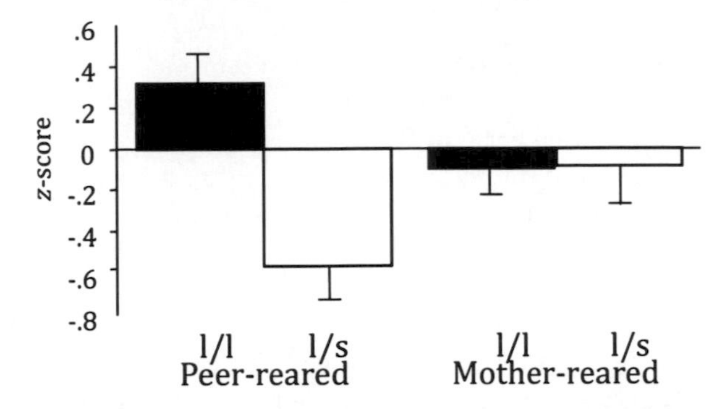

Fig. 10.1 Effect of rh5-HTT genotype (l/l and l/s) and early-rearing environment (parent-reared, l/s and peer-reared, on mean CSF 5-HIAA concentrations) (picomoles per ml of 5-HIAA) with raw values transformed to standard scores for each individual within its cohort and age at sample collection. ANOVA yielded a significant interaction between rearing environment and genotype on CSF 5-HIAA concentrations. Mother-reared monkeys' CSF 5-HIAA concentrations were not affected by genotype, but peer-reared monkeys with the short allele had lower CSF 5-HIAA concentrations than their homozygous, long allele counterparts (see Bennett et al. 2002 for more information)

the effects of varied environments on genotype-mediated phenotypic outcomes. To extend the example, among monkeys, alcohol consumption is about twice as high when monkeys drink alone and males drink more than females when drinking alone, but males drink at rates similar to females when drinking in social groups. Most important to the above discussion, the serotonin transporter gene has no effect on consumption when animals drink in social groups, but augments intake significantly when subjects drink alone (Higley, unpublished data).

This gene by environment interaction (G × E) is further illustrated by two recent studies showing the effects of differing environments on genotype-mediated outcomes. Bennett et al. (2002) tested rhesus monkey reared without adults in peer-only groups or in control groups with their mothers, approximating normative conditions. In mother-reared subjects, the 5-HTT genotype had no effects on CSF 5-HIAA concentrations. Unlike the mother-reared subjects, the peer-reared subjects with the short allele showed significantly lower CSF 5-HIAA concentrations; whereas the peer-reared which were homozygous for the long allele were undifferentiated from the mother-reared subjects (see Fig. 10.1).

Caspi et al. (2003) assessed similar G × E effects for the serotonin gene, investigating people homozygous or heterozygous for short allele, or homozygous for the long allele. They were further characterized according to the levels of severe stress in their lives, and whether they developed depression. While there were no genotype-mediated differences in rates of depression for people with the long or short allele in the no-major-stress or minimal-major-distress groups, individuals with the short allele showed modest rates of depression, but only if there were three stressful events and they found high rates of depression if there were more than three major stressful events over the past year. The relevance of these studies is particularly strong when considering the effects of genes, or for that matter the environment, on temperament and personality. While personality researchers have previously considered average genetic effects across environments, these G×E discoveries show that in one environment, the effects of a gene on an individual's personality may be negligible, whereas in another environment, that same gene may have substantial effects.

Subsequent studies in young rhesus monkeys have confirmed these original G×E findings. Neonates with the short allele were not different from those homozygous for the long allele when tested in a battery of neonatal tests that assess central functioning and emotionality, but when the subjects were characterized by rearing, possession of the short allele was shown to have a profound effect. Infants with the short allele that were reared in the nursery by humans exhibited lower scores on orientation, motor maturity, and consolability than mother-reared infants (Champoux et al. 2002).

Capitanio et al. (2008) found that aggression rates were on average higher in rhesus macaques with the rh5-HTT short allele. However, when the subjects' rates of aggression were divided according to stable or unstable environments, subjects with at least one copy of the short allele exhibited high aggression in unstable but not stable environments. Rates of aggression for subjects homozygous for the long allele were equal in both stable and unstable environments. In another study, juvenile mother-reared monkeys who seldom engage in aggression were tested. Those that exhibited aggression were no more likely to have the short allele than those subjects that never engaged in aggression. On the other hand, juvenile peer-reared monkeys with the short allele engaged in much higher levels of aggression than the long allele peer-reared and both groups of mother-reared monkeys (Barr et al. 2003). This study is particularly intriguing because a potential mechanism for the G×E interaction and resulting lower aggression was included in the analyses. Monkeys with low levels of play were more likely to act aggressively later in life (Suomi 1982; Hirsch et al. 1986; Bush et al. 1987).

In an intriguing just completed study, 150 monkeys of various ages were tested in their home cage for their response to a same age and sex intruder. The intruder was placed in a holding cage and rolled to the side of the monkeys' home cage pen. The residents were tested for various measures of violent aggression (bites, slaps, hits, etc.). Subjects with a copy of the rh5-HTT short allele were more than three times as likely to engage in violence than those homozygous for the long allele, but the effect also interacted with the genotype of the victim or the intruder in a G×E fashion. Most importantly, when both the intruder and the resident possessed the short allele, there was nearly a sixfold increase in the rate of violent contact aggression between the two monkeys, significantly more than all other types of pairings (see Fig. 10.2).

A genetic variant of MAOa, one of the principal enzymes responsible for breaking down serotonin, increases aggression. Manuck et al. (2000), for example, showed that a high-activity MAOa genetic variant was associated with impulsivity, hostility, and lifetime aggression history in men. In a ground-breaking study, Caspi and colleagues showed that the effect of the gene on aggression and antisocial

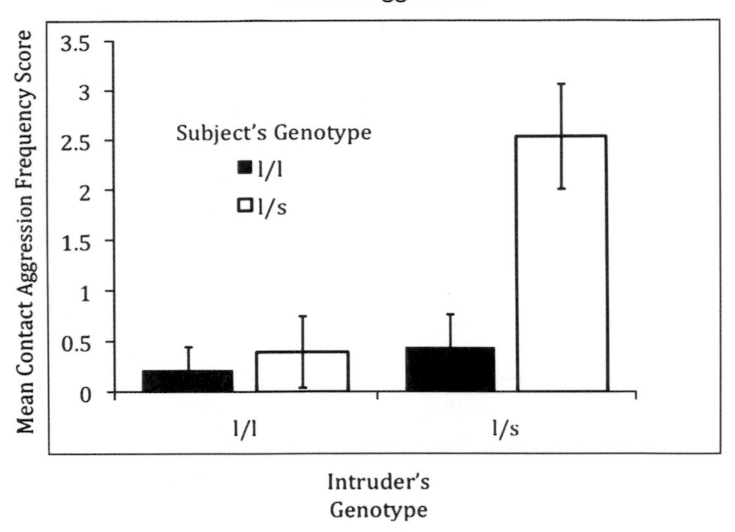

Fig. 10.2 The effect of the genotypes of both the subject and intruder ($F=6.522$, $p=0.012$) on contact aggression. Contact aggression escalated significantly when l/s genotype subjects were paired with an l/s genotype intruder (nearly sixfold greater than all other types of pairings). None of the other pairings differed significantly

behavior was only present in men exposed to early adversity, a G×E interaction. Without early adversity, the gene had little effect on rates of aggression (Caspi et al. 2002). There is also evidence for this effect in rhesus monkeys. Mother-reared rhesus macaque infants with MAOa genotypes that differed in functional effectiveness exhibited varying rates of aggression, although in rhesus macaques unlike humans, the genetic effect was with the low functioning genotype (Newman et al. 2005).

Most studies of the serotonin transporter in nonhuman primates have been performed with rhesus macaques since their high base rates of aggression are well suited for study of the relationship between the serotonin transporter genotype and aggression. Wendland et al. (2006a) characterized six other macaque species for comparison to the rhesus for the 5-HTT and MAOa genes (Wendland et al. 2006b). Since rhesus monkeys have a high frequency of polymorphic variation relative to humans, it was surprising that the other macaque species displayed no allelic variability in the promoter region of the 5-HTT gene. All of the other species had a monomorphic structure. Moreover, the other species had only a few variants in the MAOa gene. Specifically, for the serotonin transporter gene, most of the species were identical for a specific repeat number: Pigtail (*Macaca nemestrina*), stumptail (*Macaca arctoides*), Tibetan (*Macaca thibetana*), and crab-eating (*Macaca fasicularis*) macaques – all were homozygous for the long rh5-HTT allele. The Barbary macaques (*Macaca silvanus*) all had an extra-long version of this gene and the seldom studied Tibetan macaques (*Macaca thibetana*) possessed an unusual extra-short (even fewer repeats) 5-HTT promoter region. Results were similar for the MAOa gene, with rhesus macaques showing the most and the Barbary, Tibetan, and stumptail macaques showing the least allelic variability (Wendland et al. 2006b). This is interesting and parallels the dissimilar behaviors across different macaque species. Intriguingly, with the exception of the Tibetan macaque, there may be a systematic positive relationship between a high number of polymorphic repeats in the serotonin transporter and MAOa gene, and the relative aggressivity and intolerance of social companions at the species level. The Barbary macaques show the highest degree of social tolerance, exhibiting a more fluid and relaxed dominance style, a high degree of reconciliation following conflict, and the least amount of aggression. For the other species studied, those who showed no 5-HTT repeat variants and only limited MAOa variability were intermediate between the Barbary and rhesus macaques (Wendland et al. 2006b). While the Tibetan macaque appears to be an exception, a recent study of Tibetan macaque suggests that it may be more aggressive than previously thought (Berman et al. 2004), suggesting that the species-mediated gene–personality relationship may hold up in the Tibetan macaques as well.

Of all macaque species, the rhesus has been the most successful; their range and numbers are among the highest of the nonhuman primates. This adaptive success is often attributed to their aggressive nature and their temperamental hardiness. Their success as a species may also be related to their high genetic variability and its corresponding effect on adaptiveness. Harlow remarked that of all the species he studied, the rhesus, although very aggressive, seemed to adapt and flourish best in the laboratory (Harlow 2008). The Chinese-derived rhesus, possibly a subspecies of rhesus, is even more aggressive and intolerant of others within their social groups than their Indian-derived species (Champoux et al. 1994, 1997). Interestingly, they are significantly more likely to possess the rh5-HTT short allele (Higley, unpublished analysis). This rhesus variant is particularly interesting because of their geographic challenge of crossing the Himalayas to migrate to China, a feat that may be evidence of their hardiness, as well as their risk-taking and aggressive nature. The aggressive and less anxious personality of the rhesus may have led to evolutionary success that paid off in adapting to a wide variety of environments. Suomi (2006) further postulated that the increased genetic variability in the genes that modulate personality may confer advantages that allow more evolutionary adaptation across a wider variety of environments than in species that only have limited genetic variability in personality-related genes.

One interpretation of G × E interactions on personality is that the long 5-HTT allele and other normative genotypes confer protection or resiliency to early adverse environmental input and trauma. An alternative interpretation is that through evolutionary processes, normal maternal input or "safe" environments confer or protect individuals from the potential deleterious effects of less functional genetic influences that might otherwise increase their risk for adverse developmental outcomes ("maternal buffering"). These two interpretations are not mutually exclusive, but the difference in their implications for developing prevention and even intervention strategies to prevent psychopathological outcomes is considerable. From an evolutionary perspective, the appropriate input, operating at the right time may be necessary for normative brain development, and while the genetic input lays the pathways for neuronal development, the connections that survive are a result of appropriate maternal input at the correct developmental period. It is also clear that the effects of a gene on personality depend upon the environmental context in which it is expressed. Conversely, the effect of the environment on personality is dependent on the genetic background of the individual. Genetic and environmental effects are not simply additive in their interacting effects, but at least in the cases cited, genes and the environment interact in a qualitative fashion to produce, in some cases, an "either-or-effect" on personality.

10.7 Conclusions

Since most research on genes and personality in monkeys (and apes) has been focused on Old World species, additional research on New World species is desirable. This will allow us to search the evolutionary mirror for reflected commonalities. In addition, research on New World monkeys will illuminate differences that may explain the effect of ecological demands and our uniqueness. The physiology of most of the advanced nonhuman primate species is highly similar to that of humans. Like humans, they typically live in complex social groups. While rhesus monkeys share between 93 and 96% of their DNA structure with humans (Gibbs et al. 2007; Harris et al. 2007; Lee et al. 2008), some nonhuman primates such as the great apes share more than 98% of their genetic material with humans. Nevertheless, that 2–8% disparity is important (no nonhuman primate will obtain a college education), and, in generalizing, this genetic difference should be remembered.

With this caveat in mind, our findings suggest several possible similarities between human and nonhuman primates. We have described temperamental traits in rhesus monkeys that are related to how individuals respond to challenging and competitive situations. Consistent with personality theory and paralleling human traits, these traits are highly stable across time and situations. While it is difficult to directly measure central mechanisms of personality in human children, our close cousins offer important insights into the development and systems that underlie personality. Ultimately, the elucidation of etiological mechanisms and outcome of early temperaments will be the result.

References

Apter A, van Praag HM, Plutchik R et al. (1990) Interrelationships among anxiety, aggression, impulsivity, and mood: A serotonergically linked cluster? Psychiatry Res 32:191–199

Archer J (1991) The influence of testosterone on human aggression. Br J Psychol 82:1–28

Barr CS, Newman TK, Becker ML et al. (2003) The utility of the non-human primate; model for studying gene by environment interactions in behavioral research. Genes Brain Behav 2:336–340

Barr CS, Newman TK, Lindell S et al. (2004a) Early experience and sex interact to influence limbic-hypothalamic-pituitary-adrenal-axis function after acute alcohol administration in rhesus macaques (*Macaca mulatta*). Alcohol Clin Exp Res 28:1114–1119

Barr CS, Newman TK, Lindell S et al. (2004b) Interaction between serotonin transporter gene variation and rearing condition in alcohol preference and consumption in female primates. Arch Gen Psychiatry 61:1146–1152

Barr CS, Newman TK, Schwandt M et al. (2004c) Sexual dichotomy of an interaction between early adversity and the serotonin transporter gene promoter variant in rhesus macaques. Proc Natl Acad Sci USA 101:12358–12363

Barr CS, Newman TK, Shannon C et al. (2004d) Rearing condition and rh5-HTTLPR interact to influence limbic-hypothalamic-pituitary-adrenal axis response to stress in infant macaques. Biol Psychiatry 55:733–738

Bastian ML, Sponberg AC, Suomi SJ et al. (2003) Long-term effects of infant rearing condition on the acquisition of dominance rank in juvenile and adult rhesus macaques (*Macaca mulatta*). Dev Psychobiol 42:44–51

Bennett AJ, Lesch KP, Heils A et al. (2002) Early experience and serotonin transporter gene variation interact to influence primate CNS function. Mol Psychiatry 7:118–122

Bennett AJ, Tsai T, Pierre PJ et al. (1998) Behavioral response to novel objects varies with CSF monoamine concentrations in rhesus monkeys. Soc Neurosci Abstr 24:954

Berman CM, Ionica CS, Jin-Hua LI (2004) Dominance style among *Macaca thibetana* on Mt. Huangshan, China. Int J Primatol 25:214–227

Bernstein IS (1981) Dominance: The baby and the bathwater. Behav Brain Sci 4:419–457

Bernstein IS, Rose RM, Gordon TP (1977) Behavioural and hormonal responses of male rhesus monkeys introduced to females in the breeding and non-breeding seasons. Anim Behav 25:609–614

Bernstein IS, Rose RM, Gordon TP et al. (1979) Agonistic rank, aggression, social context, and testosterone in male pigtailed monkeys. Aggress Behav 5:329–339

Bertilsson L, Tybring G, Braithwaite R et al. (1982) Urinary excretion of 5-hydroxyindoleacetic acid – no relationship to the level in cerebrospinal fluid. Acta Psychiatr Scand 66:190–198

Bolig R, Price CS, O'Neill PL et al. (1992) Subjective assessment of reactivity level and personality traits of rhesus monkeys. Int J Primatol 13:287–306

Booth A, Shelley G, Mazur A et al. (1989) Testosterone, and winning and losing in human competition. Horm Behav 23:556–571

Botchin MB, Kaplan JR, Manuck SB et al. (1993) Low versus high prolactin responders to fenfluramine challenge: Marker of behavioral differences in adult male cynomolgus macaques. Neuropsychopharmacology 9:93–99

Brown GL, Ebert MH, Goyer PF et al. (1982a) Aggression, suicide, and serotonin: Relationships to CSF amine metabolites. Am J Psychiatry 139:741–746

Brown GL, Goodwin FK, Ballenger JC et al. (1979) Aggression in humans correlates with cerebrospinal fluid amine metabolites. Psychiatry Res 1:131–139

Brown GL, Goodwin FK, Bunney WEJ (1982b) Human aggression and suicide: Their relationship to neuropsychiatric diagnoses and serotonin metabolism. Adv Biochem Psychopharmacol 34:287–307

Buchanan CM, Eccles JS, Becker JB (1992) Are adolescents the victims of raging hormones: Evidence for activational effects of hormones on moods and behavior at adolescence. Psychol Bull 111:62–107

Bush DS, Steffen SL, Higley JD et al. (1987) Continuity of social separation responses in rhesus monkeys (*Macaca mulatta*) reared under different conditions. Am J Primatol 18:138

Capitanio JP, Abel K, Mendoza SP et al. (2008) Personality and serotonin transporter genotype interact with social context to affect immunity and viral set-point in simian immunodeficiency virus disease. Brain Behav Immun 22:679–689

Capitanio JP, Widaman KF (2005) Confirmatory factor analysis of personality structure in adult male rhesus monkeys (*Macaca mulatta*). Am J Primatol 65:289–294

Caramaschi D, de Boer SF, Koolhaas JM (2007) Differential role of the 5-HT1A receptor in aggressive and non-aggressive mice: An across-strain comparison. Physiol Behav 90:590–601

Caspi A, McClay J, Moffitt TE et al. (2002) Role of genotype in the cycle of violence in maltreated children. Science 297:851–854

Caspi A, Sugden D, Moffitt TE et al. (2003) Influence of life stress on depression: Moderation by a polymorphism in the 5-HTT gene. Science 301:386–389

Chamberlain B, Ervin FR, Pihl RO et al. (1987) The effect of raising or lowering tryptophan levels on aggression in vervet monkeys. Pharmacol Biochem Behav 28:503–510

Chamove AS, Eysenck HJ, Harlow HF (1972) Personality in monkeys: Factor analyses of rhesus social behaviour. Q J Exp Psychol 24:496–504

Champoux M, Bennett A, Shannon C et al. (2002) Serotonin transporter gene polymorphism, differential early rearing, and behavior in rhesus monkey neonates. Mol Psychiatry 7:1058–1063

Champoux M, Higley JD, Suomi SJ (1997) Behavioral and physiological characteristics of Indian and Chinese-Indian hybrid rhesus macaque infants. Dev Psychobiol 31:49–63

Champoux M, Suomi SJ, Schneider ML (1994) Temperament differences between captive Indian and Chinese-Indian hybrid rhesus macaque neonates. Lab Anim Sci 44:351–357

Chapais B (1983) Matriline membership and male rhesus reaching high ranks in the natal troop. In: Hinde RA (ed) Primate social relationships: An integrated approach. Sinauer Associates, Inc., Sunderland

Chapais B (1986) Why do male and female rhesus monkeys affiliate during the birth season? In: Rawlins RG, Kessler M (eds) The Cayo Santiago macaques. SUNY Press, Chicago

Chapais B (1988) Rank maintenance in female Japanese macaques: Experimental evidence for social dependency. Behaviour 102:41–59

Christiansen K, Knussmann R (1987a) Androgen levels and components of aggressive behavior in men. Horm Behav 21:170–180

Christiansen K, Knussmann R (1987b) Sex hormones and cognitive functioning in men. Neuropsychobiology 18:27–36

Cleveland A, Westergaard GC, Trenkle MK et al. (2004) Physiological predictors of reproductive outcome and mother-infant behaviors in captive rhesus macaque females (*Macaca mulatta*). Neuropsychopharmacology 29:901–910

Cloninger CR (1986) A unified biosocial theory of personality and its role in the development of anxiety states. Psychiatr Dev 4:167–226

Critchley HD, Simmons A, Daly EM et al. (2000) Prefrontal and medial temporal correlates of repetitive violence to self and others. Biol Psychiatry 47:928–934

Crockett CM, Pope T (1988) Inferring patterns of aggression from red howler monkey injuries. Am J Primatol 15:289–308

Dabbs JMJ, Frady RL, Carr TS et al. (1987) Saliva testosterone and criminal violence in young adult prison inmates. Psychosom Med 49:174–182

Eaton GG, Resko JA (1974) Plasma testosterone and male dominance in a Japanese macaque (*Macaca fuscata*) troop compared with repeated measures of testosterone in laboratory males. Horm Behav 5:251–259

Eberhart JA, Yodyingyuad U, Keverne EB (1985) Subordination in male talapoin monkeys lowers sexual behaviour in the absence of dominants. Physiol Behav 35:673–677

Ehrenkranz J, Bliss E, Sheard MH (1974) Plasma testosterone: Correlation with aggressive behavior and social dominance in man. Psychosom Med 36:469–475

Erickson K, Lindell S, Champoux M et al. (2001) Relationships between behavior and neurochemical changes in rhesus macaques during a separation paradigm. Soc Neurosci Abstr 27, program 572:14

Eysenck SB, Eysenck HJ (1971) A comparative study of criminals and matched controls on three dimensions of personality. Br J Soc Clin Psychol 10:362–366

Fairbanks LA, Fontenot MB, Phillips-Conroy JE et al. (1999) CSF monoamines, age and impulsivity in wild grivet monkeys (*Cercopithecus aethiops aethiops*). Brain Behav Evol 53:305–312

Fairbanks LA and Jorgensen MJ (2011) Objective behavioral tests of temperament in nonhuman primates. In: Weiss A, King JE, Murray L (eds) Personality and temperament in nonhuman primates. Springer, New York

Fairbanks LA, Melega WP, Jorgensen MJ et al. (2001) Social impulsivity inversely associated with CSF 5-HIAA and fluoxetine exposure in vervet monkeys. Neuropsychopharmacology 24:370–378

Ferrari PF, Palanza P, Parmigiani S et al. (2005) Serotonin and aggressive behavior in rodents and nonhuman primates: Predispositions and plasticity. Eur J Pharmacol 526:259–273

Gibbs RA, Rogers J, Katze MG et al. (2007) Evolutionary and biomedical insights from the rhesus macaque genome. Science 316:222–234

Gill K, Amit Z (1989) Serotonin uptake blockers and voluntary alcohol consumption. A review of recent studies. Recent Dev Alcohol 7:225–248

Gleeson S, Ahlers ST, Mansbach RS et al. (1989) Behavioral studies with anxiolytic drugs. VI. Effects on punished responding of drugs interacting with serotonin receptor subtypes. J Pharmacol Exp Ther 250:809–817

Gordon TP, Bernstein IS, Rose RM (1978) Social and seasonal influences on testosterone secretion in the male rhesus monkey. Physiol Behav 21:623–637

Gordon TP, Rose RM, Bernstein IS (1976) Seasonal rhythm in plasma testosterone levels in the rhesus monkey (*Macaca mulatta*): A three year study. Horm Behav 7:229–243

Harlow HF (2008) The monkey as a psychological subject. Integr Psychol Behav Sci 42: 336–347

Harris RA, Rogers J, Milosavljevic A (2007) Human-specific changes of genome structure detected by genomic triangulation. Science 316:235–237

Higley JD (2003) Aggression in Old World primates: Causes, cures, and functions. In: Maestripieri D (ed) Primate psychology. The mind and behavior of human and nonhuman primates. Harvard University Press, Cambridge

Higley JD, Barr CS (2008) Neurochemistry and behavior. In: Burbacher TM, Sackett GP, Grant KS (eds) Nonhuman primate models of children's health and developmental disabilities. Academic, New York

Higley JD, Hasert MF, Dodson A et al. (1992a) Treatment of excessive alcohol consumption using the serotonin reuptake inhibitor Sertraline in a nonhuman primate model of alcohol abuse. Paper presented at the meeting of the Research Society on Alcoholism, San Diego, June 13–18

Higley JD, Hasert MF, Suomi SJ et al. (1998) The serotonin reuptake inhibitor sertraline reduces excessive alcohol consumption in nonhuman primates: Effect of stress. Neuropsychopharmacology 18:431–443

Higley JD, King ST, Hasert MF et al. (1996a) Stability of interindividual differences in serotonin function and its relationship to severe aggression and competent social behavior in rhesus macaque females. Neuropsychopharmacology 14:67–76

Higley JD, Linnoila M, Suomi SJ (1994) Ethological contributions: Experiential and genetic contributions to the expression and inhibition of aggression in primates. In: Hersen M, Ammerman RT, Sisson L (eds) Handbook of aggressive and destructive behavior in psychiatric patients. Plenum Press, New York

Higley JD, Mehlman PT, Higley SB et al. (1996b) Excessive mortality in young free-ranging male nonhuman primates with low cerebrospinal fluid 5-hydroxyindoleacetic acid concentrations. Arch Gen Psychiatry 53:537–543

Higley JD, Mehlman PT, Poland RE et al. (1996c) CSF testosterone and 5-HIAA correlate with different types of aggressive behaviors. Biol Psychiatry 40:1067–1082

Higley JD, Mehlman PT, Taub DM et al. (1992b) Cerebrospinal fluid monoamine and adrenal correlates of aggression in free-ranging rhesus monkeys. Arch Gen Psychiatry 49:436–441

Higley JD, Suomi SJ (1989) Temperamental reactivity in non-human primates. In: Kohnstamm GA, Bates JE, Rothbart MK (eds) Temperament in childhood. Wiley, New York

Higley JD, Suomi SJ (1996) Effect of reactivity and social competence on individual responses to severe stress in children: Investigations using nonhuman primates. In: Pfeffer CR (ed) Intense stress and mental disturbance in children. American Psychiatric Press, Inc., Washington

Higley JD, Suomi SJ, Linnoila M (1991) CSF monoamine metabolite concentrations vary according to age, rearing, and sex, and are influenced by the stressor of social separation in rhesus monkeys. Psychopharmacology 103:551–556

Higley JD, Suomi SJ, Linnoila M (1992c) A longitudinal assessment of CSF monoamine metabolite and plasma cortisol concentrations in young rhesus monkeys. Biol Psychiatry 32: 127–145

Higley JD, Suomi SJ, Linnoila M (1996d) A nonhuman primate model of type II alcoholism? Part 2. Diminished social competence and excessive aggression correlates with low cerebrospinal fluid 5-hydroxyindoleacetic acid concentrations. Alcohol Clin Exp Res 20:643–650

Higley JD, Suomi SJ, Linnoila M (1996e) A nonhuman primate model of type II excessive alcohol consumption? Part 1. Low cerebrospinal fluid 5-hydroxyindoleacetic acid concentrations and diminished social competence correlate with excessive alcohol consumption. Alcohol Clin Exp Res 20:629–642

Higley JD, Thompson WW, Champoux M et al. (1993) Paternal and maternal genetic and environmental contributions to cerebrospinal fluid monoamine metabolites in rhesus monkeys (*Macaca mulatta*). Arch Gen Psychiatry 50:615–623

Hirsch RM, Higley JD, Suomi SJ (1986) Growing-up without adults: The effect of peer-only rearing on daily behaviors in rhesus monkeys. Paper presented at the meeting of the International Society for Developmental Psychobiology, Annapolis, MD

Howell S, Westergaard G, Hoos B et al. (2007) Serotonergic influences on life-history outcomes in free-ranging male rhesus macaques. Am J Primatol 69:851–865

Izquierdo A, Newman TK, Higley JD et al. (2007) Genetic modulation of cognitive flexibility and socioemotional behavior in rhesus monkeys. Proc Natl Acad Sci USA 104:14128–14133

Jarrell H, Hoffman JB, Kaplan JR et al. (2008) Polymorphisms in the serotonin reuptake transporter gene modify the consequences of social status on metabolic health in female rhesus monkeys. Physiol Behav 93:807–819

Jedema HP, Gianaros PJ, Greer PJ et al. (2010) Cognitive impact of genetic variation of the serotonin transporter in primates is associated with differences in brain morphology rather than serotonin neurotransmission. Mol Psychiatry 15:512–522, 446

Kaufman JH (1967) Social relations of adult males in a free-ranging band of rhesus monkeys. In: Altmann SA (ed) Social communication among primates. University of Chicago Press, Chicago

Keverne EB, Eberhart JA, Meller RE (1983) Plasma testosterone, sexual and aggressive behavior in social groups of talapoin monkeys. In: Steklis HD, King AS (eds) Hormones, drugs and social behavior in primates. Spectrum, New York

Kraemer GW, Ebert MH, Schmidt DE et al. (1989) A longitudinal study of the effect of different social rearing conditions on cerebrospinal fluid norepinephrine and biogenic amine metabolites in rhesus monkeys. Neuropsychopharmacology 2:175–189

Kruesi MJ, Rapoport JL, Hamburger S et al. (1990) Cerebrospinal fluid monoamine metabolites, aggression, and impulsivity in disruptive behavior disorders of children and adolescents. Arch Gen Psychiatry 47:419–426

Kuester J, Paul A (1992) Influence of male competition and female mate choice on male mating success in Barbary macaques (*Macaca sylvanus*). Behaviour 120:192–217

Kyes RC, Botchin MB, Kaplan JR et al. (1995) Aggression and brain serotonergic responsivity: Response to slides in male macaques. Physiol Behav 57:205–208

Lee AS, Gutierrez-Arcelus M, Perry GH et al. (2008) Analysis of copy number variation in the rhesus macaque genome identifies candidate loci for evolutionary and human disease studies. Hum Mol Genet 17:1127–1136

Lesch KP (2007) Linking emotion to the social brain. The role of the serotonin transporter in human social behaviour. EMBO Rep 8 Spec No:S24–S29

Lidberg L, Tuck JR, Åsberg M et al. (1985) Homicide, suicide and CSF 5-HIAA. Acta Psychiatr Scand 71:230–236

Limson R, Goldman D, Roy A et al. (1991) Personality and cerebrospinal fluid monoamine metabolites in alcoholics and controls. Arch Gen Psychiatry 48:437–441

Lindman R, Järvinen P, Vidjeskog J (1987) Verbal interactions of aggressively and nonaggressively predisposed males in a drinking situation. Aggress Behav 13:187–196

Linnoila M, Virkkunen M, George T et al. (1993) Impulse control disorders. Int Clin Psychopharmacol 8, Supplement 1:53–56

Linnoila M, Virkkunen M, Scheinin M et al. (1983) Low cerebrospinal fluid 5-hydroxyindoleacetic acid concentration differentiates impulsive from nonimpulsive violent behavior. Life Sci 33:2609–2614

Linsky AS, Colby JP, Jr., Straus MA (1986) Drinking norms and alcohol-related problems in the United States. J Stud Alcohol 47:384–393
Lorenz JG, Long JC, Linnoila M et al. (2006) Genetic and other contributions to alcohol intake in rhesus macaques (*Macaca mulatta*). Alcohol Clin Exp Res 30:389–398
Mandell W, Eaton WW, Anthony JC et al. (1992) Alcoholism and occupations: A review and analysis of 104 occupations. Alcohol Clin Exp Res 16:734–746
Manuck SB, Flory JD, Ferrell RE et al. (2000) A regulatory polymorphism of the monoamine oxidase-A gene may be associated with variability in aggression, impulsivity, and central nervous system serotonergic responsivity. Psychiatry Res 95:9–23
Manuck SB, Kaplan JR, Rymeski BA et al. (2003) Approach to a social stranger is associated with low central nervous system serotonergic responsivity in female cynomolgus monkeys (*Macaca fascicularis*). Am J Primatol 61:187–194
Martensz ND, Vellucci SV, Fuller LM et al. (1987) Relation between aggressive behaviour and circadian rhythms in cortisol and testosterone in social groups of talapoin monkeys. J Endocrinol 115:107–120
Mattsson A, Schalling D, Olweus D et al. (1980) Plasma testosterone, aggressive behavior, and personality dimensions in young male delinquents. J Am Acad Child Psychiatry 19:476–490
Mazur A (1983) Hormones, aggression, and dominance in humans. In: Svare BB (ed) Hormones and aggressive behavior. Plenum Press, New York
McBride WJ, Murphy JM, Lumeng L et al. (1989) Serotonin and ethanol preference. Recent Dev Alcohol 7:187–209
McCormack K, Newman TK, Higley JD et al. (2009) Serotonin transporter gene variation, infant abuse, and responsiveness to stress in rhesus macaque mothers and infants. Horm Behav 55:538–547
McGuire MT, Raleigh MJ, Pollack DB (1994) Personality factors in vervet monkeys: The effects of sex, age, social status, and group composition. Am J Primatol 33:1–13
Mehlman PT, Higley JD, Faucher I et al. (1994) Low CSF 5-HIAA concentrations and severe aggression and impaired impulse control in nonhuman primates. Am J Psychiatry 151: 1485–1491
Mehlman PT, Higley JD, Faucher I et al. (1995) Correlation of CSF 5-HIAA concentration with sociality and the timing of emigration in free-ranging primates. Am J Psychiatry 152:907–913
Mehlman PT, Higley JD, Fernald BJ et al. (1997) CSF 5-HIAA, testosterone, and sociosexual behaviors in free-ranging male rhesus macaques in the mating season. Psychiatry Res 72:89–102
Mehlman PT, Westergaard GC, Hoos BJ et al. (2000) CSF 5-HIAA and nighttime activity in free-ranging primates. Neuropsychopharmacology 22:210–218
Miczek KA, Donat P (1990) Brain 5-HT system and inhibition of aggressive behavior. In: Archer T, Bevan P, Cools A (eds) Behavioral pharmacology of 5-HT. Lawrence Erlbaum Associates, Inc., Hillsdale
Miczek KA, Mos J, Olivier B (1989) Brain 5-HT and inhibition of aggressive behavior in animals: 5-HIAA and receptor subtypes. Psychopharmacol Bull 25:399–403
Miller-Butterworth CM, Kaplan JR, Barmada MM et al. (2007) The serotonin transporter: Sequence variation in *Macaca fascicularis* and its relationship to dominance. Behav Genet 37:678–696
Mischel W (1968) Personality and assessment. Wiley, New York
Namboodiri MA, Sugden D, Klein DC et al. (1985) Serum melatonin and pineal indoleamine metabolism in a species with a small day/night N-acetyltransferase rhythm. Comp Biochem Physiol B Comp Biochem 80:731–736
Newman ME, Shapira B, Lerer B (1998) Evaluation of central serotonergic function in affective and related disorders by the fenfluramine challenge test: A critical review. Int J Neuropsychopharmacol 1:49–69
Newman TK, Syagailo YV, Barr CS et al. (2005) Monoamine oxidase A gene promoter variation and rearing experience influences aggressive behavior in rhesus monkeys. Biol Psychiatry 57:167–172

Nikulina EM, Avgustinovich DF, Popova NK (1992) Role of 5HT1A receptors in a variety of kinds of aggressive behavior in wild rats and counterparts selected for low defensiveness to man. Aggress Behav 18:357–364

Olivier B, Mos J (1990) Serenics, serotonin and aggression. Prog Clin Biol Res 361:203–230

Olivier B, Mos J, Tulp M et al. (1990) Modulatory action of serotonin in aggressive behavior. In: Archer T, Bevan P, Cools A (eds) Behavioral pharmacology of 5-HT. Lawrence Erlbaum Associates, Inc., Hillsdale

Olweus D, Mattsson A, Schalling D et al. (1980) Testosterone, aggression, physical, and personality dimensions in normal adolescent males. Psychosom Med 42:253–269

Olweus D, Mattsson A, Schalling D et al. (1988) Circulating testosterone levels and aggression in adolescent males: A causal analysis. Psychosom Med 50:261–272

Packer C (1979) Male dominance and reproductive activity in *Papio anubis*. Anim Behav 27: 37–45

Paul A (1989) Determinants of male mating success in a large group of Barbary macaques (*Macaca sylvanus*) at Affenberg Salem. Primates 30:344–349

Popova NK (2006) From genes to aggressive behavior: The role of serotonergic system. Bioessays 28:495–503

Popova NK, Kulikov AV, Nikulina EM et al. (1991a) Serotonin metabolism and serotonergic receptors in Norway rats selected for low aggressiveness towards man. Aggress Behav 17: 207–213

Popova NK, Voitenko NN, Kulikov AV et al. (1991b) Evidence for the involvement of central serotonin in mechanism of domestication of silver foxes. Pharmacol Biochem Behav 40: 751–756

Raleigh MJ, Brammer GL, McGuire MT (1983) Male dominance, serotonergic systems, and the behavioral and physiological effects of drugs in vervet monkeys (*Cercopithecus aethiops sabaeus*). In: Miczek KA (ed) Ethopharmacology: Primate models of neuropsychiatric disorders. Alan R. Liss, New York

Raleigh MJ, Brammer GL, McGuire MT et al. (1992) Individual differences in basal cisternal cerebrospinal fluid 5-HIAA and HVA in monkeys. The effects of gender, age, physical characteristics, and matrilineal influences. Neuropsychopharmacology 7:295–304

Raleigh MJ, Brammer GL, McGuire MT et al. (1985) Dominant social status facilitates the behavioral effects of serotonergic agonists. Brain Res 348:274–282

Raleigh MJ, Brammer GL, Ritvo ER et al. (1986) Effects of chronic fenfluramine on blood serotonin, cerebrospinal fluid metabolites, and behavior in monkeys. Psychopharmacology 90:503–508

Raleigh MJ, Brammer GL, Yuwiler A et al. (1980) Serotonergic influences on the social behavior of vervet monkeys (*Cercopithecus aethiops sabaeus*). Exp Neurol 68:322–334

Raleigh MJ, McGuire MT (1986) Animal analogues of ostracism: Biological mechanisms and social consequences. Ethology and Sociobiology 7:53–66

Raleigh MJ, McGuire MT (1990) Social influences on endocrine function in male vervet monkeys. In: Ziegler TE, Bercovitch FB (eds) Socioendocrinology of primate reproduction. Wiley-Liss, New York

Raleigh MJ, McGuire MT (1991) Bidirectional relationships between tryptophan and social behavior in vervet monkeys. Adv Exp Med Biol 294:289–298

Raleigh MJ, McGuire MT (1994) Serotonin, aggression, and violence in vervet monkeys. In: Masters RD, McGuire MT (eds) The neurotransmitter revolution. Southern Illinois University Press, Carbondale

Raleigh MJ, McGuire MT, Brammer GL (1989) Subjective assessment of behavioral style: Links to overt behavior and physiology in vervet monkeys. Am J Primatol 18:161–162

Raleigh MJ, McGuire MT, Brammer GL et al. (1991) Serotonergic mechanisms promote dominance acquisition in adult male vervet monkeys. Brain Res 559:181–190

Rejeski WJ, Brubaker PH, Herb RA et al. (1988) The role of anabolic steroids on baseline and stress heart rate in cynomolgus monkeys. Health Psychol 7:299–307

Rejeski WJ, Gregg E, Kaplan JR et al. (1990) Anabolic–androgenic steroids: Effects on social behavior and baseline heart rate. Health Psychol 9:774–791

Riddle MA, Anderson GM, McIntosh S et al. (1986) Cerebrospinal fluid monoamine precursor and metabolite levels in children treated for leukemia: Age and sex effects and individual variability. Biol Psychiatry 21:69–83

Rose RM, Bernstein IS, Gordon TP (1975) Consequences of social conflict on plasma testosterone levels in rhesus monkeys. Psychosom Med 37:50–61

Rose RM, Gordon TP, Bernstein IS (1972) Plasma testosterone levels in the male rhesus: Influences of sexual and social stimuli. Science 178:643–645

Rose RM, Gordon TP, Bernstein IS (1978) Diurnal variation in plasma testosterone and cortisol in rhesus monkeys living in social groups. J Endocrinol 76:67–74

Rose RM, Holaday JW, Bernstein IS (1971) Plasma testosterone, dominance rank and aggressive behaviour in male rhesus monkeys. Nature 231:366–368

Roy A, Virkkunen M, Linnoila M (1988) Monoamines, glucose metabolism, aggression towards self and others. Int J Neurosci 41:261–264

Sapolsky RM (1983) Individual differences in cortisol secretory patterns in the wild baboon: Role of negative feedback sensitivity. Endocrinology 113:2263–2267

Sapolsky RM (1992) Stress, the aging brain, and the mechanisms of neuron death. MIT Press, Cambridge

Scaramella TJ, Brown WA (1978) Serum testosterone and aggressiveness in hockey players. Psychosom Med 40:262–265

Shannon C, Schwandt ML, Champoux M et al. (2005) Maternal absence and stability of individual differences in CSF 5-HIAA concentrations in rhesus monkey infants. Am J Psychiatry 162: 1658–1664

Shively CA, Fontenot MB, Kaplan JR (1995) Social status, behavior, and central serotonergic responsivity in female cynomolgus monkeys. Am J Primatol 37:333–340

Smuts BB (1987) Gender, aggression and influence. In: Smuts BB, Cheney DL, Seyfarth RM, Wrangham RW, Struhsaker TT (eds) Primate societies. University of Chicago Press, Chicago

Soubrié P (1986) Reconciling the role of central serotonin neurons in human and animal behavior. Behav Brain Sci 9:319–364

Steklis HD, Brammer GL, Raleigh MJ et al. (1985) Serum testosterone, male dominance, and aggression in captive groups of vervet monkeys (*Cercopithecus aethiops sabaeus*). Horm Behav 19:154–163

Suomi SJ (1982) Abnormal behavior and primate models of psychopathology. In: Fobes JL, King JE (eds) Primate behavior. Academic, New York

Suomi SJ (2006) Risk, resilience, and gene × environment interactions in rhesus monkeys. Ann N Y Acad Sci 1994:52–62

Suomi SJ, Ripp C (1983) A history of mother-less mother monkey mothering at the university of wisconsin primate laboratory. In: Reite M, Caine N (eds) Child abuse: The nonhuman primate data. Alan R. Liss, New York

Träskman-Bendz L, Åsberg M, Bertilsson L et al. (1984) CSF monoamine metabolites of depressed patients during illness and after recovery. Acta Psychiatr Scand 69(Supplementum): 333–342

Trefilov A, Berard J, Krawczak M et al. (2000) Natal dispersal in rhesus macaques is related to serotonin transporter gene promoter variation. Behav Genet 30:295–301

Virkkunen M, De Jong J, Bartko J et al. (1989) Relationship of psychobiological variables to recidivism in violent offenders and impulsive fire setters. A follow-up study. Arch Gen Psychiatry 46:600–603

Virkkunen M, Nuutila A, Goodwin FK et al. (1987) Cerebrospinal fluid monoamine metabolite levels in male arsonists. Arch Gen Psychiatry 44:241–217

Walters JR, Seyfarth RM (1987) Conflict and cooperation. In: Smuts BB, Cheney DL, Seyfarth RM, Wrangham RW, Struhsaker TT (eds) Primate societies. University of Chicago Press, Chicago

Wendland JR, Hampe M, Newman TK et al. (2006a) Structural variation of the monoamine oxidase A gene promoter repeat polymorphism in nonhuman primates. Genes Brain Behav 5:40–45

Wendland JR, Lesch KP, Newman TK et al. (2006b) Differential functional variability of serotonin transporter and monoamine oxidase A genes in macaque species displaying contrasting levels of aggression-related behavior. Behav Genet 36:163–172

Westergaard GC, Cleveland A, Trenkle MK et al. (2003a) CSF 5-HIAA concentration as an early screening tool for predicting significant life history outcomes in female specific-pathogen-free (SPF) rhesus macaques (*Macaca mulatta*) maintained in captive breeding groups. J Med Primatol 32:95–104

Westergaard GC, Izard MK, Drake JH et al. (1999a) Rhesus macaque (*Macaca mulatta*) group formation and housing: wounding and reproduction in a specific pathogen free (SPF) colony. Am J Primatol 49:339–347

Westergaard GC, Suomi SJ, Chavanne TJ et al. (2003b) Physiological correlates of aggression and impulsivity in free-ranging female primates. Neuropsychopharmacology 28:1045–1055

Westergaard GC, Suomi SJ, Higley JD et al. (1999b) CSF 5-HIAA and aggression in female macaque monkeys: Species and interindividual differences. Psychopharmacology 146: 440–446

Wickings EJ, Dixson AF (1992) Testicular function, secondary sexual development, and social status in male mandrills (*Mandrillus sphinx*). Physiol Behav 52:909–916

Woermann FG, van Elst LT, Koepp MJ et al. (2000) Reduction of frontal neocortical grey matter associated with affective aggression in patients with temporal lobe epilepsy: an objective voxel by voxel analysis of automatically segmented MRI. J Neurol Neurosurg Psychiatry 68:162–169

Yodyingyuad U, Eberhart JA, Keverne EB (1982) Effects of rank and novel females on behaviour and hormones in male talapoin monkeys. Physiol Behav 28:995–1005

Zajicek KB, Higley JD, Suomi SJ et al. (1997) Rhesus macaques with high CSF 5-HIAA concentrations exhibit early sleep onset. Psychiatry Res 73:15–25

Chapter 11
Reactivity and Behavioral Inhibition as Personality Traits in Nonhuman Primates

Stephen J. Suomi, Andrew C. Chaffin, and J. Dee Higley

Abstract While the history of the study of personality dates back to the early 1900s, most animal research, particularly on nonhuman primates, is much more recent. That personality in animals reflects our common evolutionary history is not surprising, and given our close genetic relatedness, should be expected. The personality trait that has received the most research in nonhuman primates is what we have called elsewhere, reactivity (others have referred to it as fearfulness, timidity, shyness, etc.). While several methods have been used to study it (including personality rating scales), generally, reactivity in nonhuman primates is most often measured using behavior codings. Two paradigms have received the most research: social separations and the human intruder paradigm. Individual differences in reactivity are stable across time and situations. Reactivity can also predict multiple behavioral outcomes, including enduring anxiety, low social dominance rank and submissiveness, high alcohol intake, and other forms of affective psychopathology. One major advantage of using nonhuman primates to model personality is that the underlying physiology and central nervous system foundations can be more readily studied than in humans. These studies show the importance of the amygdala and frontal cortex, as well as the HPA Axis, central norepinephrine, and serotonin in regulating reactivity. Studies also show the importance of early parental influence and genes on reactivity. Recent studies using molecular genetics show that the serotonin transporter and corticotrophin releasing hormone genes probably play important roles in its etiology but interact with early rearing history and situations to modulate reactivity.

S.J. Suomi (✉)
Laboratory of Comparative Ethology, National Institute of Child Health and Human Development, 6105 Rockledge Drive Suite 8030, MSC 7971, Bethesda, MD 20892-7971, USA
e-mail: suomis@lce.nichd.nih.gov

A. Weiss et al. (eds.), *Personality and Temperament in Nonhuman Primates*, Developments in Primatology: Progress and Prospects, DOI 10.1007/978-1-4614-0176-6_11,

11.1 Introduction

The study of personality or temperament (as it is often called when studying animals) is a major field of human psychology and, as the other chapters in this volume show, more recently a major field of study in animals as well. Since both temperament and personality are commonly used when studying personality in nonhuman primates, we will use the terms interchangeably in this chapter.

Paramount in influence was Eysenck (Eysenck and Eysenck 1976) and his theoretical model of personality. In Eysenck's original model of personality, three main independent personality traits were proffered: Introversion–Extraversion, Neuroticism–Emotional Stability, and Psychoticism. While the names of the first two traits have endured and become a part of the common lexicon, Psychoticism has a somewhat different meaning today than when Eysenck first proposed it. For Eysenck, individuals rated high on Psychoticism were likely to develop psychotic states of schizophrenia or other mental illnesses (Eysenck and Eysenck 1976). The traits he described as comprising Psychoticism are today referred to as impulsivity – mercurial, flighty, quick to act without foresight. Individuals high in impulsivity often act in an apparently thoughtless manner, to their long-term disadvantage, oblivious to or ignoring the potential risk for injury or negative social consequences. While his trait Psychoticism has not been useful for predicting mental illness, Eysenck's genius is shown in the endurance of his original proposal. While details may differ, all of his original "big-three" have withstood the test of time, with subsequent researchers adding one or two additional traits leading to Five-Factor Models. Eysenck's personality model was largely atheoretical, based on data from large groups. Subsequent personality theorists have used other means to group behaviors as traits, but generally arrive at similar personality categories. Cloninger (1986), to illustrate, was to define traits according to common biochemical groupings, showing three larger traits that each had homogenous biochemical profiles.

Temperament, or personality, is a constellation of co-occurring behaviors or traits that are mediated by a common underlying hypothetical construct. Traits are behaviors that show stability across time and setting and are the basic components of personality. Figure 11.1 shows our description of traits such as reactivity with overlapping Venn diagrams, with each circle representing impulsive-like behaviors, and a central core that incorporates the actual hypothetical construct, such as impulsivity.

The central core of the overlapping circles in the Venn diagram would be the trait controlling its behavioral expressions (e.g., fearfulness or impulsivity). This central core is intrinsic to individuals and affects how they respond to reinforcing or desired stimuli. For example, the diagram implies that impulsive or impulse-dysregulated monkeys are quick to approach potential reinforcement without regard to potentially negative consequences ("quick to approach" here is a central core mediated trait). However, some monkeys may approach reinforcing stimuli because they are low in anxiety, more highly motivated by reinforcement, lack the cognitive capability to understand the dangers involved, or are unafraid of the potential unknowns

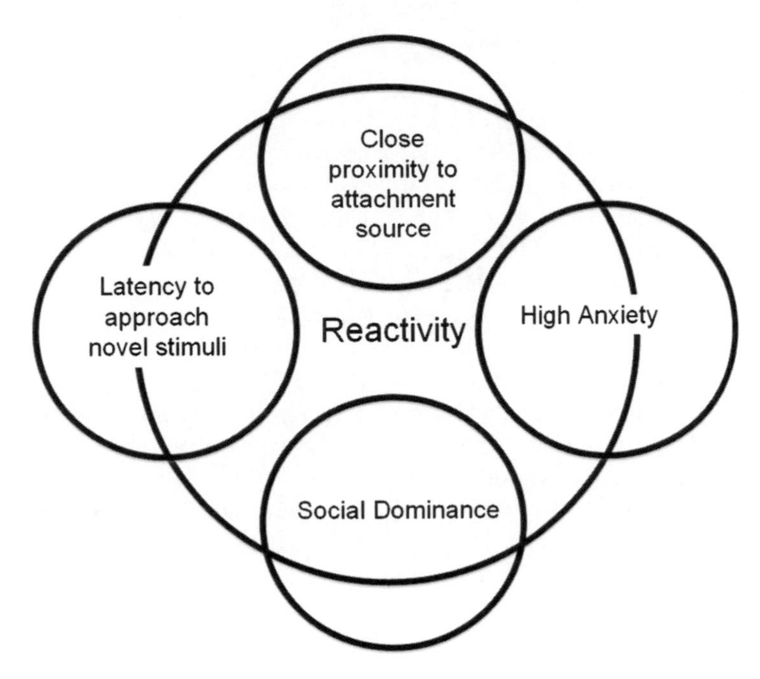

Fig. 11.1 Illustrates the variables that may underlie the temperamental trait, reactivity (see text for discussion)

(overlapping peripheral traits). Situational variations (e.g., predator presence, rain, or hunger) would also influence the activation of different systems. Thus other potentially overlapping traits or situational aspects may also affect the trait. Returning to the Venn diagram, because fear may partially modulate the quickness to react in dangerous situations, low fearfulness may only partially overlap with impulsivity. While the central core construct is the major influence for a trait-like impulsivity, impulsive-like behavior is partially mediated by other variables, and it would not correlate perfectly with fearfulness. Another example is novelty seeking, as measured by the speed to approach novel or highly arousing stimuli, traits that overlapp with several other traits such as reward seeking, curiosity, impulsivity, or low sympathetic nervous system arousal. Similarly, aggressiveness overlaps with social dominance rank, but not all competitive or aggressive individuals would rank high. This is because other variables also mediate social dominance rank – reactive monkeys may flee as a result of high fear and anxiety. Social support probably also plays an important role in social dominance rank. Moreover, violence at times can result in high social dominance rank as bullies sometimes win such encounters (Howell et al. 2007). Thus while personality traits can be described, predicting whether one personality trait will affect behavior will depend on situational variables and whether other systems are activated.

11.2 Personality in Animals

When personality was first outlined for study, animal and human studies often appeared in the same journal, but with the advent of the Second World War, unlike human personality research, the study of personality in animals waned. Since then, personality has been widely studied in humans, but the reemergence of animal personality research has been more recent. In early studies with rodent models temperament was most widely used to describe traits in animals tested in open-field tests. Animals described as fearful were slow to enter open areas, froze, and spent the bulk of the testing period in the shadows and edges of the testing chambers, and showed high levels of defecation (Scott and Fuller 1965; Blizard 1981; Gray 1987). Strains of rats and mice were selectively bred for levels of reactivity, as measured by exploration and defecation during an open-field test. Later, temperament was described in a variety of animals: dogs, horses, monkeys, apes, and more recently in wild animals such as silver foxes (Scott and Fuller 1965; Francois et al. 1990; Popova et al. 1976, 1991a, b; van Oortmerssen and Bakker 1981; Sandnabba 1996; Popova 2006).

One advantage of using animals to investigate personality was that it allowed researchers to assess the underlying biochemical processes that modulate specific traits. Reactivity, aggression, and impulsivity were linked to neurotransmitters such as norepinephrine (NE). Selection for high levels of NE led to strains that were fearful and anxious (Liang and Blizard 1978; Insel and Hill 1987; Blizard 1989; Scott et al. 1996; Weiss et al. 2008). Other studies showed that treatments early in life could augment or reduce levels of reactivity (Higley et al. 1991a; Holmes et al. 2005), demonstrating experimentally both genetic and environment components to reactivity, something not possible to experimentally manipulate in humans.

In his autobiography, Eysenck would in fact remark that a criterion for a personality dimension is that it "...should be apparent not only in humans, but also in animals" (Eysenck and Eysenck 1985). Using common terminology recognizes both our shared heritage, and facilitates a better understanding of our own underpinnings. This is particularly true when assessing nonhuman primates. As research on animal personality increased, nonhuman primate research in personality emerged with increasing frequency (for reviews see Clarke and Boinski 1995; Capitanio 2004; Weinstein et al. 2008). Initially, personality studies of nonhuman primates focused on macaques, particularly the rhesus macaque. Remarkably, Eysenck's foresight is shown again as he was a collaborator on an early study of 168 macaques – a very large number for that period (Chamove et al. 1972). In this study, behavioral codings were factor analyzed and as in Eysenck's study of humans, three personality traits emerged: Affiliative, Hostile, and Fearful – traits that the authors indicated were markedly similar to Extraversion, Psychoticism, and Neuroticism found in humans (Chamove et al. 1972). Similarly, in lion-tailed macaques, behavioral codings revealed three components: Extraversion, Aggressiveness, and Bold-Cautious (Rouff et al. 2005). While not a formal study of personality, Harlow's description of peer-reared monkeys was an apt description of reactivity and fearfulness, indicating early in the research of personality using monkeys, that environmental factors modulate some nonhuman primate traits (Harlow and Harlow 1965; Chamove et al. 1973). Still other studies showed species differences in the response to more-or-less identical environments. For example, pigtail macaques were more aggressive than bonnet macaques or squirrel monkeys, and bonnet macaques were more friendly, sitting in and maintaining close social proximity, than pigtail macaques (Rosenblum and Kaufman 1967). Stevenson-Hinde and colleagues (Stevenson-Hinde and Zunz 1978; Stevenson-Hinde and Simpson 1980; Stevenson-Hinde et al. 1980) measured individual differences in emotional reactivity in rhesus monkeys who lived in a large group.

As with humans, the macaque neocortex makes up a substantial portion of the brain. While there are general patterns of rhesus social behavior, there are also striking individual differences. These individual differences are so stable, that recognition of individuals is largely done by looking for habitual patterns of behavior such as gregariousness or social isolation. Such differences are a basis for modeling human temperament and personality, particularly for researchers doing genetic studies. For example, genetic mutations that affect temperament and personality in humans, such as the serotonin transporter and MAOa genotypes, have orthologues (genetic polymorphisms that produce the same outcome but may not be identical in location or precise sequence) in the rhesus macaque, but they are often not found in rodents (Bennett et al. 2002; Newman et al. 2005). Such variability and homogeneity in genetic sequence of genes related to personality provides important information about the evolution of temperament and personality and the underlying pressure to maintain certain traits. Also as with humans, personality modulates a wide variety of rhesus monkey characteristics including behavior, health, social status, and general well-being (Capitanio 2008; Higley et al. 1996a, b; Zajicek et al. 1997; Westergaard et al. 1999a; Howell et al. 2007; Weinstein et al. 2008).

11.3 Reactivity

Reactivity has been alternatively labeled as timidity, fearfulness, anxiety, introversion, shyness, and more commonly as behavioral inhibition. Reactivity reflects an affective and behavioral predisposition to respond to novel or challenging stimuli. Individuals rated high in reactivity would be less likely to approach new stimuli and do so with a strong emotional component. They are more anxious, more socially inhibited, and less likely to face social challenges (see Kagan 1992; Schwartz et al. 2003).

Several different paradigms and behaviors have been used to measure fearfulness/reactivity, as well as aggressiveness. These include measurements of anxiety or fearfulness during novel or possible threatening situations or behavioral withdrawal during social interactions. Table 11.1 is a listing of common methods that have been used to measure reactivity in monkeys.

Reactivity is most often measured during stressful events based on the severity of the reaction to the stressors involved. Two commonly used methods to measure reactivity are social separation and the human intruder paradigm.

Table 11.1 Methods used to measure reactivity in nonhuman primates, as well as the variables measured

Methods and variables used to measure reactivity	Reference
Freezing, and distress vocals to a provocative human intruder	Fairbanks (2001); Kalin and Shelton (1989)
Withdrawal and fear to a threatening same aged and sex unfamiliar monkey intruder	Erickson et al. (2001); Fairbanks (2001); Habib et al. (2000); Higley (1985); Higley et al. (1991a)
Distress during social separation	Kalin et al. (1992)
Subjective ratings of fearfulness or emotionality while being handled by an experimenter	Schneider and Suomi (1992), Suomi et al. (1981)
Subjective ratings of fearfulness or emotionality multiple home-cage observations	Bolig et al. (1992); Capitanio and Widaman (2005); Stevenson-Hinde and Simpson (1980); Stevenson-Hinde and Zunz (1978)
Measurements of infants maintaining proximity to an attachment source	Suomi (1983)
Latency to approach stimuli in a novel setting and seldom initiating social interactions with unfamiliar individuals	Thompson et al. (1986)
Physiological measures of reactivity such as high levels of cortisol, heart rate, or central amines – norepinephrine, serotonin, etc.	Barr et al. (2004a); Erickson et al. (2005); Fahlke et al. (2000); Higley et al. (1992c, 1982)

The second column is the reference for each of the methods described

11.3.1 Social Separation

The response to social separation was first described and later elaborated upon by Harlow and colleagues (Seay and Harlow 1965; Suomi et al. 1970; McKinney et al. 1972). While the settings may vary, the response of separated animals is predictable, with a relatively universal acute or protest phase (Bowlby 1973), which varies considerably between individuals. The acute phase is followed by what Harlow and Bowlby called a despair phase. Because despair is not universal, more recently it has been described as the chronic separation phase (Higley 1985; Anderson et al. 2002). Some distress during the initial separation or acute phase is virtually universal. Bowlby (1973) and others (e.g., Higley 1985) have suggested that distress and vocalizations during separation from one's mother and social group are evolutionarily adaptive. In the natural environment the behaviors of the acute phase lead to an increased potential for reunion with family or group members, thus increasing the chances of survival. As evidence of this potential for being reunited with the social group, when a separated monkey hears other animals answer its distress calls, the frequency of calling is much higher than when the monkey is alone and its cries are not reciprocated (Bayart et al. 1990; Wiener et al. 1990; Kalin et al. 1992).

Distress vocalizations vary from a few to more than one per second during the first few minutes of separation (Suomi et al. 1976; Kalin et al. 1988; Becker et al. 2004; Barr et al. 2008b). Monkeys who exhibit a high rate of distress vocalizations during separation are considered to be reactive or more fearful, whereas monkeys that seldom vocalize would be considered as less reactive. This relationship varies with context. In the presence of a potential intruder, distress vocalizations are lower than baseline (Bayart et al. 1990; Kalin et al. 1992). During social separation, reactive monkeys show increased anxiety-like behavior such as self-orality, hand-wringing and self-clasping, immobility, and huddling (Mineka and Suomi 1978; Mineka et al. 1981).

Anxiety and fear are typically associated with separation (Suomi et al. 1970, 1983; Mineka and Suomi 1978; Mineka et al. 1981; Higley 1985). While this paradigm is widely used, it is not standardized. As noted earlier, for example, monkeys separated from their group are more likely to vocalize when they can hear other group members than when they are isolated from their social group and separation studies are quite variable in controlling for whether the monkeys are isolated or placed where they can hear other monkeys. Length of separation also affects both behavioral and physiological measures of reactivity. For example, monkeys separated for 1 h show lower cortisol levels than monkeys separated for 2 h, and cortisol levels are much lower by the next morning (Higley 1985; Higley et al. 1992c). The variability in the length of time separated makes measures across separation studies difficult to interpret. Adding to the problems, in some cases the separated animals are left in their social group and the attachment source is removed. In such cases, the infant may or may not be "adopted" or cared for by other group members. For the animal moved to a new setting, there is the added stressor of a novel environment and possibly unfamiliar animals in nearby cages. Thus, when interpreting social separation data we must consider the homogenous vs. heterogenous nature of the separation paradigm across studies.

11.3.2 The Human Intruder Paradigm

The human intruder paradigm has been widely used (Kalin and Shelton 1989; Kalin et al. 1991; Kalin 1993, 1999; Bethea et al. 2004). This paradigm has three phases, each eliciting a different set of behaviors. The advantage of the human intruder paradigm is that it has a clearly defined set of procedures, making it one of the more standardized tests used to measure emotional behavior in nonhuman primates. Not only are the components in identical sequence each time, but also the behavior of the human intruder is clearly defined. Because the stimulus is a human, the response of the stimulus can be controlled and be more-or-less identical each time.

The initial phase of the test, lasting a few minutes, is a brief social separation. Subjects are separated from other members of the group thereby eliciting distress vocalizations and agitation, the classic protest of a social separation described by Bowlby (1973). In the second phase, the human intruder enters the room, stands in profile, at a right angle to the cage, looking at the wall, and not making eye contact with the subject, a context meant to parallel the presence of a predator. Because the intruder is not looking at the subject, the ecological context is one of avoiding detection. With the experimenter looking away from the cage, the monkey typically acts as if it is hidden in the bushes, freezing, remaining motionless, and quiet. Finally in the third phase of the experiment, the human intruder turns, facing the frozen subject, and makes eye contact. Here the initial response is momentary freezing, but then the subject acts as if it is detected and exhibits flight to the back of the cage. Because escape is not possible, a series of threats and lunges typically results. Again, as with the initial phase, there are wide individual differences in the intensity of the response, with some subjects persisting in freezing even into the third phase, while others are quick to adjust to the new ecological demands with rapid flight and threats (Kalin et al. 1998b, 1991; Kalin and Shelton 2003; Rogers et al. 2008). Nevertheless, even in this paradigm there is some variability in methodology, with the proximity to the monkey and time in each phase varying between studies.

An advantage of this paradigm is that separate biochemical processes have been shown to underlie each of the different experimental phases and each phase has been linked to different ecological demands (see below – Kalin et al. 1998b, 2000, 2005). This is important because different temperamental traits are activated in each phase of the intruder paradigm. This paradigm has explanatory power, because as in humans (Cloninger 1994; Stallings et al. 1996; Ruchkin et al. 1998), differing aspects of personality are modulated by different biochemical systems (Higley et al. 1991a; Kalin 1993, 1999; Higley and Linnoila 1997a; Fahlke et al. 2000; Kalin and Shelton 2003; Erickson et al. 2005).

11.3.3 Other Methods

Most studies have focused on reactivity during introduction of a major stressor. Other studies have looked at reactivity in response to the day-to-day vicissitudes and demands that occur in normative social challenges and interactions with the environment (Stevenson-Hinde and Simpson 1980; Stevenson-Hinde et al. 1980; Higley 1985; Champoux 1988b). Because anxiety and fearfulness inhibit play (see, e.g., Higley 1985), some studies have used levels of play in normal settings as an assessment of reactivity (Higley 1985). There is also evidence that reactivity is related to other forms of social behavior. In one early study the least fearful monkeys were often the most social (Stevenson-Hinde and Zunz 1978), possibly because they sought social support to palliate their anxiety. In day-to-day social interactions, subjects that were less likely to show leadership skills, such as high position in the group's social dominance hierarchy, were later classified as more reactive, based on how they responded to the stressor of social separation: those lacking leadership and social competence showed increased hand-wringing, self-clasping, and distress vocalizations during a subsequent social separation (Scanlan 1988). Thus, reactive monkeys are more likely to acquiesce to other monkeys during social interactions or competition for resources.

Behaviors characteristic of reactivity show age-related changes. In infancy ventral clinging and nipple contact, self-orality, hand-wringing and self-clasping, distress vocalizations, and social withdrawal occur in all animals, but more often and in more situations in highly reactive animals. By adolescence, in response to challenge, increased social withdrawal and immobility or flight become the most prevalent behaviors, especially in reactive monkeys. In addition, behavioral regression may occur with repeated exposure to stressors (Suomi et al. 1970, 1981; Scanlan 1984; Higley 1985).

We measured reactivity by using temperament ratings and behavioral observations during exposure to a novel room. To obtain monkeys who differed in levels of reactivity, we selectively bred mothers with males who had previously produced offspring that were high or low in reactivity. During the first month of life, neonatal rating scales of fearfulness and consolability were independently developed by two researchers. There was a high positive correlation between these two measures (Scanlan; Schneider; unpublished data). When these same subjects were 4 months old they were placed with their respective inanimate surrogate mothers into a novel playroom with interesting objects, subjects that were previously rated as high in reactivity took significantly longer to approach the objects and explore the room, and spent longer in close proximity to their surrogate mother. Also, consistent with Kagan's findings (Kagan et al. 1987; Biederman et al. 1993), highly reactive infants spent more time observing their partner manipulating the novel objects and exploring the unfamiliar room, and after watching its partner in this, often the highly reactive monkey would approach the same object that its partner had just manipulated and use it in the same fashion (Thompson et al. 1986). Kagan interpreted this in human children as evidence that the inhibited children were wary approaching unfamiliar stimuli, and particularly watchful of others to determine the relative safety before approaching novel objects and settings.

11.4 Continuity and Predictive Markers

Individual differences in reactive temperament have long-term stability, often beginning early in life. Distress vocalizations have intraindividual stability in the social separation paradigm beginning early in life with vocalization rates in 15-day-old infants showing a positive correlation with rates of vocalizations 6 months later (Becker et al. 2004). In the Thompson et al. (1986) study described above, distress vocalizations showed a within session correlation of $r=0.60$ when measured across 5-min segments. Furthermore, distress vocalizations in separated 6-month-old monkeys predicted distress vocalizations 1 and 3 years later (Higley 1985). In addition, Scanlan and Suomi (1988) found that irritability and distress vocalizations in the first month of life predicted self-directed behavior during a prolonged social separation at 6 months of life, a measure often considered to be despair during chronic social separation stress.

Other measures of fearfulness and timidity also show intraindividual stability across time and situations. Neonatal temperament ratings predicted individual differences in affective state at 6 months of age during the stressor of a social separation. Specifically, subjective rating scales of consolability and fearfulness predicted levels of despair during a 6-month social separation (Becker et al. 1984)

Other studies have found continuity across longer periods. Since fearfulness and anxiety are likely to be manifested in response to the day-to-day social challenges and changes in the home environment, some studies have assessed levels of reactivity in the animals' home-cages. Higley (1985) found that behaviors exhibited during home-cage conspecific challenges such as self-directed orality, clasping, and immobility showed strong continuity from infancy (month 9) to childhood (month 18), and early adolescence (month 30). Subsequent studies have replicated these findings in monkeys reared and tested under different conditions (Bush et al. 1987; Higley et al. 1992a, 1994, 1996c, d), and with different species studied in different laboratories (Capitanio et al. 1986). In addition, levels of play in infancy were inhibited by anxiety and fearfulness and were negatively correlated with an aggregate measure of day-to-day home-cage reactivity (defined as immobility while self-clasping and self-mouthing) during subsequent developmental periods (Higley 1985).

If physiological processes underlie stable traits, those physiological processes should also show temporal stability. There is good evidence that this is true. Neonatal temperament ratings of fearfulness and anxiety predicted individual differences in sympathetic arousal, as measured by heart rate during exposure to a novel room, with highly reactive monkeys having higher heart rates (Thompson et al. 1986). In a separate study, monkeys that showed large heart-rate changes to a conditioned stimulus signaling a burst of white noise at 22 days of age demonstrated significantly more anxiety and fearfulness in each quarter of the succeeding 12 months (Suomi et al. 1981). In a follow-up study of an independent group of subjects, heart-rate change at 22 days of age was negatively correlated with activity during a short-term social separation from their social group at 18 months of age (Suomi 1983).

11.5 Reactivity and Psychopathology

Among humans, high anxiety or fearfulness in infancy is a risk for childhood and adult anxiety disorders and alcohol abuse in young adults (Rosenbaum et al. 1991; Hill et al. 1999; Biederman et al. 2001). Similarly, fearful temperament early in life is related to subsequent anxiety and despair during later social separations. Higley (1985) measured levels of anxiety exhibited in the home-cage 2 weeks prior to a series of social separations. Anxiety, as measured by hand-wringing, self-clasping, and self-mouthing during undisturbed home-cage activity, was strongly correlated with behaviors characteristic of despair during social separation. Furthermore, fearfulness during infancy was strongly correlated with behaviors characteristic of anxiety and despair displayed during social separations in adolescence. Harlow's early studies (Harlow and Harlow 1965) indicated that play only occurs when fear, arousal, and anxiety are low. Not only is social play negatively correlated with future home-cage reactivity, but also home-cage play in infancy is also negatively correlated with levels of anxiety during social separation in adolescence (Higley 1985). These studies indicate that as with humans, anxiety and fearfulness may predispose individual monkeys to future affective problems.

11.6 Reactivity and the Fear Circuit of the Brain

An overactive amygdala exerts a strong influence on fear and anxiety (Kalin et al. 2004, 2005, 2007). Bilateral removal of the amygdala reduces socially mediated anxiety, leading to more approaches and close proximity to unfamiliar social partners (Machado et al. 2008). Perhaps due to their lack of anxiety, these amygdalectomized animals receive and exhibit higher rates of aggression and overall exhibit decreased positive social interactions with their companions (Machado et al. 2008). When compared to intact controls, initially the amygdalectomized monkeys are more likely to approach ambiguous and fear provoking stimuli, but as the controls become familiar with the situation, these differences disappear (Mason et al. 2006). Thus, temperamental, fear-mediated inhibitions only occur when situations have cues of ambiguity and danger. During other periods, amygdala-mediated differences in anxious temperament are largely absent.

Behaviors indicative of anxiety and fear vary in intensity and do not always covary. Moreover, reactivity may not be mediated by only one neurotransmitter or brain circuit and some fearful responses may be specifically related to specific contextual demands, and as a consequence, dependent on different neurotransmitter systems (Keverne et al. 1989). Brain responses to challenging, stressful stimuli may vary across brain areas and neurotransmitter systems according to ecological demands. Crying for maternal retrieval, for example, produces high activation in the frontal cortex, but damps the response of the amygdala (Fox et al. 2005). Individual differences in the amount of crying by an infant when it is separated from its mother are highly correlated with this inverse pattern of response in the frontal cortex and amygdala (Fox et al. 2005). Furthermore, individual differences remain stable, even within groups of bilateral amygdalectomized monkeys, suggesting that additional areas of the brain may contribute to the long-term individual reactive response pattern of fearful and bold monkeys (Mason et al. 2006).

Further evidence of anxiety/fearful response fractionation of systems is found in the effects of the serotonin transporter gene on behavior. While the less functional serotonin transporter genotype affects socially-mediated reactivity, it does not affect the response to a potential predator (Kalin et al. 2008; Rogers et al. 2008). In fact, even within systems, measures may not covary. For example, in many years of simultaneously measuring plasma cortisol and its stimulating hormone ACTH, we virtually never found a significant positive correlation between the two.

11.7 Effect of Early Rearing History

Early rearing experiences affect reactivity. The strongest evidence comes from the studies of infants in peer-reared groups without adults. Peer-reared monkeys are usually housed for the first 30 days in a neonatal nursery, sometimes with a terry-cloth surrogate mother. Afterward, the infants are placed together without adults, usually in groups of four and allowed unfettered daily social interactions. Peer-only rearing largely ameliorates the behavior deficits of subjects raised without companions in social isolation, mainly because peers provide each other with contact comfort, warmth, and contingent daily social interactions. Thus, they show the full panoply of rhesus behaviors exhibiting social competence and successful reproductive skills, and in the case of females, competent maternal behavior (Harlow 1969; Suomi et al. 1974; Novak and Harlow 1975).

Nevertheless, following the experience of growing up without adults, peer-reared monkeys are more fearful and anxious than their mother-reared counterparts. To illustrate, during the first month of life, infants of mother-reared monkeys are seldom are out of ventral contact with their mother. After the first month, close ventral contact diminishes largely because infants spend increasing amounts of time in exploration and social play. Over the first 6 months of life, peer-only-reared monkeys display more infant-like social clinging, behavioral withdrawal, and less exploration of their environment than mother-reared monkeys (Harlow 1969; Chamove et al. 1973; Higley et al. 1992a, 1996c, d). Mother-reared monkeys seldom engage in such intimate contact after the ninth month of life, and hardly ever after the first year (Hinde and Simpson 1967). Peer-reared subjects continue this immature behavior into the second and third years of life (Higley et al. 1991a, 1992a, 1996c, d). The differences are most dramatic under mild challenges such as exposure to novel sounds or the appearance of a stranger in their housing room.

Additional evidence comes from a study in which peer-reared monkeys were exposed to an analogue of the Ainsworth Strange Situation procedure (Ainsworth et al. 1978) designed for rhesus monkey infants. With interesting toys and objects to explore, peer-reared monkeys remained closer to their attachment source (a favorite peer) and explored the toys less than mother-reared monkeys (Higley and Danner 1988). Moreover, the effects of anxiety-inducing, peer-rearing have long-term consequences. Even as juveniles and adolescents, peer-reared subjects are more anxious in a novel room or during social separation (Higley et al. 1992a, 1996c, d).

We have suggested that the deficits of peer-reared subjects may be due to nonsecure attachment relations (Higley et al. 1992a, 1994). Besides acting as a secure base to reduce arousal, mothers also exhibit facial emotions that infants use as social cues to elicit information (known as social referents) when environmental situations are ambiguous. A mother showing fear to unknown or ambiguous stimuli can quickly elicit fear in her infant (Mineka et al. 1984). Conversely, approach by an infant's mother to an ambiguous stimulus or social partner reduces fear in her infant. It is as if the infant uses the mother to gain information about the potential danger of ambiguous stimuli. In the absence of the cues that a mother provides, peer-reared monkeys are left in a chronic state of uncertainty since they can only orient to each other and get the same cues of uncertainty or even fear from one another.

Many of the deficits in peer-reared monkeys are related to deficits in the central nervous system. We suggest that the behavioral deficits occur because in the absence of mothers who provide a secure base to reduce anxiety, infants fail to develop the capacity to become self-reliant in reducing their own arousal and anxiety. We suggest that under normal conditions, during development, maternal input may induce the survival of important synapses, leading to the appropriate connections between the frontal cortex and the amygdala. This frontal connection to the amygdala in turn allows the infant to bring the arousal under voluntary control. In the absence of maternal input, the selective association of the frontal cortex with the amygdala will not develop appropriately. Hence, the capacity to bring the amygdala under voluntary control of the frontal cortex does not occur.

There is evidence for this. Serotonin tracts between the frontal cortex and amygdala modulate and produce inhibitory control of the amygdala, reducing amygdala activity and fear (Porrino et al. 1981; Raine et al. 1998; Critchley et al. 2000; Kalin and Shelton 2003; Kalin et al. 2004, 2005, 2007; Urry et al. 2006). The effect of rearing on the serotonin system thus becomes relevant. As discussed below, peer-rearing produces impaired CNS serotonin functioning (Higley and Linnoila 1997a, b). PET and SPECT studies using rhesus monkeys show reduced serotonin functioning in the orbital frontal area of the frontal cortex of peer-reared monkeys (Doudet et al. 1995; Ichise et al. 2006), suggesting that frontal deficits underlie the peer-reared monkeys' temperamental inhibition.

11.8 Biological Correlates of Reactivity

In humans and in monkeys, right frontal brain activation is a potent predictor of inhibited or reactive temperament (Davidson and Fox 1989; Kalin et al. 1998a, 2000; Buss et al. 2003). One of the most potent central anxiety-provoking neuropeptides is corticotrophin releasing hormone (CRH). CSF levels of CRH are positively correlated with extreme right brain, frontal cortex activation (Kalin et al. 1998a, 2000). CRH activates the amygdala to induce freezing and immobility (Kalin et al. 1983; Strome et al. 2002), but not vocalizations (Kalin et al. 1989). Injection of CRH into the ventricles of an adult rhesus monkey induces an almost instantaneous behavioral withdrawal and depression (Kalin et al. 1989; Strome et al. 2002). A potent selective CRH receptor antagonist, antalarmin, on the other hand, reverses freezing and anxiety in rhesus adults during stressful situations (Habib et al. 2000).

In contrast to CRH, neuropeptide Y (NPY) is an anxiety reducing brain peptide that is reciprocally related to CRH, at least in the rat (Wahlestedt et al. 1993; Heilig et al. 1994). Consistent with findings in rodents, there is a negative correlation between CSF NPY and CRH concentrations during the stress of a social separation. As shown in Fig. 11.2, during both baseline and repeated social separations, both groups show increases in CRH but the more anxious peer-reared monkeys exhibit higher levels of CSF CRH when compared to the less anxious mother-reared monkeys (Gabry et al. unpublished manuscript). CRH genetic variants in rhesus macaques lead to differences in the activity of the CRH system in the brain and to differences in anxiety-like behaviors (Barr et al. 2008a). This study, and those discussed above, suggest a provocative proposal that temperamental differences in the disposition for behavior inhibition may be in part due to experiential and heritable influences on the CRH system (Kalin et al. 2004).

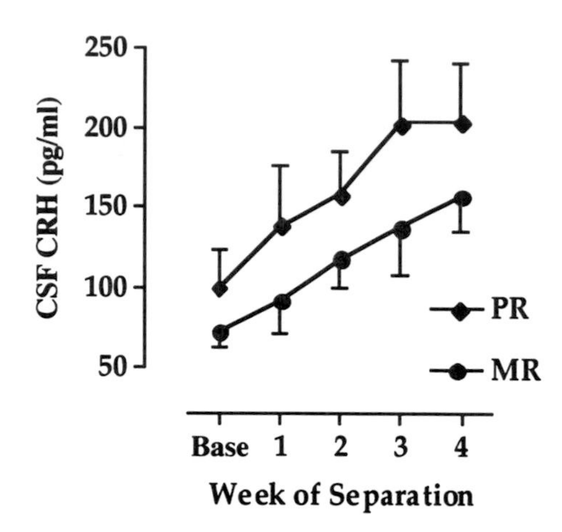

Fig. 11.2 Peer-reared and mother-reared CSF concentrations of the neurohormone corticotropin-releasing hormone (CRH-values are in pg/mL). The peer-reared monkeys are illustrated with diamond-shaped data points and the mother-reared monkeys are illustrated by the circles. Samples were obtained at baseline (base), and over four separations (1–4). ANOVA shows a main effect, with the peer-reared monkeys showing higher average CRH concentrations

While CRH plays an important if not paramount role in behavior inhibition, several studies have shown that sympathetic arousal is also positively correlated with levels of behavioral reactivity. For example, two groups of peer-reared monkeys were raised either in a large noisy room or in a small quiet room. When both groups were exposed to an experimental cage with interesting but novel toys, the subjects reared in the large noisy room took longer to adapt to the new cage and had a higher, less variable heart rate (Byrne et al. 1988). Capitanio et al. (1986) found both cross time stability for heart rate and a bank of stress-related behaviors, heart rate being positively correlated with the level of stress-related behaviors.

For monkeys, one of the most studied indices of sympathetic arousal is the output of the adrenocortical system (cortisol and ACTH). Mean cortisol and ACTH is increased in stress sensitive monkeys. For example, monkeys that show behavioral withdrawal and freezing exhibit higher plasma cortisol levels (Erickson et al. 2005). Peer-reared monkeys have higher cortisol and ACTH than mother-reared monkeys (Suomi 1983; Higley et al. 1991a, 1992b; Fahlke et al. 2000; Kinnally et al. 2008). Low ranking rhesus monkeys have higher cortisol levels than high ranking animals (Suomi et al. 1989). In a comprehensive meta-analysis, Abbott and colleagues (2003) found high cortisol levels in low ranking Old World monkeys particularly when the threat of aggression was high. In natural settings, monkeys that are frequently wounded (as measured by high numbers of scars and wounds) have higher plasma cortisol and ACTH even when they are not currently wounded (Higley et al. 1992b).

While there are sometimes failures to find significant associations between stress hormones and behavior, positive correlations have been reported between reactivity and cortisol. In 24-month-old peer-reared monkeys, we found that levels of cortisol were positively correlated with an aggregate measure of distress which included immobility, self-directed orality, self-clasping, and huddling (Higley 1985). Scanlan and Suomi found that the pituitary hormone responsible for the release of cortisol, ACTH, was positively correlated with measures of reactivity such as self-directed behavior and distress vocalizations (Scanlan 1987; Scanlan and Suomi 1988). In addition, high social dominance rank is negatively correlated with ACTH (Scanlan and Suomi 1986). It should be noted, however, that not all laboratories have found such covariance between adrenal cortical measures and behavior (see, e.g., Levine et al. 1987). In our experience as well as others (Levine and Wiener 1988; Norcross and Newman 1999), hypothalamic–pituitary–adrenal (HPA) axis responses do not reliably correlate with concurrent stress-related behaviors. The behaviors that sometimes but not always covary with HPA axis response are distress vocalizations and coos (Lyons et al. 1999; Becker et al. 2004; Cross and Rogers 2006).

Individual differences in the cortisol response are stable across time. Several studies from our laboratory have found that cortisol levels during the stress of social separation are stable over extended periods (Champoux 1988a, b; Scanlan 1988; Scanlan and Suomi 1988; Higley et al. 1992c; Fahlke et al. 2000). For example, in a study of ten rhesus monkeys, Suomi (1983) found that cortisol levels under stressful conditions at 22 days of age had a correlation of $r=0.56$ with cortisol levels at 18 months of age during a brief social separation.

11.9 Reactivity and Neurotransmitters

Cloninger (1986) outlined three general personality systems, each controlled by a different neurotransmitter system. Individuals with high harm avoidance and chronic anxiety have frequent anticipatory worries based on specific cues, high sedation thresholds, and easy fatigability, a personality style similar to that of reactivity (Cloninger 1986) postulated that reactivity was controlled by the NE system. Activity of the adrenocortical system is partly controlled by the neurotransmitters NE and serotonin in adults (Southwick et al. 2005; van Stegeren 2008) and children (Kagan et al. 1987, 1988; Kagan 1994). Kraemer (1997) also described the NE system as underlying reactivity in nonhuman primates. In a now classic review, Soubrié (1986) postulated that serotonin acted as a general inhibitory neurotransmitter, with high levels of serotonin correlating positively with levels of behavior inhibition. Neurotransmitters are developmentally stable from infancy to adulthood (Kraemer et al. 1989; Higley et al. 1990, 1991b, 1992c, 1996c; Clarke et al. 1996; Shannon et al. 2005; Howell et al. 2007). The trait-like stability of the neurotransmitter system is important to the extent that these neurotransmitters underlie stable personality traits, showing that individual stability in neurotransmitter activity may explain the enduring long-term stability of traits across time and situation.

11.10 Early Rearing and Neurotransmitter Functioning

Of particular note is the effect of parental absence on the development of the serotonin system. Neonates, 2-year-olds, and adults reared without parents in peer-reared groups showed significantly lower CSF levels of the serotonin metabolite (Shannon et al. 2005). In fact, one of the most replicated findings from our laboratory is that early parental absence (i.e., peer-rearing) leads to long-term, impaired CNS serotonin (Higley et al. 1991a, 1992c, 1993, 1996c; Bennett et al. 2002; Barr et al. 2004b; Shannon et al. 2005; Ichise et al. 2006). The effect of early maternal absence on the serotonin system has been replicated in other laboratories using different methods (Kinnally et al. 2008). On the other hand, while separation is highly stressful for both peer-reared monkeys and mother-reared monkeys, serotonin transporter mRNA expression was not affected in the nursery-peer-reared monkeys when they were separated from their peers (Kinnally et al. 2008).

11.11 Species Differences in Reactivity

Different species of nonhuman primates display different levels of reactivity. For example, when gibbons and rhesus monkeys were exposed to novel stimuli, gibbons were more likely to approach and explore the stimuli. Similarly, when exposed to a large open caging area, gibbons were more likely to explore, whereas rhesus monkeys were more likely to remain immobile. Gibbons were also more likely than rhesus monkeys to approach a stranger of their own species and interact with that stranger (Bernstein et al. 1963). Under similar test conditions, langurs were more likely to investigate novel objects and less likely to urinate and defecate than rhesus monkeys (Singh 1992).

Other studies have shown that even closely related species may vary in reactivity. Clarke et al. (1988) compared rhesus, crab-eating, and bonnet macaques, in tests of emotionality. Across tasks, the crab-eating macaques were the most reactive, showing less exploration and manipulation of novel stimuli, and higher levels of distress cries and behavioral immobility. Rhesus monkeys were more active and explorative and showed more aggression; bonnet macaques fell in-between. When levels of cortisol and heart rate were compared across the different tasks, crab-eaters had the highest levels, rhesus the lowest, and bonnets fell in-between, consistent with their behavioral levels of reactivity.

There are also substantial differences in aggressiveness within the macaque genus. While rhesus are aggressive, pigtail macaques are more friendly and placid (Westergaard et al. 1999b). Chinese-derived rhesus macaques show significantly more reactivity and aggression than the Indian-derived rhesus. These differences are present early in life and are consistent with differences in CSF 5-HIAA concentrations (Champoux et al. 1997, 1994).

11.12 rh5-HTT, Other Genes and Reactive Temperament

The discovery of two genetic variants in the human serotonin transporter gene (hereafter 5-HTT) led to the discovery of a similar bivariate set of alleles in rhesus macaques (Bennett et al. 2002). This repeat polymorphism is orthologous to that found in humans, with two variants, one with a short and the other a long allele (hereafter rh5-HTT). The short variant exhibits reduced decreased rh5-HTT expression. As a consequence, monkeys with the short allele are likely to exhibit low CSF 5-HIAA concentrations (Bennett et al. 2002), and a variety of serotonin mediated temperamental and stress-axis hormonal differences (Champoux et al. 2002; Barr et al. 2004a, c, d), although many of these genotype effects are exaggerated or only present after specific rearing conditions or only in some environments. In one study, researchers found low transporter binding in subjects with the short allele (Lopez and Higley 2002), but the effect was limited to the peer-reared monkeys. A recent study suggests that the differences in the serotonin system between subjects with the rh5-HTT short and homozygous long allele variants are not simply the immediate consequence of how the gene modulates neurotransmission, but instead the transporter gene acts developmentally to modify the brain structurally during development, with reductions in size in the left amygdala frontal, orbital frontal, anterior cingulate, and medial prefrontal of animals with the short allele (Jedema et al. 2010).

Following the discovery of the human 5-HTT genotypes, several studies showed a relationship between anxiety and possession of the short allele in humans. A meta-analysis showed that the short allele is strongly associated with neuroticism (Sen et al. 2004). A more recent comprehensive meta-analysis of research on the human serotonin transporter gene showed that possession of the 5-HTT short allele is strongly associated with irritable and anxious temperaments that lead to depression (Gonda et al. 2009).

The linkage of the 5-HTT short allele to anxious temperaments in humans may be true for the rh5-HTT genotypes in nonhuman primates. Monkeys homozygous for the rh5-HTT short allele were more likely to exhibit submissive "lipsmacks" to a threatening human intruder. During exposure to an interesting playroom, they spent less time away from their mother exploring, and they were particularly fearful when a novel remotely controlled toy car moved toward them and then away (Bethea et al. 2004). An independent study showed that during the stress of group formation (a potent stressor for rhesus monkeys), subjects with the rh5-HTT short allele were more stress-susceptible showing more weight loss during group formation than those homozygous for the long allele (Jarrell et al. 2008). In a study of mother–infant interactions, among both the mothers and their infants, possession of the short allele was associated with higher basal levels of cortisol and greater hormonal responses to stress in the infants (McCormack et al. 2009). The short variant of the rh5-HTT genotype also leads to increased vulnerability to the effects of psychosocial stress that are associated with low social dominance rank. Animals homozygous for the long allele, on the other hand, were somewhat buffered from the deleterious effects of low social dominance rank (Jarrell et al. 2008).

A recent study suggests that social anxiety may be partly mediated by how monkeys with the short allele variant process faces and facial emotional expressions. When given the opportunity to view monkey faces, rhesus with the rh5-HTT short allele show little interest. In fact, they prefer looking at faces in which the eyes, nose, and mouth are scrambled more than the normal rhesus faces (Watson et al. 2009). More consistent with reactive temperament, when they were shown faces of monkeys high in social status, all monkeys minimized time looking at the eyes of the high ranking faces, but when the monkeys did look at faces of high ranking animals, the rhesus with the short allele exhibited greater pupil dilation, a measure of high sympathetic arousal.

Differences in emotionality among monkeys based on serotonin transporter variants may be mediated by the amygdala and related neuronal networks. Consistent with earlier studies showing a positive relationship between amygdala activity and high fearfulness, humans with the short allele exhibited an overly active amygdala response when shown provocative, fear provoking stimuli (Hariri et al. 2002). Subsequent meta-analysis confirms this finding that the amygdala is overactive in subjects with one or two copies of the rh5-HTT short allele (Munafo et al. 2008). In rhesus monkeys, carriers of the rh5-HTT short allele displayed high amygdala and frontal cortex activity during relocation to a new environment (Kalin et al. 2008). In response to a threatening human, young monkeys with the rh5-HTT short allele displayed increased activity in the orbitofrontal cortex and bed nucleus of the stria terminalis (Kalin et al. 2008), a network linked to the amygdala complex and important in regulating anxiety.

The rh5-HTT gene also mediates the function of the serotonin system in additional areas of the brain. When compared to subjects with the long allele, subjects with the short allele had increased 5-HT2a mRNA levels, and decreased 5-HTT binding in the prefrontal cortex (Lopez et al. 2001). This pattern of brain functioning was first described as a pattern in suicide victims, particularly those whose psychological autopsies showed an affective disorder (Arango et al. 2001, 2002). The latter effect was stronger when rearing was also considered (Lopez et al. 2001), with peer-reared monkeys that possessed the short allele having decreased 5-HTT binding and increased 5HT2a binding.

11.13 Conclusions

In this chapter we have described a temperamental trait, reactivity, in rhesus monkeys that is related to how individuals respond to stressful situations. As in impulsivity and aggression, reactivity is stable across time and situations. Likewise, individual differences in this trait reflect the neurophysiology of individuals and are products of genes, early rearing environments, and their interaction.

As we noted in our previous chapter, we strongly urge the study of the genetic, environmental, and physiological underpinnings of personality in New World monkeys. With respect to reactivity, such studies would enable us to better understand the evolutionary, ecological, and social bases of similarities and differences in a personality trait that is important to and present in humans as well as a wide range of primate and non-primate species.

The strong genetic similarities between humans and rhesus macaques (Gibbs et al. 2007; Harris et al. 2007; Hernandez et al. 2007; Lee et al. 2008) as well as the common evolutionary challenges of living in complex social groups arguably makes a case for using rhesus macaques as a model of human personality. We thus cautiously recommend that future research of this sort focuses on individual differences in other traits related to human experiences, including higher-order traits such as "transcendence". While measuring these traits may be not be as straightforward as measuring traits with obvious behavioral manifestations, studies of chimpanzees (King and Figueredo 1997) and even rhesus macaques (Weiss et al. 2011) have identified analogues named "Openness".

We think that such studies may lead to discoveries about those aspects of human nature that go beyond how individuals react when facing challenges or potential competitors. Indeed, these studies will offer a glimpse into the evolution of individual differences in how we process our thoughts and experiences, which, for the most part, has been the realm of philosophers.

References

Abbott DH, Keverne EB, Bercovitch FB et al. (2003) Are subordinates always stressed? A comparative analysis of rank differences in cortisol levels among primates. Horm Behav 43:67–82

Ainsworth MDS, Blehar MC, Waters E et al. (1978) Patterns of attachment: A psychological study of the strange situation. Lawrence Erlbaum Associates, Hillsdale

Anderson GM, Bennett AJ, Weld KP et al. (2002) Serotonin in cisternal cerebrospinal fluid of rhesus monkeys: Basal levels and effects of sertraline administration. Psychopharmacology 161:95–99

Arango V, Underwood MD, Boldrini M et al. (2001) Serotonin 1A receptors, serotonin transporter binding and serotonin transporter mRNA expression in the brainstem of depressed suicide victims. Neuropsychopharmacology 25:892–903

Arango V, Underwood MD, Mann JJ (2002) Serotonin brain circuits involved in major depression and suicide. Prog Brain Res 136:443–453

Barr CS, Dvoskin RL, Yuan Q et al. (2008a) CRH haplotype as a factor influencing cerebrospinal fluid levels of corticotropin-releasing hormone, hypothalamic-pituitary-adrenal axis activity, temperament, and alcohol consumption in rhesus macaques. Arch Gen Psychiatry 65: 934–944

Barr CS, Newman TK, Lindell S et al. (2004a) Early experience and sex interact to influence limbic-hypothalamic-pituitary-adrenal-axis function after acute alcohol administration in rhesus macaques (*Macaca mulatta*). Alcohol Clin Exp Res 28:1114–1119

Barr CS, Newman TK, Lindell S et al. (2004b) Interaction between serotonin transporter gene variation and rearing condition in alcohol preference and consumption in female primates. Arch Gen Psychiatry 61:1146–1152

Barr CS, Newman TK, Schwandt M et al. (2004c) Sexual dichotomy of an interaction between early adversity and the serotonin transporter gene promoter variant in rhesus macaques. Proc Natl Acad Sci USA 101:12358–12363

Barr CS, Newman TK, Shannon C et al. (2004d) Rearing condition and rh5-HTTLPR interact to influence limbic-hypothalamic-pituitary-adrenal axis response to stress in infant macaques. Biol Psychiatry 55:733–738

Barr CS, Schwandt ML, Lindell SG et al. (2008b) Variation at the mu-opioid receptor gene (OPRM1) influences attachment behavior in infant primates. Proc Natl Acad Sci USA 105:5277–5281

Bayart F, Hayashi KT, Faull KF et al. (1990) Influence of maternal proximity on behavioral and physiological responses to separation in infant rhesus monkeys (*Macaca mulatta*). Behav Neurosci 104:98–107

Becker ML, Bernhards DE, Chisholm KL et al. (2004) Calling rate as a measure of stress reactivity in mother- and nursery-reared rhesus infants (*Macaca mulatta*). Am J Primatol 62:60

Becker MS, Suomi SJ, Marra L et al. (1984) Developmental data as predictors of depression in infant rhesus monkeys. Infant Behav Dev 7:26

Bennett AJ, Lesch KP, Heils A et al. (2002) Early experience and serotonin transporter gene variation interact to influence primate CNS function. Mol Psychiatry 7:118–122

Bernstein IS, Schusterman RJ, Sharpe LG (1963) A comparison of rhesus monkey and gibbon responses to unfamiliar situations. J Comp Physiol Psychol 56:914–916

Bethea CL, Streicher JM, Coleman K et al. (2004) Anxious behavior and fenfluramine-induced prolactin secretion in young rhesus macaques with different alleles of the serotonin reuptake transporter polymorphism (5HTTLPR). Behav Genet 34:295–307

Biederman J, Hirshfeld-Becker DR, Rosenbaum JF et al. (2001) Further evidence of association between behavioral inhibition and social anxiety in children. Am J Psychiatry 158:1673–1679

Biederman J, Rosenbaum JF, Bolduc-Murphy EA et al. (1993) A 3-year follow-up of children with and without behavioral inhibition. J Am Acad Child Adolesc Psychiatry 32:814–821

Blizard DA (1981) The Maudsley reactive and nonreactive strains: a North American perspective. Behav Genet 11:469–489

Blizard DA (1989) Analysis of stress susceptibility using the Maudsley reactive and non-reactive strains. In: Palermo DS (ed) Coping with uncertainty: Behavioral and developmental perspectives. Lawrence Erlbaum Associates, Hillsdale

Bolig R, Price CS, O'Neill PL et al. (1992) Subjective assessment of reactivity level and personality traits of rhesus monkeys. Int J Primatol 13:287–306

Bowlby J (1973) Attachment and loss: Separation. Basic Books, Inc., New York

Bush DS, Steffen SL, Higley JD et al. (1987) Continuity of social separation responses in rhesus monkeys (*Macaca mulatta*) reared under different conditions. Am J Primatol 18:138

Buss KA, Schumacher JR, Dolski I et al. (2003) Right frontal brain activity, cortisol, and withdrawal behavior in 6-month-old infants. Behav Neurosci 117:11–20

Byrne EA, DiGregorio G, Thompson WW (1988) The effect of rearing environment on infant rhesus monkeys' physiological and behavioral adaptation to novelty. Infant Behav Dev 11:43

Capitanio JP (2004) Personality factors between and within species. In: Thierry B, Singh M, Kaumanns W (eds) Macaque societies: A model for the study of social organizations. Cambridge University Press, Cambridge

Capitanio JP (2008) Personality and disease. Brain Behav Immun 22:647–650

Capitanio JP, Rasmussen KL, Snyder DS et al. (1986) Long-term follow-up of previously separated pigtail macaques: Group and individual differences in response to novel situations. J Child Psychol Psychiatry 27:531–538

Capitanio JP, Widaman KF (2005) Confirmatory factor analysis of personality structure in adult male rhesus monkeys (*Macaca mulatta*). Am J Primatol 65:289–294

Chamove AS, Eysenck HJ, Harlow HF (1972) Personality in monkeys: Factor analyses of rhesus social behaviour. Q J Exp Psychol 24:496–504

Chamove AS, Rosenblum LA, Harlow HF (1973) Monkeys (*Macaca mulatta*) raised with only peers. A pilot study. Anim Behav 21:316–325

Champoux M (1988a) Behavioral development and temporal stability of reactivity to stressors in mother-reared and nursery/peer-reared rhesus macaques. University of Wisconsin, Madison

Champoux M (1988b) Behavioral development of nursery-reared rhesus monkeys (*Macaca mulatta*) neonates. Infant Behav Dev 11:367–371

Champoux M, Bennett A, Shannon C et al. (2002) Serotonin transporter gene polymorphism, differential early rearing, and behavior in rhesus monkey neonates. Mol Psychiatry 7:1058–1063

Champoux M, Higley JD, Suomi SJ (1997) Behavioral and physiological characteristics of Indian and Chinese-Indian hybrid rhesus macaque infants. Dev Psychobiol 31:49–63
Champoux M, Suomi SJ, Schneider ML (1994) Temperament differences between captive Indian and Chinese-Indian hybrid rhesus macaque neonates. Lab Anim Sci 44:351–357
Clarke AS, Boinski S (1995) Temperament in nonhuman primates. Am J Primatol 37:103–125
Clarke AS, Hedeker DR, Ebert MH et al. (1996) Rearing experience and biogenic amine activity in infant rhesus monkeys. Biol Psychiatry 40:338–352
Clarke AS, Mason WA, Moberg GP (1988) Interspecific contrasts in responses of macaques to transport cage training. Lab Anim Sci 38:305–309
Cloninger CR (1986) A unified biosocial theory of personality and its role in the development of anxiety states. Psychiatr Dev 3:167–226
Cloninger CR (1994) Temperament and personality. Curr Opin Neurobiol 4:266–273
Critchley HD, Simmons A, Daly EM et al. (2000) Prefrontal and medial temporal correlates of repetitive violence to self and others. Biol Psychiatry 47:928–934
Cross N, Rogers LJ (2006) Mobbing vocalizations as a coping response in the common marmoset. Horm Behav 49:237–245
Davidson RJ, Fox NA (1989) Frontal brain asymmetry predicts infants' response to maternal separation. J Abnorm Psychol 98:127–131
Doudet D, Hommer D, Higley JD et al. (1995) Cerebral glucose metabolism, CSF 5-HIAA levels, and aggressive behavior in rhesus monkeys. Am J Psychiatry 152:1782–1787
Erickson K, Gabry KE, Lindell S et al. (2005) Social withdrawal behaviors in nonhuman primates and changes in neuroendocrine and monoamine concentrations during a separation paradigm. Dev Psychobiol 46:331–339
Erickson K, Lindell S, Champoux M et al. (2001) Relationships between behavior and neurochemical changes in rhesus macaques during a separation paradigm. Soc Neurosci Abstr 27, program 572:14
Eysenck HJ, Eysenck MW (1985) Personality and individual differences. A natural science approach. Plenum, New York
Eysenck HJ, Eysenck SBG (1976) Psychoticism as a dimension of personality. Hodder and Stoughton, London
Fahlke C, Lorenz JG, Long J et al. (2000) Rearing experiences and stress-induced plasma cortisol as early risk factors for excessive alcohol consumption in nonhuman primates. Alcohol Clin Exp Res 24:644–650
Fairbanks LA (2001) Individual differences in response to a stranger: Social impulsivity as a dimension of temperament in vervet monkeys (*Cercopithecus aethiops sabaeus*). J Comp Psychol 115:22–28
Fox AS, Oakes TR, Shelton SE et al. (2005) Calling for help is independently modulated by brain systems underlying goal-directed behavior and threat perception. Proc Natl Acad Sci USA 102:4176–4179
Francois MH, Nosten BM, Roubertoux PL et al. (1990) Opponent strain effect on eliciting attacks in NZB mice: Physiological correlates. Physiol Behav 47:1181–1185
Gabry KE, Erickson K, Champoux M et al. (unpublished manuscript) Increased CSF corticotropin-releasing hormone and decreased neuropeptide Y concentrations in response to psychosocial stress
Gibbs RA, Rogers J, Katze MG et al. (2007) Evolutionary and biomedical insights from the rhesus macaque genome. Science 316:222–234
Gonda X, Fountoulakis KN, Juhasz G et al. (2009) Association of the s allele of the 5-HTTLPR with neuroticism-related traits and temperaments in a psychiatrically healthy population. Eur Arch Psychiatry Clin Neurosci 259:106–113
Gray JA (1987) The psychology of fear and stress. Cambridge University Press, New York
Habib KE, Weld KP, Rice KC et al. (2000) Oral administration of a corticotropin-releasing hormone receptor antagonist significantly attenuates behavioral, neuroendocrine, and autonomic responses to stress in primates. Proc Natl Acad Sci USA 97:6079–6084
Hariri AR, Mattay VS, Tessitore A et al. (2002) Serotonin transporter genetic variation and the response of the human amygdala. Science 297:400–403

Harlow HF (1969) Age-mate or peer affectional system. Adv Study Behav 2:333–383
Harlow HF, Harlow MK (1965) The affectional systems. In: Schrier AM, Harlow HF, Stollinitz F (eds) Behavior of nonhuman primates. Academic, New York
Harris RA, Rogers J, Milosavljevic A (2007) Human-specific changes of genome structure detected by genomic triangulation. Science 316:235–237
Heilig M, Koob GF, Ekman R et al. (1994) Corticotropin-releasing factor and neuropeptide Y: Role in emotional integration. Trends Neurosci 17:80–85
Hernandez RD, Hubisz MJ, Wheeler DA et al. (2007) Demographic histories and patterns of linkage disequilibrium in Chinese and Indian rhesus macaques. Science 316:240–243
Higley JD (1985) Continuity of social separation behaviors in rhesus monkeys from infancy to adolescence. University of Wisconsin, Madison
Higley JD, Danner GR (1988) Attachment in rhesus monkeys reared either with only peers or with their mothers as assessed by the Ainsworth Strange Situation procedure. Infant Behav Dev 11:139
Higley JD, Hasert MF, Suomi SJ et al. (1991a) Nonhuman primate model of alcohol abuse: Effects of early experience, personality, and stress on alcohol consumption. Proc Natl Acad Sci USA 88:7261–7265
Higley JD, Hopkins WD, Thompson WW et al. (1992a) Peers as primary attachment sources in yearling rhesus monkeys (*Macaca mulatta*). Dev Psychol 28:1163–1171
Higley JD, Linnoila M (1997a) Low central nervous system serotonergic activity is traitlike and correlates with impulsive behavior. A nonhuman primate model investigating genetic and environmental influences on neurotransmission. Ann N Y Acad Sci 836:39–56
Higley JD, Linnoila M (1997b) A nonhuman primate model of excessive alcohol intake: Personality and neurobiological parallels of Type I- and Type II-like alcoholism. Recent Dev Alcohol 13:192–219
Higley JD, Linnoila M, Suomi SJ (1994) Ethological contributions: Experiential and genetic contributions to the expression and inhibition of aggression in primates. In: Hersen M, Ammerman RT, Sisson L (eds) Handbook of aggressive and destructive behavior in psychiatric patients. Plenum Press, New York
Higley JD, Mehlman PT, Higley SB et al. (1996a) Excessive mortality in young free-ranging male nonhuman primates with low cerebrospinal fluid 5-hydroxyindoleacetic acid. Arch Gen Psychiatry 53:537–543
Higley JD, Mehlman PT, Poland RE et al. (1996b) CSF testosterone and 5-HIAA correlate with different types of aggressive behaviors. Biol Psychiatry 40:1067–1082
Higley JD, Mehlman PT, Taub DM et al. (1992b) Cerebrospinal fluid monoamine and adrenal correlates of aggression in free-ranging rhesus monkeys. Arch Gen Psychiatry 49:436–441
Higley JD, Suomi SJ, Linnoila M (1990) Parallels in aggression and serotonin: Consideration of development, rearing history, and sex differences. In: van Praag HM, Plutchik R, Apter A (eds) Violence and suicidality: Perspectives in clinical and psychobiological research. Brunner/Mazel, New York
Higley JD, Suomi SJ, Linnoila M (1991b) CSF monoamine metabolite concentrations vary according to age, rearing, and sex, and are influenced by the stressor of social separation in rhesus monkeys. Psychopharmacology 103:551–556
Higley JD, Suomi SJ, Linnoila M (1992c) A longitudinal assessment of CSF monoamine metabolite and plasma cortisol concentrations in young rhesus monkeys. Biol Psychiatry 32:127–145
Higley JD, Suomi SJ, Linnoila M (1996c) A nonhuman primate model of type II alcoholism? Part 2. Diminished social competence and excessive aggression correlates with low cerebrospinal fluid 5-hydroxyindoleacetic acid concentrations. Alcohol Clin Exp Res 20:643–650
Higley JD, Suomi SJ, Linnoila M (1996d) A nonhuman primate model of type II excessive alcohol consumption? Part 1. Low cerebrospinal fluid 5-hydroxyindoleacetic acid concentrations and diminished social competence correlate with excessive alcohol consumption. Alcohol Clin Exp Res 20:629–642
Higley JD, Suomi SJ, Scanlan JM et al. (1982) Plasma cortisol as a predictor of individual depressive behavior in rhesus monkeys (*Macaca mulatta*). Paper presented at the meeting of Society for Neuroscience, Minneapolis, MN

Higley JD, Thompson WW, Champoux M et al. (1993) Paternal and maternal genetic and environmental contributions to cerebrospinal fluid monoamine metabolites in rhesus monkeys (*Macaca mulatta*). Arch Gen Psychiatry 50:615–623
Hill SY, Lowers L, Locke J et al. (1999) Behavioral inhibition in children from families at high risk for developing alcoholism. J Am Acad Child Adolesc Psychiatry 38:410–417
Hinde RA, Simpson MJA (1967) Qualities of mother-infant relationships in monkeys. Ciba Found Symp 33:39–68
Holmes A, le Guisquet AM, Vogel E et al. (2005) Early life genetic, epigenetic and environmental factors shaping emotionality in rodents. Neurosci Biobehav Rev 29:1335–1346
Howell S, Westergaard G, Hoos B et al. (2007) Serotonergic influences on life-history outcomes in free-ranging male rhesus macaques. Am J Primatol 69:851–865
Ichise M, Vines DC, Gura T et al. (2006) Effects of early life stress on [11 C]DASB positron emission tomography imaging of serotonin transporters in adolescent peer- and mother-reared rhesus monkeys. J Neurosci 26:4638–4643
Insel TR, Hill JL (1987) Infant separation distress in genetically fearful rats. Biol Psychiatry 22: 786–789
Jarrell H, Hoffman JB, Kaplan JR et al. (2008) Polymorphisms in the serotonin reuptake transporter gene modify the consequences of social status on metabolic health in female rhesus monkeys. Physiol Behav 93:807–819
Jedema HP, Gianaros PJ, Greer PJ et al. (2010) Cognitive impact of genetic variation of the serotonin transporter in primates is associated with differences in brain morphology rather than serotonin neurotransmission. Mol Psychiatry 15:512–522, 446
Kagan J (1992) Behavior, biology, and the meanings of temperamental constructs. Pediatrics 90:510–513
Kagan J (1994) On the nature of emotion. Monogr Soc Res Child Dev 59:7–24
Kagan J, Reznick JS, Snidman N (1987) The physiology and psychology of behavioral inhibition in children. Child Dev 58:1459–1473
Kagan J, Reznick JS, Snidman N (1988) Biological bases of childhood shyness. Science 240: 167–171
Kalin NH (1993) The neurobiology of fear. Sci Am 268:94–101
Kalin NH (1999) Primate models to understand human aggression. J Clin Psychiatry 60, Supplement 15:29–32
Kalin NH, Larson C, Shelton SE et al. (1998a) Asymmetric frontal brain activity, cortisol, and behavior associated with fearful temperament in rhesus monkeys. Behav Neurosci 112:286–292
Kalin NH, Shelton SE (1989) Defensive behaviors in infant rhesus monkeys: Environmental cues and neurochemical regulation. Science 243:1718–1721
Kalin NH, Shelton SE (2003) Nonhuman primate models to study anxiety, emotion regulation, and psychopathology. Ann N Y Acad Sci 1008:189–200
Kalin NH, Shelton SE, Barksdale CM (1988) Opiate modulation of separation-induced distress in non-human primates. Brain Res 440:285–292
Kalin NH, Shelton SE, Barksdale CM (1989) Behavioral and physiologic effects of CRH administered to infant primates undergoing maternal separation. Neuropsychopharmacology 2:97–104
Kalin NH, Shelton SE, Davidson RJ (2000) Cerebrospinal fluid corticotropin-releasing hormone levels are elevated in monkeys with patterns of brain activity associated with fearful temperament. Biol Psychiatry 47:579–585
Kalin NH, Shelton SE, Davidson RJ (2004) The role of the central nucleus of the amygdala in mediating fear and anxiety in the primate. J Neurosci 24:5506–5515
Kalin NH, Shelton SE, Davidson RJ (2007) Role of the primate orbitofrontal cortex in mediating anxious temperament. Biol Psychiatry 62:1134–1139
Kalin NH, Shelton SE, Fox AS et al. (2005) Brain regions associated with the expression and contextual regulation of anxiety in primates. Biol Psychiatry 58:796–804
Kalin NH, Shelton SE, Fox AS et al. (2008) The serotonin transporter genotype is associated with intermediate brain phenotypes that depend on the context of eliciting stressor. Mol Psychiatry 13:1021–1027

Kalin NH, Shelton SE, Kraemer GW et al. (1983) Corticotropin-releasing factor administered intraventricularly to rhesus monkeys. Peptides 4:217–220
Kalin NH, Shelton SE, Rickman M et al. (1998b) Individual differences in freezing and cortisol in infant and mother rhesus monkeys. Behav Neurosci 112:251–254
Kalin NH, Shelton SE, Snowdon CT (1992) Affiliative vocalizations in infant rhesus macaques (*Macaca mulatta*). J Comp Psychol 106:254–261
Kalin NH, Shelton SE, Takahashi LK (1991) Defensive behaviors in infant rhesus monkeys: Ontogeny and context-dependent selective expression. Child Dev 62:1175–1183
Keverne EB, Martensz ND, Tuite B (1989) Beta-endorphin concentrations in cerebrospinal fluid of monkeys are influenced by grooming relationships. Psychoneuroendocrinology 14: 155–161
King JE, Figueredo AJ (1997) The Five-Factor Model plus Dominance in chimpanzee personality. J Res Pers 31:271–271
Kinnally EL, Lyons LA, Abel K et al. (2008) Effects of early experience and genotype on serotonin transporter regulation in infant rhesus macaques. Genes Brain Behav 7:481–486
Kraemer GW (1997) Psychobiology of early social attachment in rhesus monkeys. Clinical implications. Ann N Y Acad Sci 807:401–418
Kraemer GW, Ebert MH, Schmidt DE et al. (1989) A longitudinal study of the effect of different social rearing conditions on cerebrospinal fluid norepinephrine and biogenic amine metabolites in rhesus monkeys. Neuropsychopharmacology 2:175–189
Lee AS, Gutierrez-Arcelus M, Perry GH et al. (2008) Analysis of copy number variation in the rhesus macaque genome identifies candidate loci for evolutionary and human disease studies. Hum Mol Genet 17:1127–1136
Levine S, Wiener SG (1988) Psychoendocrine aspects of mother-infant relationships in nonhuman primates. Psychoneuroendocrinology 13:143–154
Levine S, Wiener SG, Coe CL et al. (1987) Primate vocalization: A psychobiological approach. Child Dev 58:1408–1419
Liang B, Blizard DA (1978) Central and peripheral norepinephrine concentrations in rat strains selectively bred for differences in response to stress: Confirmation and extension. Pharmacol Biochem Behav 8:75–80
Lopez JF, Higley JD (2002) The effect of early experience on brain corticosteroid and serotonin receptors in rhesus monkeys. Biol Psychiatry 51:100S–100S
Lopez JF, Vazquez DM, Zimmer CA et al. (2001) Chronic unpredictable stress and antidepressant modulation of mineralocorticoid, and glucocorticoid receptors. Soc Neurosci Abstr 27, program 352
Lyons DM, Martel FL, Levine S et al. (1999) Postnatal experiences and genetic effects on squirrel monkey social affinities and emotional distress. Horm Behav 36:266–275
Machado CJ, Emery NJ, Capitanio JP et al. (2008) Bilateral neurotoxic amygdala lesions in rhesus monkeys (*Macaca mulatta*): Consistent pattern of behavior across different social contexts. Behav Neurosci 122:251–266
Mason WA, Capitanio JP, Machado CJ et al. (2006) Amygdalectomy and responsiveness to novelty in rhesus monkeys (*Macaca mulatta*): Generality and individual consistency of effects. Emotion 6:73–81
McCormack K, Newman TK, Higley JD et al. (2009) Serotonin transporter gene variation, infant abuse, and responsiveness to stress in rhesus macaque mothers and infants. Horm Behav 55:538–547
McKinney WT, Jr., Suomi SJ, Harlow HF (1972) Repetitive peer separations of juvenile-age rhesus monkeys. Arch Gen Psychiatry 27:200–203
Mineka S, Davidson M, Cook M et al. (1984) Observational conditioning of snake fear in rhesus monkeys. J Abnorm Psychol 93:355–372
Mineka S, Suomi SJ (1978) Social separation in monkeys. Psychol Bull 85:1376–1400
Mineka S, Suomi SJ, DeLizio R (1981) Multiple separations in adolescent monkeys: An opponent-process interpretation. J Exp Psychol Gen 110:56–85
Munafo MR, Brown SM, Hariri AR (2008) Serotonin transporter (5-HTTLPR) genotype and amygdala activation: A meta-analysis. Biol Psychiatry 63:852–857

Newman TK, Syagailo YV, Barr CS et al. (2005) Monoamine oxidase A gene promoter variation and rearing experience influences aggressive behavior in rhesus monkeys. Biol Psychiatry 57:167–172

Norcross JL, Newman JD (1999) Effects of separation and novelty on distress vocalizations and cortisol in the common marmoset (*Callithrix jacchus*). Am J Primatol 47:209–222

Novak MA, Harlow HF (1975) Social recovery of monkeys isolated for the first year of life: 1. Rehabilitation and therapy. Dev Psychol 11:453–465

Popova NK (2006) From genes to aggressive behavior: The role of serotonergic system. Bioessays 28:495–503

Popova NK, Kulikov AV, Nikulina EM et al. (1991a) Serotonin metabolism and serotonergic receptors in Norway rats selected for low aggressiveness towards man. Aggress Behav 17:207–213

Popova NK, Voitenko NN, Kulikov AV et al. (1991b) Evidence for the involvement of central serotonin in mechanism of domestication of silver foxes. Pharmacol Biochem Behav 40: 751–756

Popova NK, Voitenko NN, Trut LN (1976) Changes in the content of serotonin and 5-hydroxyindoleacetic acid in the brain in the selection of silver foxes according to behavior. Neurosci Behav Physiol 7:72–74

Porrino L, Crane AM, Goldman-Rakic PS (1981) Direct and indirect pathways from the amygdala to the frontal lobe in rhesus monkeys. J Comp Neurol 198:121–136

Raine A, Meloy JR, Bihrle S et al. (1998) Reduced prefrontal and increased subcortical brain functioning assessed using positron emission tomography in predatory and affective murderers. Behav Sci Law 16:319–332

Rogers J, Shelton SE, Shelledy W et al. (2008) Genetic influences on behavioral inhibition and anxiety in juvenile rhesus macaques. Genes Brain Behav 7:463–469

Rosenbaum JF, Biederman J, Hirshfeld DR et al. (1991) Further evidence of an association between behavioral inhibition and anxiety disorders: Results from a family study of children from a non-clinical sample. J Psychiatr Res 25:49–65

Rosenblum LA, Kaufman IC (1967) Laboratory observations of early mother-infant relations in pigtail and bonnet macaques. In: Altmann SA (ed) Social communication among primates. University of Chicago Press, Chicago

Rouff JH, Sussman RW, Strube MJ (2005) Personality traits in captive lion-tailed macaques (*Macaca silenus*). Am J Primatol 67:177–198

Ruchkin VV, Eisemann M, Hagglof B et al. (1998) Interrelations between temperament, character, and parental rearing in male delinquent adolescents in northern Russia. Compr Psychiatry 39:225–230

Sandnabba NK (1996) Selective breeding for isolation-induced intermale aggression in mice: Associated responses and environmental influences. Behav Genet 26:477–488

Scanlan JM (1984) Adrenocortical and behavioral responses to acute novel and stressful conditions: The influence of gonadal status, time course of response, age, and motor activity. University of Wisconsin, Madison

Scanlan JM (1987) Social dominance as a predictor of behavioral and pituitary-adrenal response to social separation in rhesus monkey infants. Society for Research in Child Development, Baltimore.

Scanlan JM (1988) Continuity of stress responsivity in infant rhesus monkeys (*Macaca mulatta*): State, hormonal, dominance, and genetic influences. University of Wisconsin, Madison.

Scanlan JM, Suomi SJ (1986) Social dominance as a predictor of behavioral and pituitary-adrenal response to social separation in rhesus monkey infants. International Society for Developmental Psychobiology, Annapolis

Scanlan JM, Suomi SJ (1988) Neonatal predictors of separation response in rhesus monkeys. Paper presented at the meeting of the American Society of Primatologists, New Orleans, LA

Schneider ML, Suomi SJ (1992) Neurobehavioral assessment in rhesus monkey neonates (*Macaca mulatta*): Developmental changes, behavioral stability, and early experiences. Infant Behav Dev 15:155–177

Schwartz CE, Wright CI, Shin LM et al. (2003) Inhibited and uninhibited infants "grown up": Adult amygdalar response to novelty. Science 300:1952–1953

Scott JP, Fuller JL (1965) Genetics and the social behavior of the dog. University of Chicago Press, Chicago

Scott PA, Cierpial MA, Kilts CD et al. (1996) Susceptibility and resistance of rats to stress-induced decreases in swim-test activity: a selective breeding study. Brain Res 725:217–230

Seay B, Harlow HF (1965) Maternal separation in the rhesus monkey. J Nerv Ment Dis 140:434–444

Sen S, Burmeister M, Ghosh D (2004) Meta-analysis of the association between a serotonin transporter promoter polymorphism (5-HTTLPR) and anxiety-related personality traits. Am J Med Genet B Neuropsychiatr Genet 127B:85–89

Shannon C, Schwandt ML, Champoux M et al. (2005) Maternal absence and stability of individual differences in CSF 5-HIAA concentrations in rhesus monkey infants. Am J Psychiatry 162:1658–1664

Singh R (1992) Ontogeny of behavioral dynamics among free-ranging rhesus monkeys (*Macaca mulatta*) theory and observation. In: Seth PK, Seth S (eds) Perspectives in primate biology. Today & Tomorrow's Printers and Publisher's, New Delhi

Soubrié P (1986) Reconciling the role of central serotonin neurons in human and animal behavior. Behav Brain Sci 9:319–364

Southwick SM, Vythilingam M, Charney DS (2005) The psychobiology of depression and resilience to stress: Implications for prevention and treatment. Ann Rev Clin Psychol 1:255–291

Stallings MC, Hewitt JK, Cloninger CR et al. (1996) Genetic and environmental structure of the Tridimensional Personality Questionnaire: Three or four temperament dimensions? J Pers Soc Psychol 70:127–140

Stevenson-Hinde J, Simpson MJA (1980) Subjective assessment of rhesus monkeys over four successive years. Primates 21:66–82

Stevenson-Hinde J, Stillwell-Barnes R, Zunz M (1980) Individual differences in young rhesus monkeys: Consistency and change. Primates 21:498–509

Stevenson-Hinde J, Zunz M (1978) Subjective assessment of individual rhesus monkeys. Primates 19:473–482

Strome EM, Wheler GH, Higley JD et al. (2002) Intracerebroventricular corticotropin-releasing factor increases limbic glucose metabolism and has social context-dependent behavioral effects in nonhuman primates. Proc Natl Acad Sci USA 99:15749–15754

Suomi SJ (1983) Social development in rhesus monkeys: Consideration of individual differences. In: Oliverio A, Zappella M (eds) The behavior of human infants. Plenum Press, New York

Suomi SJ, Collins ML, Harlow HF et al. (1976) Effects of maternal and peer separations on young monkeys. J Child Psychol Psychiatry 17:101–112

Suomi SJ, Harlow HF, Domek CJ (1970) Effect of repetitive infant-infant separation of young monkeys. J Abnorm Psychol 76:161–172

Suomi SJ, Harlow HF, Novak MA (1974) Reversal of social deficits produced by isolation rearing in monkeys. J Hum Evol 3:527–534

Suomi SJ, Kraemer GW, Baysinger CM et al. (1981) Inherited and experiential factors associated with individual differences in anxious behavior displayed by rhesus monkeys. In: Klein DF, Rabkin J (eds) Anxiety: New research and changing concepts. Raven Press, New York

Suomi SJ, Mineka S, DeLizio RD (1983) Short- and long-term effects of repetitive mother-infant separations on social development in rhesus monkeys. Dev Psychol 19:770–786

Suomi SJ, Scanlan JM, Rasmussen KL et al. (1989) Pituitary – adrenal response to capture in Cayo Santiago – derived group M rhesus monkeys. P R Health Sci J 8:171–176

Thompson WW, Higley JD, Byrne EA et al. (1986) Behavioral inhibition in nonhuman primates: Psychobiological correlates and continuity over time. Paper presented at the meeting of the International Society for Developmental Psychobiology, Annapolis, MD

Urry HL, van Reekum CM, Johnstone T et al. (2006) Amygdala and ventromedial prefrontal cortex are inversely coupled during regulation of negative affect and predict the diurnal pattern of cortisol secretion among older adults. J Neurosci 26:4415–4425

van Oortmerssen GA, Bakker TC (1981) Artificial selection for short and long attack latencies in wild *Mus musculus domesticus*. Behav Genet 11:115–126
van Stegeren AH (2008) The role of the noradrenergic system in emotional memory. Acta Psychol (Amst) 127:532–541
Wahlestedt C, Pich EM, Koob GF et al. (1993) Modulation of anxiety and neuropeptide Y-Y1 receptors by antisense oligodeoxynucleotides. Science 259:528–531
Watson KK, Ghodasra JH, Platt ML (2009) Serotonin transporter genotype modulates social reward and punishment in rhesus macaques. PLoS ONE 4:e4156
Weinstein TAR, Capitanio JP, Gosling SD (2008) Personality in animals. In: John OP, Robins RW, Pervin LA (eds) Handbook of personality: Theory and research. Guilford Press, New York
Weiss A, Adams MJ, Widdig A, Gerald MS (2011) Rhesus macaques (*Macaca mulatta*) as living fossils of hominoid personality and subjective well-being. J Comp Psychol 125:72–83
Weiss JM, West CH, Emery MS et al. (2008) Rats selectively-bred for behavior related to affective disorders: proclivity for intake of alcohol and drugs of abuse, and measures of brain monoamines. Biochem Pharmacol 75:134–159
Westergaard GC, Izard MK, Drake JH et al. (1999a) Rhesus macaque (*Macaca mulatta*) group formation and housing: Wounding and reproduction in a specific pathogen free (SPF) colony. Am J Primatol 49:339–347
Westergaard GC, Mehlman PT, Shoaf SE et al. (1999b) CSF 5-HIAA and aggression in female macaque monkeys: Species and interindividual differences. Psychopharmacology 146:440–446
Wiener SG, Bayart F, Faull KF et al. (1990) Behavioral and physiological responses to maternal separation in squirrel monkeys (*Saimiri sciureus*). Behav Neurosci 104:108–115
Zajicek KB, Higley JD, Suomi SJ et al. (1997) Rhesus macaques with high CSF 5-HIAA concentrations exhibit early sleep onset. Psychiatry Res 73:15–25

Chapter 12
Behavioral Syndromes: A Behavioral Ecologist's View on the Evolutionary and Ecological Implications of Animal Personalities

Andrew Sih

Abstract Recent years have seen an explosion of interest in animal personalities (also known as behavioral syndromes) in a broad range of taxa, where individuals exhibit within-individual behavioral consistency (i.e., individuals have a behavioral type, BT) and different individuals differ consistently in BT across a range of situations or contexts. Most of the study on animal personalities has focused on individual variation in boldness, aggressiveness, activity, or exploratory tendency, with a growing interest in sociability. Because behavioral syndromes clearly influence fitness, often in complex, context-dependent ways, they can have important ecological and evolutionary implications. Perhaps most intriguingly, some studies show that having a BT can be associated with limited plasticity and suboptimal behavior. Here, I report on three frontline areas of interest in the study of behavioral syndromes. First, I review literature on proximate and ultimate approaches to understanding patterns about the existence, stability and structure of animal personalities. I present a new conceptual framework that integrates proximate and ultimate approaches, emphasizing the importance of positive feedback loops between proximate mechanisms and adaptive behavior. Second, I discuss emerging ideas on the social ecology of behavioral syndromes – on how social interactions and social situations influence the expression of BTs, and in turn, how the mix of BTs in a social group influences social dynamics. Finally, I discuss the ecological implications of behavioral syndromes – how individual or species' BTs affect ecological interactions, how variation in BTs or BTs of particular keystone individuals affect population/community level patterns, and most interestingly, how behavioral correlations affect population/community level phenomena. A particularly intriguing example of the latter involves the importance of BT-dependent dispersal in ecological invasions. I conclude with a list of suggestions on exciting areas for future study.

A. Sih (✉)
Department of Environmental Science and Policy, University of California,
One Shields Avenue, Davis, CA 95616, USA
e-mail: asih@ucdavis.edu

A. Weiss et al. (eds.), *Personality and Temperament in Nonhuman Primates*,
Developments in Primatology: Progress and Prospects, DOI 10.1007/978-1-4614-0176-6_12,

12.1 Introduction

While animal behaviorists have long been interested in individual variation in the "personalities" of primates (Yerkes 1939; Stevenson-Hinde et al. 1980), laboratory rodents (Koolhaas et al. 1999), and a few other domesticated animals (Hessing et al. 1993), recent years have seen a rapid growth of interest in studying individual variation in behavioral tendencies (animal personalities, temperament, behavioral syndromes, etc.) in a broad range of other animals (Gosling 2001) including undomesticated mammals (e.g., Réale et al. 2000, 2007; Réale and Festa-Bianchet 2003), birds (e.g., Dingemanse et al. 2003; Duckworth 2006), reptiles (e.g., Stapley and Keogh 2005; Cote and Clobert 2007), amphibians (e.g., Sih et al. 2003), fish (e.g., Huntingford 1976; Bell and Sih 2007), crustaceans (e.g., Pintor et al. 2009), insects (e.g., Hedrick 2000; Sih and Watters 2005; Kortet and Hedrick 2007), spiders (Riechert and Hedrick 1993; Johnson and Sih 2005), and various nonarthropod invertebrates (Sinn et al. 2006, 2008). Although much of the earlier work on "model systems" focused on genetic and neuroendocrine correlates of personality, much of the recent work on other taxa has been from an evolutionary or ecological view. Indeed, three conceptual overview papers emphasizing evolutionary or ecological aspects of personality have each been cited approximately 200 or more times (as of October 2009) and are thus among the most highly cited animal behavior papers of the last few years (Wilson et al. 1994; Sih et al. 2004a, b).

We coined the term "behavioral syndrome" in earlier work (Sih et al. 2004a, b). Since then, many have asked – how is a behavioral syndrome different from an animal personality? Because the term "animal personality" has been used in many different ways (thus it is difficult to pin down a consensus on the exact meaning of the term), it is difficult to precisely pin down the difference between the two terms; however, I use a broadly inclusive definition of the term "behavioral syndrome." A behavioral syndrome involves behavioral consistency, both within and between individuals. Within-individual consistency occurs when individuals behave in a consistent way through time or across situations, i.e., individuals have a behavioral type (BT). Between-individual consistency occurs when individuals differ in their BT, which would be reflected statistically as a behavioral correlation (e.g., a significant rank order correlation across time or across contexts). That is, a behavioral syndrome can be defined and quantified by a behavioral correlation across situations or contexts. In an earlier study (Sih and Bell 2008), we clarified that behavioral syndromes and BTs do *not* have to: (1) be stable over a lifetime, or even over a large proportion of a lifetime; (2) involve a genetic basis; (3) involve both multiple contexts and multiple situations; (4) be independent of social status or condition; (5) involve a dichotomy of BTs (as opposed to a continuous range of BTs), and (6) be associated with suboptimal behavior. They certainly do not require animals to show little or no behavioral plasticity. Although a behavioral syndrome might be more interesting or more important if it has a strong genetic basis, is stable over a lifetime, carries over across multiple contexts, and results in suboptimal behavior, these are not part of the definition of the concept.

While the study of personalities in humans and primates has often focused on the Big Five or analogs of the Big Five (Gosling 2001; Freeman and Gosling 2010; King and Weiss 2011), studies on behavioral syndromes have most often examined boldness, aggressiveness, activity, and exploratory tendencies, with a few studies on sociability (Réale et al. 2007). The notion is that individuals often appear to exhibit behavioral carryovers across contexts in general boldness or aggressiveness, where e.g., some individuals are more bold than others in both the presence and absence of predators (Sih et al. 2003), or some individuals are more aggressive than others in a competitive contest situation, but also in foraging, mating, or unfortunately in parental care contexts. Most intriguingly perhaps, boldness with predators is often correlated with aggressiveness with conspecifics (Huntingford 1976; Riechert and Hedrick 1993; Bell 2005; Johnson and Sih 2007; Stevenson-Hinde and Hinde 2011).

Behavioral ecologists have been interested in these behavioral axes in part because they seem clearly related to evolutionary fitness (survival and reproductive success) that underlies many evolutionary and ecological outcomes. For example, activity, aggressiveness, and boldness are generally thought to be often associated with better competitive ability, but also with a cost of greater exposure to predation and other risks (Stamps 2007; Biro and Stamps 2008). In fact, relatively few studies have actually quantified the connection between BT and fitness, and those that have quantified this have found mixed results (Dingemanse and Réale 2005; Smith and Blumstein 2008). Similarly, while exploratory tendency appears to be associated with an array of ecologically important outcomes (survival, within-pair and extra-pair reproductive success, dispersal), these associations can be gender and environment-dependent (Dingemanse et al. 2003).

Given that behavioral syndromes can have important albeit complex effects on fitness, key issues of interest from an ecological and evolutionary perspective include: (1) using a blend of proximate and ultimate approaches to understand why behavioral syndromes are the way they are; (2) understanding the social ecology of BTs, how BTs relate to social dynamics, and (3) addressing the ecological implications of behavioral syndromes. Here, I provide a progress report on these issues with some reference to their possible applications to primates. It should be noted that I am not a primatologist. My explicit role in this book is to provide a behavioral ecologist's view from outside of primatology.

12.2 Explaining the "Mysteries" of Behavioral Syndromes

The heart of the concept of a behavioral syndrome includes two key points. First, individuals exhibit within-individual consistency in behavior. This at least sometimes implies limited (as compared with unlimited, optimal) behavioral plasticity. Rather than being "free" to exhibit the optimal behavior in all circumstances, an individual's BT limits, at least somewhat, the range of behaviors expressed by that individual. Instead of exhibiting both very bold or very shy behavior depending on what is best for the situation, some individuals tend to be more bold, while others tend to be more shy. Second, behaviors are correlated across contexts; e.g., a tendency to be aggressive with conspecifics might be correlated to a tendency to be bold even when predators are present. In this context, one of the most intriguing insights to emerge from recent work on behavioral syndromes is the possibility that individuals sometimes exhibit apparently suboptimal behavior in particular contexts, in part because of their BT (Sih et al. 2004a, b). Relatively bold individuals are sometimes inappropriately bold in dangerous situations, while shy individuals sometimes miss out on important opportunities that they could have had if they were not so shy (e.g., Sih et al. 2003). Similarly, aggressive individuals that out-compete others are sometimes inappropriately aggressive in contexts where aggressive behavior is not favored (e.g., Johnson and Sih 2005; Duckworth 2006).

The notion that aggressive or bold behavior can generate tradeoffs is familiar and well accepted; however, our thinking on this tradeoff often derives from a clear constraint – e.g., a time budget constraint – that is physically difficult to get around. If, for example, feeding and hiding are mutually exclusive behaviors, then increased time spent feeding (e.g., associated with being bold and active) must mean less time spent hiding and thus more exposure to risks. Sih et al. (2004a, b) called this a negative behavioral correlation due to a time budget conflict. It is easy to understand why animals cannot maximize both their feeding rates and safety at the same time. In contrast, the existence of BTs and behavioral syndromes seem to be associated with suboptimal behavior that is not due to a simple constraint like a time budget conflict. The fact that bolder individuals that are more active (than others) when predators are absent also tend to be inappropriately bold when predators are present is not obviously due to an unavoidable time budget constraint. There is no law of physics that says that an animal could not be highly active when predators are absent, but then shut down as necessary when predators appear. Similarly, the fact that animals that are more aggressive in social interactions also often tend to be bolder at a later time when predators are present is not obviously due to a simple time budget constraint.

In this context, a fascinating question is – why should individuals have a personality at all? If an individual's BT can spillover to cause suboptimal behavior, why have a BT? Why not adjust behavior to be optimal in all circumstances? To emphasize, we are not asking why is a particular individual more bold or shy than others? And, we are not asking what maintains variation in BTs in a population? We are asking the more fundamental question of why do individuals have a BT at all? Do individuals always have a personality, or are some so behaviorally flexible that it would be fair to say that they have no personality? If that is the case, can we explain why individuals differ in their tendency to have a BT?

A related question involves the stability of personality over a lifetime – the strength of correlations between personality at different ages over a lifetime. Again, although, as humans, we are accustomed to the idea that personalities are reasonably stable (Caspi et al. 2005), unless the same BT is favored in all life history stages, having a stable BT can result in suboptimal behavior in some parts of the life history. Most intriguingly, species or populations appear to differ in the stability of their BTs. Some species show evidence of a statistically stable BT over an entire lifetime (e.g., Johnson and Sih 2005), while others appear to exhibit periods of instability, perhaps associated with the onset of maturity (e.g., Bell and Stamps 2004; Carere et al. 2005; Caspi et al. 2005; Roberts et al. 2006). How might we explain this variation in the stability of BTs?

Another related, somewhat more complex question is – what explains the structure of personality? (Uher 2011). What explains which behaviors are correlated and which are not? In humans, the structure of personality can be captured by the 5-Factor Model or the Big Five (Digman 1990). An active area of study in nonhumans focuses on quantifying whether analogs of the Big Five dimensions can be identified in other species (Gosling 2001). What might explain differences among species on whether they exhibit all aspects of the Big Five (Gosling 2001; King and Weiss 2011; Fairbanks and Jorgensen 2011). Some studies suggest that consistency over time in the same behavior is stronger than consistency across contexts in different behaviors (Uher 2011). Why might this be so?

As noted earlier, rather than studying the human-derived Big Five, ecologists often measure activity, boldness, aggressiveness, exploratory tendency, and sociability. Similar issues arise, however, about the structure of behavioral correlations. These BT axes are sometimes correlated; in particular, boldness and aggressiveness are often correlated (see earlier references), but sometimes they are not correlated (e.g., Bell 2005; Bell and Sih 2007; Dingemanse et al. 2007; Moretz et al. 2007). Interestingly, behavioral correlations are sometimes stable over ontogeny, but in other cases, the correlation structure changes over ontogeny (Bell and Stamps 2004). To emphasize, this is a distinctly different issue than the question of whether BTs are stable over ontogeny. Even if individuals change their personalities over time, the population can still retain the same behavioral correlations (the same behavioral syndrome). Given that behavioral correlations can be ecologically important, a key issue is to develop a conceptual framework to explain variation in behavioral correlations.

More broadly, we seek a unified framework that yields insights on all of the above fundamental questions about personality. At their heart, they all involve carryovers (across time or contexts) that imply some mechanism that limits behavioral plasticity. Thus the key is to elucidate a general framework for explaining both the existence of, and variation in behavioral consistency and behavioral correlations. Next, I summarize published ideas and present new ideas on such a framework – some of which emphasize proximate mechanisms (e.g., genetic, neuroendocrinological, or physiological mechanisms), some of which emphasize adaptive, cost–benefit explanations, and finally, one that attempts to blend the two.

12.3 Proximate Approaches

My particular interest here is on how proximate mechanisms might explain the behavioral consistency (across time, across contexts) that lies at the heart of the behavioral syndrome concept. How might an individual's physiology or neuroendocrine profile explain why its behavior at one time in one context carries over to other times and/or other contexts? Logic suggests that one type of mechanism that could limit behavioral plasticity is the connection of a behavior to a proximate "anchor" that inherently exhibits limited plasticity. If an individual's proximate state cannot change or at least, cannot change quickly, then this proximate mechanism is a good candidate for explaining limited behavioral plasticity. Recent literature suggests two, perhaps interrelated, types of proximate mechanisms that might play this role in explaining behavioral syndromes.

The dominant proximate framework for explaining human and animal personalities revolves around hormonal and neuroendocrine mechanisms and related genes. For example, variation in BTs has been associated with variation in levels of dopamine, serotonin, and other neurochemicals (Koolhaas et al. 1999, 2006; Carere et al. 2001; Kalin 2003; Veenema et al. 2006; Overli et al. 2007; Bell et al. 2007) that, in turn, have been associated with variation in specific candidate genes (e.g., serotonin transporters or dopamine receptors; e.g., Reif and Lesch 2003; Munafo et al. 2008). Studies on primates by Capitanio, Sapolsky, Suomi, and others have played a leading role in developing this field (Suomi 1987, 2004; Sapolsky and Ray 1989; Sapolsky 1994; Suomi et al. 1996; Capitanio et al. 1998; Capitanio 1999, 2011; Higley et al. 2011). Relationships between hormones and behavior are complex, with interactions and feedbacks involving multiple neurochemicals, multiple receptors, binding globulins, tissue specificity, etc. In the context of behavioral syndromes, a key point is that neuroendocrine systems are highly plastic. When individuals move from one ecological situation or context to another, hormone levels can respond rapidly (e.g., Wingfield et al. 1990; Bell et al. 2007). To understand how neuroendocrine mechanisms might explain long-term stability of BTs, we need more study on the deeper anchors that limit neuroendocrine plasticity.

An alternative framework for explaining variation in BTs focuses on variation in metabolic physiology (Biro and Stamps 2008; Careau et al. 2008). The assumption is that individuals with higher (basal and active) metabolic rates have both higher energy demands and the capacity to process more resources, and/or compete well for resources. These high metabolic rate individuals should then be both bold and aggressive as two behavioral means for feeding this "geared up" physiology. Being bold and aggressive, however, is also often a risky lifestyle both in terms of exposure to danger and in terms of a higher risk of energy shortfall and starvation. An alternative, perhaps equally successful lifestyle might be a low risk, low return lifestyle (shy and unaggressive) which is associated with lower metabolic rates (Stamps 2007; Biro and Stamps 2008). The metabolic and hormonal approaches are interrelated in the sense that metabolic rates are controlled, in part, by hormones.

A particularly intriguing aspect of the metabolic rate framework is the possibility that variation in metabolic rate is controlled, in part, by variation in the size of metabolic organs (e.g., heart, liver, lungs). Unlike hormone levels or metabolic rate per se, both of which can change rapidly, organ size can only change relatively slowly. Morphological traits such as organ size thus seem *ab initio* to be good candidates for serving as an anchor to explain behavioral consistency. In particular, metabolic organ sizes might be part of a positive feedback loop where the morphology and associated physiology favor a BT that, in turn, maintains differences among individuals in morphology and physiology. Individuals with large metabolic organs have high energy demands and high energy capacities that lead them to be bold and aggressive which garners the high energy intake needed to support the large metabolic organs and associated fast lifestyle. In contrast, individuals with smaller metabolic organs might have lower energy needs (thus less need to be bold and aggressive) and lower ability to compete well for energy (thus less benefit for being bold and aggressive). If low metabolic rate individuals are then shy and unaggressive, they will likely be unable to bring in enough energy to grow larger metabolic organs. To my knowledge, the prediction that bolder, more aggressive individuals should tend to have larger metabolic organs and higher metabolic rates has not yet been examined in primates.

12.4 Adaptive Approaches

An alternative or complementary approach for understanding variation in behavioral syndromes is the cost–benefit adaptive approach. At first, this may seem paradoxical (Neff and Sherman 2004; Sih et al. 2004c). How can it be adaptive to exhibit a stable, consistent BT if in some circumstances that can result in suboptimal performance? Recent theory suggests several main hypotheses all involving either benefits of specialization (on a particular BT), costs, or constraints relative to switching specializations, or both. Here I provide a summary of the main ideas with some reference to primates.

Some attempts to explain behavioral syndromes have emphasized the benefits of behavioral consistency per se. One set of ideas involves the benefits of behavioral consistency (i.e., of being predictable) in systems with repeated social interactions (cooperative or competitive). Being predictable has costs – predictable individuals can be easily cheated; however, in the right social circumstances, where credible threats or promises are favorable enough, behavioral consistency can be favored (Dall et al. 2004; McNamara et al. 2009). In aggressive situations, individuals clearly benefit if they can win contests by simply using threats without engaging in escalated fights. Game theory suggests that threats should work if the threat is backed by superior fighting ability (Maynard Smith 2003). In principle, however, a weaker, but irrationally aggressive individual could cause a rational, stronger competitor to back down (to avoid the costs of a fight) by threatening to escalate fights despite the costs. Such a threat, however, should only work if it is credible; that is, if the over-aggressive individual is indeed predictably aggressive. Similarly, in situations involving reciprocal altruism, individuals should be more likely to do favors for others who are dependably cooperative (who make credible promises to reciprocate). Many primates likely have the repeated social interactions with individual recognition and reputational effects (knowledge of who is aggressive or cooperative) required to fit the above scenarios. If social predictability is indeed a major factor in explaining the existence of behavioral syndromes, species that better fit the required social scenario should exhibit "more personality" than less social species – more behavioral consistency, particularly for BT axes that relate to social interactions. An example might be the difference in personality structure of chimpanzees and orangutans where in chimpanzees the strongest personality factor is dominance, while in orangutans, dominance is only the second strongest factor behind extraversion (King and Figueredo 1997; Weiss et al. 2006; Weiss, personal communication). Further quantitative study of the social benefits of behavioral consistency in primates would be exciting.

A rather different mechanism favoring behavioral consistency per se involves the apparent physiological benefits of maintaining a stable growth rate (Stamps 2007). That is, in some systems, including apparently humans, there are physiological costs and inefficiencies associated with alternating periods of rapid growth with slow growth. Moderation is better than alternating highs and lows. This is an example of what Sih and Bell (2008) called "*status quo selection*" – selection favoring consistency per se. Multiple consistent BTs can then persist in a population as alternative solutions to a growth/mortality tradeoff where bold–aggressive behavior supports faster growth but with the cost of higher mortality risk (Stamps 2007). The idea is that, under the right conditions, high risk–high growth, intermediate risk–intermediate growth, and low risk–low growth might all yield roughly similar overall fitness.

Other hypotheses focus on the possible role of uncertainty in favoring consistency. When Sih et al. (2004a, b) first posed the paradox that having a BT might sometimes, perhaps often, result in suboptimal behavior, they noted that the existing literature on adaptive phenotypic plasticity should be a useful source of ideas. In brief, adaptive plasticity theory predicts that even in a variable environment where different traits are favored in different environments, it can be adaptive for individuals (or genotypes) to exhibit relatively fixed phenotypes if: (1) the benefits of being a specialist in some environment is strong; (2) the cost of switching phenotypes is large; and (3) individuals have poor information (e.g., either imprecise cues or poor cognitive abilities), so they sometimes, perhaps often, "choose" the wrong phenotype for a particular environment. McElreath and Strimling (2006) formally modeled this idea in an animal personality context. Their basic idea can be captured via a stock market analogy. For those of us who find the market quixotic and unpredictable, it might be better to take a fixed tactic (buy and hold, or stay out entirely) rather than be a "day trader" who actively attempts to "buy low and sell high" on a day-to-day basis. In essence, in the absence of good information, it can be unproductive to attempt to adaptively track environmental variation. Instead, pick a strategy, a BT, and stick with it. If individuals differ in a state variable (e.g. assets, energy reserves, condition), some should be better off sticking with a relatively bold BT, while others stick with a relatively shy BT (McElreath and Strimling 2006). This hypothesis predicts that for equal levels of environmental fluctuation (that induce behavioral responses), we should be more likely to see consistent BTs (where some are generally more bold or aggressive than others) in environments that are inherently more unpredictable or uncertain.

While McElreath and Strimling (2006) posited that differences in individual state might explain differences in BT, they did not explicitly consider how individuals might end up with differences in state, or how these differences in state might be maintained. Wolf et al. (2007a, b) recently provided such a model linking BT with the individual's life history state. Their specific model assumed that early in life, some individuals explore more, gain more information about high-quality habitats, and thus have good potential to enjoy high reproductive success later in life (i.e., they have high "assets"), while others explore little and thus have lower assets. Following what Clark (1994) termed the "asset protection principle," high explorers with large assets should be cautious (not bold, not aggressive) to protect their assets, while low explorers should be bolder and aggressive. In this model, life history state (assets) provides the anchor that stabilizes an individual's BT. If differences among individuals in assets are maintained for long periods, then differences in BT should also be stable.

The problem with Wolf et al.'s (2007a, b) hypothesis, however, is that asset protection is inherently a negative feedback mechanism that results in convergence in state (assets) over time, rather than maintenance of differences (Clark 1994; McElreath et al. 2007). Individuals with high assets should be cautious and unaggressive, which should cause their assets to decrease over time. Conversely, individuals with low assets should be bold and aggressive which, assuming that they survive, should increase assets over time. Wolf et al. (2007a, b) suggest that despite the negative feedback inherent in asset protection, their life history-based model can explain behavioral syndromes if: (1) behavioral consistency is only for short periods of time; or (2) behavior (being bold/shy or more vs. less aggressive) has little effect on state (e.g., if new assets are converted to reproduction immediately). While these points might apply to some examples of behavioral syndromes, they do not appear to explain long-term behavioral consistency in other systems, including stable personalities in humans and primates.

In contrast, McElreath et al. (2007) suggested that *positive* feedback mechanisms are a key for explaining behavioral syndromes (also see Wolf et al. 2008). For example, Luttbeg and Sih (2010) modeled a common, positive feedback mechanism – state-dependent safety. High state individuals that are larger, in better condition, more vigorous, or have more energy reserves should often be better able to cope with predators than low state individuals, either by fleeing faster or by being better at defending themselves (Caro 2005). Animals with higher state thus enjoy lower predation risk (than low state individuals) while being bold (i.e., while foraging actively despite the presence of predators). State-dependent safety thus allows high state individuals to be bolder (forage more actively) than low state individuals. If resources are available, this allows high state individuals to maintain their high state. In contrast, less vigorous, low state individuals suffer high risks if they forage boldly. Thus they tend to be cautious which prevents them from collecting the resources required to increase in state. Being in poor condition is not a desirable situation; however, if risk is high, particularly if resource levels are low, the best option can be to be cautious and shy – to hunker down and accept low state as making the "best of a bad situation."

Luttbeg and Sih (2010) modeled the above logic using a state-dependent dynamic programming approach (Houston and McNamara 1999; Clark and Mangel 2000) where individuals go through a series of time steps "choosing" a best behavior (relative boldness) in each time step depending on ecological conditions and their

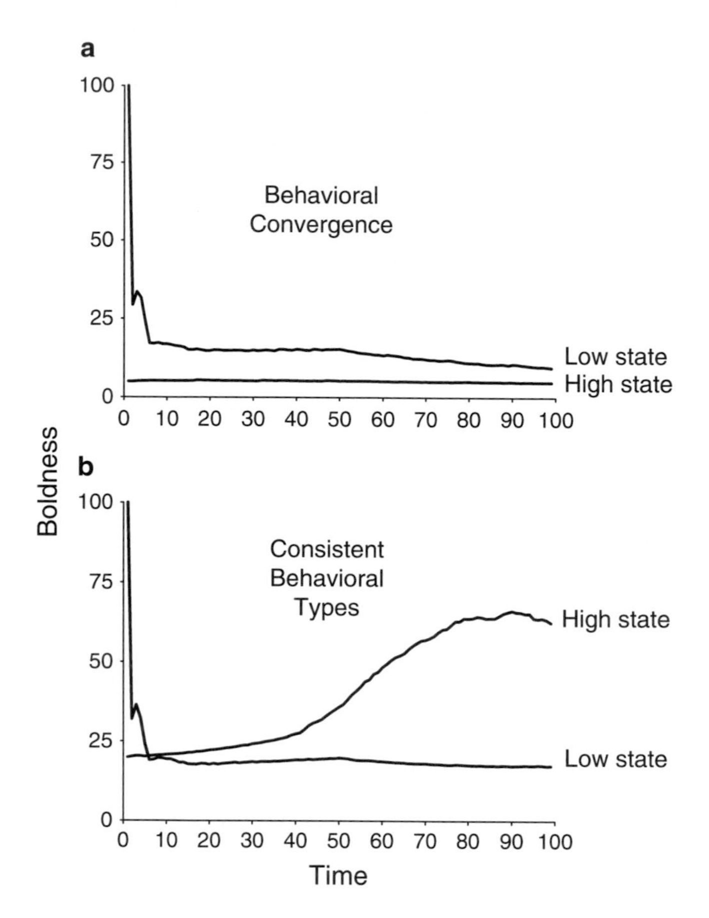

Fig. 12.1 Predicted patterns of optimal boldness over time for individuals that start with high vs. low initial state (e.g., energy reserves, condition). (**a**) With asset protection, but no state-dependent safety, individuals in low state are initially very bold (to get energy to prevent starvation) while those in high state are cautious (to protect their high assets). Over time, behavior converges; i.e., individuals do not exhibit consistent, distinct behavioral types. (**b**) With state-dependent safety, individuals diverge and maintain distinct behavioral types with high state individuals being more bold than low state individuals

current state. Higher boldness exposes animals to higher mortality risk, but also brings in more resources that contribute to future state and future reproductive success. Predation risk varies over time causing immediate changes in adaptive behavior; however, as long as differences in state persist, differences in BT (relative boldness in a given ecological situation) are maintained. The key then is the push–pull between negative feedback (asset protection) tending to cause convergence of state and behavior, and positive feedback (via state-dependent safety) tending to maintain differences in state and BT (Fig. 12.1).

Interestingly, Luttbeg and Sih (2010) found that the long-term persistence of consistent, adaptive behavioral syndromes depends on factors that affect the relative importance of negative vs. positive feedback mechanisms: on resource levels, predation risk, and variation in the strength of state-dependent safety. Long-term stable differences in BT only emerged under moderately favorable conditions where risk and resources were matched; i.e., moderate resources and moderate risk, or high resources offset by high risk, or low resources with low risk. Under highly unfavorable conditions (low resources and high risk), the potential benefits of state-dependent safety are too weak to favor a bold BT, whereas under highly favorable conditions (high resources, low risk), everyone should forage boldly. Stronger state-dependent safety increases the benefit of being bold and thus allows stable BTs under less favorable ecological conditions.

To emphasize, a key insight from the Luttbeg and Sih model is its focus on explaining *variability* in the existence of behavioral syndromes. Rather than simply identifying mechanisms that can result in behavioral syndromes, we predicted both when we should expect to see behavioral syndromes and when we might not. Empirical studies have indeed found variation in the strength and stability of behavioral syndromes (see earlier references). In particular, one study, on a coral reef fish, confirmed that behavioral syndromes were clear-cut under natural conditions (presumed to be moderately favorable), but that an experimental increase in habitat quality resulted in the loss of the syndrome (Snekser et al. 2009). Tests of the effects of habitat favorability on primate personalities would be valuable.

12.5 Integrating Proximate and Adaptive Approaches

Proximate and adaptive approaches can be integrated by using information on proximate mechanisms to define key state variables more realistically in state-dependent models. For example, Luttbeg and I (unpublished model) considered also the situation where resource intake is state-dependent. Higher state individuals have higher resource intake than lower state individuals, either by being faster foragers (in exploitative competition for food) or by winning contests for food (in interference competition). This again results in a positive feedback loop where high state begets bold or aggressive behavior that maintains high state, while low state forces an individual into being shy and unaggressive. Furthermore, Luttbeg and Sih explored the scenario where the key state variable for high resource intake is not energy reserves per se, but is a morphological trait that requires energy to build or maintain (e.g., metabolic organs). Perhaps obviously, this corresponds potentially to the metabolic mechanisms suggested by Careau et al. (2008) and Biro and Stamps (2008). The key is that in this synthetic framework, instead of linking BT and morphology by a uni-directional cause and effect pathway where morphology determines BT, here, BT and morphology respond to each other in an ongoing, adaptive, positive feedback loop. Metabolic organ size is not simply a proximate mechanism that governs BT, instead both BT and organ size are components of an overall integrated phenotype. Organ size influences BT that, in turn, brings in the energy to maintain or further increase organ size that affects future adaptive behavior and so on.

In systems with positive feedback loops, initial conditions often have major impacts on subsequent outcomes. Here, Luttbeg and Sih found that initial state (e.g., energy reserves, condition, or metabolic organ size when individuals begin foraging on their own) often has persistent effects on subsequent BTs. In particular, under a range of ecological conditions, the relationship between initial state and later BT is sigmoid. A range of low values of initial state results in a persistent shy BT. A range of high initial states results in a persistent bold BT. For an intermediate range of initial states, higher initial state results in a bolder BT.

This role of initial state and subsequent state changes then potentially provides a framework for addressing how genetics, maternal effects, and development (experience) might influence BTs. Genetics, maternal effects, and parental care determine an individual's initial state that sets the individual out on an initial trajectory. Later developmental environments along with feedback mechanisms shape further change in BT (e.g., see Fig. 12.1). All else the same, well-provisioned offspring should have high state that results in a bold, aggressive BT, while poorly provisioned offspring should develop a shy, unaggressive BT. The notion that genetics along with a window of early experience might determine an individual's stable BT then emerges not as a genetic/developmental constraint, but as an adaptive outcome of a positive feedback loop (Higley et al. 2011). If positive feedback loops are strong, BTs should be stable unless the individual either experiences a large perturbation of the key state variable, or a persistent change in the rules of state-dependent success that govern BTs. Broadly speaking, this is perhaps just a model-based statement of the obvious – that BTs are more likely to change when individuals go through major life changes – metamorphoses, major dispersal events, windfalls, or disasters. The test of the value of this general framework will come when empiricists better understand specifics about key state variables, and state-dependent cost–benefit dynamics, so modelers can make specific predictions on how genetics, early experiences (shaped often by parents), and later experiences ought to shape short- and long-term BTs.

12.6 Frontiers in the Social Ecology of Behavioral Syndromes

Interestingly, for many of the social behaviors that are among the main topics in behavioral ecology (e.g., mating behavior, parental care, cooperation, mate choice, and learning, including social learning), surprisingly few studies have quantified individual variation in behavioral styles and how they relate to success (Sih and Bell 2008). For example, although there is a large theoretical literature on cooperation/deception, few empirical studies on nonhumans have actually quantified individual variation in cooperative tendency and how that affects social outcomes (Bergmuller and Taborsky 2007; Rutte and Taborsky 2007; Wright 2007). Similarly, although there have probably been thousands of studies on mate choice, the field has only recently begun to emphasize individual variation (primarily, among females) in mate preferences. For example, in some species, some females prefer more aggressive males, while others actually shun more aggressive males (Ophir et al. 2005). Are these differences in female mate preferences associated with differences in the females' personality? Although most studies find variation in mate choosiness (e.g., some females strongly prefer some males over others, while other females are relatively unchoosy), few, if any, studies have looked for a choosiness syndrome. Are some females generally choosier about mates than others (anecdotally we think that it is true in humans), and are females that are choosier about males also choosier in other aspects of their life (e.g., diet choice, habitat choice)? How might choosiness as a BT be correlated with other aspects of BT? For example, are shy, unaggressive individuals more choosy than bold, aggressive ones? Numerous novel issues about "social BTs" or more broadly, about BTs beyond boldness, aggressiveness, and exploratory tendency (Réale et al. 2007) remain relatively unexplored (Sih and Bell 2008).

Thus a major frontier in the field of behavioral syndromes is the study of the social ecology of BTs. How might an individual's behavior and fitness depend not just on its BT, but also on the mix of BTs in its social group? How might the mix of BTs in a social group influence group outcomes such as the pattern of social interactions (the social network, Croft et al. 2008; Wey et al. 2008; Sih et al. 2009), the stability of social structure, sexual selection, or population fitness? For four decades, game theory has been a major tool for analyzing how the mix of BTs in a group influences social dynamics (Dugatkin and Reeve 1998; Maynard Smith 2003). The basic premise of game theory is that fitness is frequency-dependent, i.e., that the relative fitness of different BTs (hawks and doves, cooperators and defectors, producers and scroungers, foragers that prefer habitat A vs. habitat B) should depend on the proportion of the population utilizing each of these strategies. Despite decades of theory, to date, amazingly few empirical studies have actually experimentally manipulated the relative frequency of these BTs to quantify how the mix of BTs actually affects fitness or social dynamics. For social behavioral ecology, one of the most important insights from the recognition that individuals have BTs is that we can identify the BT of each individual, and create groups with different mixes of BTs to quantify impacts on individual and group outcomes. To date, only a few studies have done this in a behavioral syndrome context (e.g., Sih and Watters 2005). My impression talking to scientists at primate centers is that they sometimes have opportunities to create small- to mid-sized groups with known individuals. Experimentally creating different mixes of known personalities, with an eye toward testing game theory should prove insightful.

Studying how "the mix matters" (cf. Sih and Watters 2005) should also tell us more about how individuals differ in how they alter their behavior in response to variation in social conditions (i.e., how they differ in *social plasticity*). The idea here is to find a more realistic middle ground between extreme, perhaps caricatured views of classic behavioral ecology, and behavioral syndromes (Fig. 12.2). Classic behavioral ecology is built on optimality thinking which, in its simplest sense, assumes that individuals have unlimited, optimal behavioral plasticity with no individual variation in BT (Fig. 12.2a). In contrast, a simple caricature of behavioral syndromes posits a world where individuals differ in average BT (e.g., average aggressiveness), but all exhibit similarly limited behavioral plasticity (Fig. 12.2b). To emphasize, studies on behavioral syndromes do not find that animals show no behavioral plasticity – e.g., clearly, all or almost all individuals

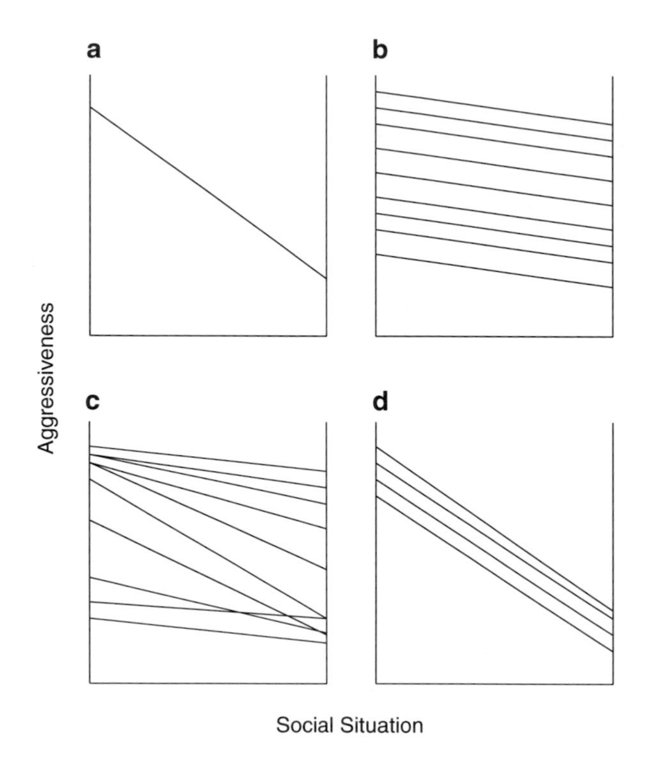

Fig. 12.2 Four hypothetical scenarios on how social situation might influence aggressiveness. Social situations might, e.g., be different social partners that differ in rank, or changes in sex ratio. The key is that optimal aggressiveness decreases as we go from *left* to *right*. (**a**) Simple, optimality-based behavioral ecology predicts that all individuals show the optimal pattern of behavioral plasticity. (**b**) A simple, behavioral syndromes view assumes that individuals differ in behavioral type (BT) where all exhibit the same amount of limited (less than optimal) plasticity, but individuals differ in average aggressiveness. (**c**) A scenario where individuals differ in both average BT and plasticity where more extreme BTs are less plastic. (**d**) A scenario where individuals differ in average BT, but the social situation plays a large role in governing expressed behaviors

change their aggressiveness depending on the social or ecological context; however, the behavioral syndromes' view focuses on the relatively consistent differences in BT above and beyond the individual plasticity (Sih et al. 2004a, b). In reality, individuals usually differ in both average BT and in their plasticity. Although the literature on coping styles looks at individual variation in both average BT and plasticity (Koolhaas et al. 1999, 2006), studies on other aspects of BT have generally not done so. How might average BT and plasticity be inter-related? A plausible hypothesis is that extreme BTs (very bold, very shy) are also low in plasticity, whereas at least some individuals with intermediate BTs might be more plastic (Fig. 12.2c). Additional empirical work is clearly needed on this issue. New empirical insights generated by these new studies should then help to guide future, more realistic game models.

Some studies have found that social situation (and social plasticity) might be more important than inherent BT in determining observed behavior (Fig. 12.2d). For example, in some systems, dominance rank might be the main determinant of a suite of social behaviors – not just contest behavior, but also foraging, mating, antipredator, and/or parental behavior (i.e., a behavioral syndrome) (King and Weiss 2011). As long as the dominance hierarchy is stable, individuals exhibit apparently consistent BTs. However, if an individual changes its rank, it can change its behavior accordingly. This begs the question of whether these individuals truly exhibit a BT. Individuals show behavioral consistency, but it is associated with their social rank. Nelson et al. (2009) suggest that this is either not a behavioral syndrome, or it is only a syndrome in a "weak sense." It is perhaps just a semantic point, but my view is that even when an individual's BT is associated with an underlying mechanism, in this case social rank, it is still a BT. In other cases, the mechanism underlying the BT might be the individual's physiology or hormonal state, or morphology, or energy or life history state. Indeed, for a mechanistic scientist, presumably in all cases, an individual's BT has an underlying mechanism. In all cases, if that mechanism changed, the BT might change. If we accept that hormonally based BTs are real, why not accept that socially governed BTs are real?

In any case, instead of arguing about whether socially governed BTs are real BTs, a more fruitful question might be to understand why in some situations, individuals exhibit strong BTs independent of the social situation (Fig. 12.2b), whereas in other cases, the social situation plays the larger role in governing behavior (Fig. 12.2d). Recent studies on humans suggest that a positive feedback between the social system and BTs might explain two main outcomes. Heine and Buchtel (2009) review the large literature contrasting social systems and personality in Western (primarily, American) and Eastern (primarily Japanese) cultures. In brief, in Western societies, people are more individualistic, and less ruled by social constraints. Social position is more fluid. Personality assays show strong signals of what I call a BT (Fig. 12.2b). These parts fit together. Weak social constraints allow individuals to express a stronger, individual BT, and the fact that people who place high value on their individualism (including their BT) keeps social constraints weak. In contrast, in Eastern societies, social constraints are stronger. People often do not view themselves as having a consistent BT. Instead, they recognize that their behavior is largely situation-dependent. Am I an aggressive person? It depends entirely on whether I am with my peers, my underlings, my boss, or my parents. Breaking the social rules is taboo. In statistical terms, their BTs are more subtle (Fig. 12.2d). Again, the parts fit together. Because social constraints are strong, individuals have weaker BTs which allow strong social constraints to be stable. Of course, I have oversimplified a much richer, nuanced contrast between these cultures and social philosophies. Nonetheless, the literature on humans suggests a potentially general framework where social structure determines the costs and benefits of behavioral consistency vs. social plasticity that shapes the structure of personality that in turn allows the maintenance of stable social structures.

12.7 Ecological Implications of Behavioral Syndromes

The term "ecological implications of behavioral syndromes" could either mean how: (1) differences among individuals in BT affect *individual* performance, e.g., fitness; (2) differences among species in average BT affect *species* interactions and other population/community-level ecological phenomena; (3) *variance in BT* or *particular BTs* within populations affects population/community-level ecological patterns; or (4) within-species *behavioral correlations* affect population/community-level phenomena. All of these topics have been surprisingly under-studied, and thus offer exciting opportunities for important progress on understanding how behavioral syndromes might influence ecological patterns and dynamics.

At the individual level, it is generally thought that more bold or aggressive individuals gain more resources, but at the cost of higher mortality risk (Stamps 2007). Interestingly, however, a meta-analysis found that while some empirical studies support this idea, others do not (Smith and Blumstein 2008). A likely explanation for the variation in results is that fitness effects of BT are often context-dependent; e.g., some field studies suggest that selection on BT depends on food levels, density, or predation risk (Dingemanse et al. 2003, 2007; Réale and Festa-Bianchet 2003). Interestingly, in contrast to the large number of species-level studies manipulating ecological conditions to quantify how ecological context influences the relative performance of different species (with different BTs), to date, few studies have experimentally manipulated ecological conditions to examine effects on individual selection on BT (e.g., Bell and Sih 2007). In addition, relatively few studies have quantified the actual behavioral mechanisms underlying the effects of BT on fitness. Nonetheless, it would be fair to say that the notion that individual BT affects individual fitness (survival and reproductive success) is part of conventional dogma.

An exciting, but relatively under-emphasized area of study on effects of BT on individual fitness involves effects of personality on disease transmission. The likelihood of getting sick depends on both exposure to disease and probability of getting sick per exposure. Both can be affected by an individual's BT. Depending on the disease (or parasite), exposure can be higher for individuals that are more bold or active (i.e., that thus frequently encounter infected individuals or high risk sites), more sociable (for diseases transmitted in affiliative interactions), or more aggressive (e.g., for diseases transmitted in fights). For sexually transmitted diseases, exposure will be presumably higher for BTs that are associated with high mating rates. BT can also affect the likelihood of getting ill per encounter via associations with hormones and the immune system. Capitanio (2011) provides an excellent review of these effects in monkeys. A fascinating feedback between BT and disease involves the fact that diseases often change host behavior; in particular, many parasites apparently manipulate host behavior in ways that facilitate further transmission (Thomas et al. 2005; Fenton and Rands 2006). Thus BT can affect the probability of getting infected which can then alter BT in ways that enhance further passing on of the infection (as well as other BT-dependent ecological and social interactions).

If individual BT affects individual fitness then by extension, the average BT of a species ought to be related to the species' ecological interactions and performance. Indeed, ecologists have long thought (without using behavioral syndrome or personality terminology) that a species' average BT can be ecologically important. Bold (active) species are thought to do well in exploitative competition, but fare poorly with predators (e.g., Sih 1987; Werner and Anholt 1993), whereas aggressive species do well in interference competition. Aggressiveness is also often associated with successful invasive species (Lodge 1993; Higley et al. 2011). In addition, recent work suggests that ability to invade new habitats is also associated with exploratory tendency or lack of neophobia. These personality traits have been associated with behavioral flexibility or innovative behavior in several taxa including primates (Reader and Laland 2002, 2003). In turn, comparative studies on numerous bird or mammal species find that taxa that are innovative (that exhibit more feeding innovations) have higher invasion success (Sol et al. 2002, 2005, 2008). A corollary idea is that exploratory tendency, lack of neophobia, behavioral flexibility, and innovativeness might be generally associated with ability to respond well to novel conditions, e.g., those caused by human-induced rapid environmental change.

Above and beyond the effects of a species' average BT on population/community-level phenomena, several recent ideas focus on ecological effects of the population's variance in BTs, or on the BT of a particular subset of individuals. Population variance among individuals in BT can be important for at least three reasons. First, individual differences in BT represent a key component of within-species niche variation that can significantly reduce intraspecific competition (Bolnick et al. 2003). For the same range of total species' niche breadth, individuals should exhibit weaker intraspecific competition if they have different, specialized BTs as opposed to being generalists all with the same BT. Second, variation among individuals in BT can be important for a species' response to environmental change in the same way that genetic variation is thought to be important for an evolutionary response. That is, even if the species' average BT is poorly adapted to a new environment, if the species has substantial variation in BTs, it might persist because it has a sufficient number of individuals that are well adapted to the new environment. An idea that in a sense combines the other two ideas involves the importance of social/ecological complementarity in enhancing a group's response to environmental change. In socially complex systems, like those of many primates, different BTs might play different complementary roles that are critical for the overall population response to environmental change, as well for population performance, in general. The notion that it is important to actively enhance variation in BTs for species conservation, reintroductions, or zoo management was discussed by Watters and Meechan (2007).

Other ecological implications of behavioral syndromes revolve around the importance of the BT of a subset of the individuals in the population. In many social systems, some individuals have particularly large effects on the rest of the group. Sih and Watters (2005) called these "keystone individuals." Examples include alpha males or females (assuming that they are not only dominant, but also interactive and influential), conflict mediators in pigtailed macaques (Flack et al. 2006), facilitators of social cohesion in dolphins (Lusseau and Newman 2004), hyper-aggressive males in water striders (Sih and Watters 2005), bridge individuals that facilitate information flow between social groups that are otherwise largely unconnected (Burt et al. 1998; Burt 2004), and super-spreaders that, by dint of having an unusual amount of social contacts, facilitate spread of disease (Newman 2003). Social scientists have long recognized, and animal behaviorists have recently begun to champion the value of using social network metrics and theory to quantify the relative importance of different individuals in these dynamics (Wey et al. 2008; Croft et al. 2009; Sih et al. 2009). In the behavioral syndrome context, a key issue that has only received limited attention is the effect of the personality of keystone individuals in shaping overall dynamics (e.g., Burt et al. 1998). A well-known example in primates is Sapolsky's observation in baboons that when highly aggressive males died (due to eating tainted meat) and were survived by more affiliative males, the "personality" and dynamics of the group for more than two decades (Sapolsky 2006).

Finally, several ideas on the ecological implications of behavioral syndromes revolve around behavioral correlations per se, a key defining point of the concept of behavioral syndromes. At a simple level, an individual or species' BT often features correlations between behaviors (expressed in different ecological situations or contexts) that generate tradeoffs (negative performance correlations across situations). For example, individuals or species that are bolder than others in both the presence and absence of predators do well in competition, but poorly with predators. This tradeoff can limit species distributions and abundances (e.g., Sih et al. 2003). While this effect of a behavioral syndrome is important, as discussed earlier, it can be considered merely a minor variation on a well-known theme in ecology – the role of tradeoffs in explaining many patterns on population and community ecology. More interesting, and less explored, is the role of behavioral correlations in generating possible *positive* performance correlations across contexts, and nonrandom filtering of BTs in a spatial context.

Contrary, in a sense, to the general idea of tradeoffs limiting species, invasive pest species appear to be "superspecies" that are good at dispersing, good at colonizing and establishing in new habitats, and have large impacts on native communities, often both by becoming very abundant and by having large per capita impacts. How do they do it all? In many cases, invasive pests appear to be bold and aggressive – thus allowing them to establish well, outcompete similar native species and have large impacts (Kolar and Lodge 2001; Higley et al. 2011), particularly if native predators are ineffective. An interesting paradox is – if invaders are so aggressive, how do they attain super-high densities and maintain high feeding rates at high densities? In some cases, particular population genetic structures appear to result in invaders that exhibit strong inter-specific aggression, but weak intraspecific aggression (Holway and Suarez 1999). In other cases, invasive species exhibit strong inter- and intraspecific aggression and yet still maintain high densities. Recent work suggests that a key might be the fact that a bold–aggressive–active syndrome allows invasive species to be aggressive and yet continue to be active and exhibit high feeding and growth rates despite the aggression (Pintor et al. 2009). In general, comparing the overall behavioral syndrome (not just BT, but the behavioral correlation structure) of invasive vs. noninvasive species should prove insightful.

Another area of growing interest in invasion ecology and spatial ecology, in general, is the possibility of BT-dependent dispersal (a correlation between a BT and dispersal tendency). In several cases, it has been noted (often, anecdotally) that invaders are more bold or aggressive in their invaded range than back home in their native range. This has been attributed to reduced predation risk in the invaded range or evolution in the invaded range; however, growing evidence suggests that it might often be due to nonrandom dispersal by BT. Recent studies have found, e.g., that in great tits, dispersers (relative to stay-at-home residents) tend to have high exploratory tendencies (Dingemanse et al. 2003), in bluebirds, dispersers tend to be more aggressive than others (Duckworth and Badyaev 2007; Duckworth 2008), in killifish (Fraser et al. 2001) and in humans (Whybrow 2005), dispersers tend to be bold, and in the common lizard, they tend to be asocial (Cote and Clobert 2007). The suggested scenario in humans is intriguing. Whybrow (2005) asks in essence – why are Americans as a culture so greedy, rebellious, risk-taking, over-consumptive, and obese? Part of the answer, he suggests, is that America is made up of wave after wave of "maverick migrants," the bold, aggressive individuals from other nations who elected to seek their fortune in America. When you invade a country with these mavericks, you get the American culture. Others are beginning to find parallel scenarios in other invasions (e.g., Duckworth and Badyaev 2007).

The role of BT-dependent dispersal in invasions is one example of the general phenomenon of "selective filtering by BT" where some mechanism (e.g., dispersal) selectively moves some BTs forward while leaving others behind. The key then is that due to behavioral correlations, the filtering, often based on one aspect of BT, carries along with it other aspects of BT. In invasions, it might be that bold individuals are more likely to have no fear of the unknown and to thus disperse. If boldness is positively correlated to aggressiveness, then we have an invasive pest. On the other hand, the prediction is that if boldness and aggressiveness are not positively correlated, the invasion is not as likely to attain pest proportions. Another obvious main filtering mechanism is death due to natural selection or human harvesting. If fishing tends to catch fish that are more bold (Biro and Post 2008), and if bold fish are also aggressive, then fishing pressure can push a fished species toward a more shy, unaggressive personality that affects its social dynamics, sexual selection, population ecology, and species interactions.

12.8 Conclusions

While the study of behavioral syndromes (a.k.a. animal personalities) in other animals has learned much and yet has much to learn from ongoing studies on personalities in humans and nonhuman primates, studies of behavioral syndromes in other animals also offer some perhaps fresh perspectives and ideas for understanding primate (and human) personalities. In particular, I suggest that exciting areas worthy of further consideration include:

1. Contrasting and integrating the role of neuroendocrine and metabolic traits in providing a proximate basis for understanding various issues about BTs and the structure of behavioral syndromes.
2. Testing ideas about how social predictability, environmental (including social) uncertainty, status quo selection, and positive feedback loops involving BTs and various state variables (e.g., energy reserves, vigor, metabolic organ size) might explain behavioral consistency over time and across different ecological contexts.
3. Integrating proximate and adaptive views by using proximate traits as the state variables in adaptive state-dependent models.
4. More study of BTs associated with major social behaviors including mating, cooperation, parental care, and social learning.
5. Testing and developing new game theory on how the mix of BTs in a social group influences individual and group outcomes, and conversely how social systems influence BTs (social plasticity).
6. More work on how variation among individuals in BT, or the BT of particular keystone individuals might influence population performance.
7. The role of BT-dependent dispersal in spatial ecology, in general, and in invasion ecology in particular.

Taken together, the above suggested studies should help further develop and refine an integrative understanding of personalities. Work on primates, valuable in itself, and as a bridge between the still largely separate fields of studying human and animal personalities, should play a critical role in this integration.

References

Bell AM (2005) Differences between individuals and populations of three spined stickleback. J Evol Biol 18:464–473

Bell AM, Backstrom T, Huntingford FA et al. (2007) Variable behavioral and neuroendocrine responses to ecologically-relevant challenges in sticklebacks. Physiol Behav 91:15–25

Bell AM, Sih A (2007) Exposure to predation generates personality in threespined sticklebacks. Ecol Lett 10:828–834

Bell AM, Stamps JA (2004) The development of behavioural differences between individuals and populations of stickleback. Anim Behav 68:1339–1348

Bergmuller R, Taborsky M (2007) Adaptive behavioural syndromes due to strategic niche specialization. BMC Ecol 7:12 doi:10.1186/1472-6785/7/12

Biro PA, Post JR (2008) Rapid depletion of genotypes with fast growth and bold personality traits from harvested fish populations. Proc Natl Acad Sci USA 105:2919–2922

Biro PA, Stamps JA (2008) Life-history productivity is linked to animal personality traits. Trends Ecol Evol 23:361–368

Bolnick DI, Svanback R, Fordyce JA et al. (2003) The ecology of individuals: Incidence and implications of individual specialization. Am Nat 161:1–28

Burt RS, Jannotta JE, Mahoney JT (1998) Personality correlates of structural holes. Soc Networks 20:63–87

Burt RS (2004) Structural holes and good ideas. Am J Sociol 110:349–399
Capitanio JP (1999) Personality dimensions in adult male rhesus macaques: prediction of behaviors across time and situation. Am J Primatol 47:299–320
Capitanio JP, Mendoza SP, Lerche NW (1998) Individual differences in peripheral blood immunological and hormonal measures in adult male rhesus macaques (*Macaca mulatta*): Evidence for temporal and situational consistency. Am J Primatol 44:29–41
Capitanio JP (2011) Nonhuman primate personality and immunity: Mechanisms of health and disease. In: Weiss A, King JE, Murray L (eds) Personality and temperament in nonhuman primates. Springer, New York
Carere C, Welink D, Drent PJ et al. (2001) Effect of social defeat in a territorial bird (*Parus major*) selected for different coping styles. Physiol Behav 73:427–433
Carere C, Drent PJ, Privutera L et al. (2005) Personalities in great tits, *Parus major*: stability and consistency. Anim Behav 70:795–805
Caro T (2005) Antipredator defenses in birds and mammals. Chicago, University of Chicago Press
Careau J, Thomas M, Humphries M et al. (2008) Metabolism and animal personality. Oikos 117:641–653
Caspi A, Roberts BW, Shiner RL (2005) Personality development: Stability and change. Annu Rev Psychol 56:453–484
Clark CW (1994) Antipredator behavior and the asset-protection principle. Behav Ecol 5:159-170
Clark CW, Mangel M (2000) Dynamic state variable models in ecology. Oxford, Oxford University Press
Cote J, Clobert J (2007) Social personalities influence natal dispersal in a lizard. Proc R Soc Lond B Biol Sci 274:383–390
Croft DP, James R, Krause J (2008) Exploring animal social networks. Princeton, Princeton University Press
Croft DP, Krause J, Darden SK et al. (2009) Behavioral trait assortment in a social network: patterns and implications. Behav Ecol Sociobiol 63:1495–1503
Dall SRX, Houston AI, McNamara JM (2004) The behavioural ecology of personality: consistent individual differences from an adaptive perspective. Ecol Lett 7:734–739
Digman JM (1990) Personality structure: emergence of the Five-Factor Model. Annu Rev Psychol 41:417–440
Dingemanse N J, Réale D (2005) Natural selection and animal personality. Behaviour 142:1159–1184
Dingemanse NJ, Both C, van Noordwijk AJ et al. (2003) Natal dispersal and personalities in great tits (*Parus major*). Proc R Soc Lond B Biol Sci 270:741–747
Dingemanse NJ, Thomas DK, Wright J et al. (2007) Behavioural syndromes differ predictably between twelve populations of three-spined stickleback. J Anim Ecol 76:1128–1138
Duckworth RA (2006) Behavioral correlations across breeding contexts provide a mechanism for a cost of aggression. Behav Ecol 17:1011–1019
Duckworth RA (2008) Adaptive dispersal strategies and the dynamics of a range expansion. Am Nat 172:S4–S17
Duckworth RA, Badyaev AV (2007) Coupling of dispersal and aggression facilitates the rapid range expansion of a passerine bird. Proc Natl Acad Sci USA 104:15017–15022
Dugatkin LA, Reeve HK (1998) Game theory and animal behavior. New York, Oxford University Press
Fairbanks LA, Jorgensen MJ (2011) Objective behavioral tests of temperament in nonhuman primates. In: Weiss A, King JE, Murray L (eds) Personality and temperament in nonhuman primates. Springer, New York
Fenton A, Rands SA (2006) The impact of parasite manipulation and predator foraging behavior on predator-prey communities. Ecology 87:2832–2841
Flack JC, Girvan M, de Waal FBM et al. (2006) Policing stabilizes construction of social niches in primates. Nature 439:426–429
Fraser DF, Gilliam JF, Daley MJ et al. (2001) Explaining leptokurtic movement distributions: Intrapopulation variation in boldness and exploration. Am Nat 158:124–135
Freeman, H, Gosling, SD (2010) Personality in nonhuman primates: A review and evaluation of past research. Am J Primatol 71:1–19

Gosling SD (2001) From mice to men: What can we learn about personality from animal research? Psychol Bull 127:45–86
Hedrick AV (2000) Crickets with extravagant mating songs compensate for predation risk with extra caution. Proc Biol Sci 267:671–675
Heine SJ, Buchtel EE (2009) Personality: the universal and the culturally specific. Annu Rev Psychol 60:369–394
Hessing MJC, Hagelso AM, Van Beek JAM et al. (1993) Individual behavioural characteristics in pigs. Appl Anim Behav Sci 37:285–295
Higley JD, Suomi SJ, Chaffin AC (2011) Impulsivity and aggression as personality traits in nonhuman primates. In: Weiss A, King JE, Murray L (eds) Personality and temperament in nonhuman primates. Springer, New York
Holway DA, Suarez AV (1999) Animal behavior: An essential component of invasion biology. Trends Ecol Evol 14:328–330
Houston AI, McNamara JM (1999) Models of adaptive behaviour: an approach based on state. Cambridge University Press, New York
Huntingford FA (1976) The relationship between anti-predator behaviour and aggression among conspecifics in the three-spined stickleback. Anim Behav 24:245–260
Johnson J, Sih A (2005) Pre-copulatory sexual cannibalism in fishing spiders (*Dolomedes triton*): A role for behavioral syndromes. Behav Ecol Sociobiol 58:390–396
Johnson JC, Sih A (2007) Fear, food, sex and parental care: A syndrome of boldness in the fishing spider, *Dolomedes triton*. Anim Behav 74:1131–1138
Kalin NH (2003) Nonhuman primate studies of fear, anxiety and temperament and the role of benzodiazepine receptors and GABA systems. J Clin Psychiatry 64:41–44
King JE, Figueredo AJ (1997) The Five-Factor Model plus Dominance in chimpanzee personality. J Res Pers 31:257–271
King JE, Weiss A (2011) Personality from the perspective of a primatologist. In: Weiss A, King JE, Murray A (eds) Personality and temperament in nonhuman primates. Springer, New York
Kolar CS, Lodge DM (2001) Progress in invasion biology: Predicting invaders. Trends Ecol Evol 16:199–204
Koolhaas JM, Korte SM, De Boer SF et al. (1999) Coping styles in animals: Current status in behavior and stress-physiology. Neurosci Biobehav Rev 23:925–935
Koolhaas JM, De Boer SF, Buwalda B (2006) Stress and adaptation. Curr Dir Psychol Sci 15:109–112
Kortet R, Hedrick A (2007) A behavioural syndrome in the field cricket *Gryllus integer*: Intrasexual aggression is correlated with activity in a novel environment. Biol J Linn Soc 91:475–482
Lodge DM (1993) Biological invasions: Lessons for ecology. Trends Ecol Evol 8:133–137
Lusseau D, Newman MEJ (2004) Identifying the role that animals play in their social networks. Proc Biol Soc (Suppl) 271:S477–S481
Luttbeg B, Sih A (2010) Risk, resources and state-dependent adaptive behavioral syndromes. Phil Trans Roy Soc 365:3977–3990
Maynard Smith J (2003) Evolution and the theory of games. Cambridge University Press, New York
McElreath R, Strimling P (2006) How noisy information and individual asymmetries can make "personality" an adaptation: A simple model. Anim Behav 72:1135–1139
McElreath R, Luttbeg B, Fogarty SP et al. (2007) Evolution of animal personalities. Nature 450:E5
McNamara JM, Stephens PA, Dall SRX et al. (2009) Evolution of trust and trustworthiness: social awareness favours personality differences. Proc Biol Soc 276:605–613
Moretz J, Martins E, Robison B (2007) Behavioral syndromes and the evolution of correlated behavior in zebrafish. Behav Ecol 18:556–562
Munafo MR, Yalcin B, Willis-Owen SA et al. (2008) Association of the dopamine D4 receptor (DRD4) gene and approach-related personality traits: Meta-analysis and new data. Biol Psychiatry 63:197–206
Neff BD, Sherman PW (2004) Behavioral syndromes versus Darwinian algorithms. Trends Ecol Evol 19:621–623

Nelson XJ, Wilson DR, Evans CS (2009) Behavioral syndromes in stable social groups: An artificat of external constraints? Ethology 114:1154–1165
Newman MEJ (2003) The structure and function of complex networks. SIAM Rev 45:167–256
Ophir AG, Persaud KN, Galef BG (2005) Avoidance of relatively aggressive male Japanese quail (*Coturnix japonica*) by sexually experienced conspecific females. J Comp Psychol 119:3–7
Overli O, Sorensen C, Pulman KGT et al. (2007) Evolutionary background for stress-coping styles: Relationships between physiological, behavioral, and cognitive traits in non-mammalian vertebrates. Neurosci Biobehav Rev 31:396–412
Pintor LM, Sih A, Kerby JL (2009) The effect of correlated behaviors and invader density on the impacts of an invasive crayfish. Ecology 90:581–587
Reader SM, Laland KN (2002) Social intelligence, innovation and enhanced brain size in primates. Proc Natl Acad Sci USA 99:4435–4441
Reader SM, Laland KN (2003) Animal innovation. Oxford University Press, New York
Réale D, Festa-Bianchet M (2003) Predator-induced natural selection on temperament in bighorn ewes. Anim Behav 65: 463–470
Réale D, Gallant BY, Leblanc M et al. (2000) Consistency of temperament in bighorn ewes and correlates with behaviour and life history. Anim Behav 60:589–597
Réale D, Martin J, Cottman DW et al. (2009) Male personality, life history strategies and reproductive success in a promiscuous mammal. J Evol Biol 22:1599–1607
Réale D, Reader SM, Sol D et al. (2007) Integrating animal temperament within ecology and evolution. Biol Rev 82:291–318.
Reif A, Lesch KP (2003) Toward a molecular architecture of personality. Behav Brain Res 139:1–20
Riechert SE, Hedrick AV (1993) A test for correlations among fitness-linked behavioural traits in the spider *Agelenopsis aperta*. Anim Behav 46:669–675
Roberts BW, Walton KE, Viechtbauer W (2006) Patterns of mean-level change in personality traits across the life course: A meta-analysis of longitudinal studies. Psychol Bull 132:1–25
Rutte C, Taborsky M (2007) General reciprocity in rats. PLoS Biol 7:1421–1425
Sapolsky RM, Ray JC (1989) Styles of dominance and their endocrine correlates among wild baboons (*Papio anubis*). Am J Primatol 18:1–13
Sapolsky RM (1994) Individual differences and the stress response. Semin Neurosci 6:261–269
Sapolsky RM (2006) Social cultures among nonhuman primates. Curr Anthropol 47:641–656
Sih A (1987) Predators and prey lifestyles: An evolutionary and ecological overview. In: Kerfoot WC, Sih A (eds) Predation: Direct and indirect impacts on aquatic communities. University Press of New England, Hanover
Sih A, Bell AM (2008) Insights for behavioral ecology from behavioral syndromes. Adv Stud Behav 38:227–281
Sih A, Bell AM, Johnson JC (2004a) Behavioral syndromes: An ecological and evolutionary overview. Trends Ecol Evol 19:372–378
Sih A, Bell AM, Johnson JC (2004) Reply to Neff and Sherman. Behavioral syndromes versus Darwinian algorithms. Trends Ecol Evol 19:622–623
Sih A, Bell AM, Johnson JC et al. (2004b) Behavioral syndromes: An integrative overview. Q Rev Biol 79:241–277
Sih A, Hanser SF, McHugh KA (2009) Social network theory: New insights and issues for behavioral ecologists. Behav Ecol Sociobiol 63:975–988
Sih A, Kats LB, Maurer EF (2003) Behavioral correlations across situations and the evolution of antipredator behaviour in a sunfish-salamander system. Anim Behav 65:29–44
Sih A, Watters JV (2005) The mix matters: Behavioural types and group dynamics in water striders. Behaviour 142:1417–1431
Sinn DL, Apiolaza LA, Moltschaniwsky NA (2006) Heritability and fitness-related consequences of squid personality traits. J Evol Biol 19:1437–1447
Sinn DL, Gosling SD, Moltschaniwsky NA (2008) Development of shy/bold behaviour in squid: Context-specific phenotypes associated with developmental plasticity. Anim Behav 75:433–442
Smith BR, Blumstein DT (2008) Fitness consequences of personality: A meta-analysis. Behav Ecol 19:448–455

Snekser JL, Leese J, Ganim A et al. (2009) Caribbean damselfish with varying territory quality: Correlated behaviors but not a syndrome. Behav Ecol 20:124–130
Sol D, Timmermans S, Lefebvre L (2002) Behavioural flexibility and invasion success in birds. Anim Behav 63:495–502
Sol D, Duncan RP, Blackburn TM et al. (2005) Big brains, enhanced cognition, and response of birds to novel environments. Proc Natl Acad Sci USA 102:5460–5464
Sol D, Bacher S, Reader SM et al. (2008) Brain size predicts the success of mammal species introduced into novel environments. Am Nat 172:S63–S71
Stamps JA (2007) Growth-mortality tradeoffs and 'personality traits' in animals. Ecol Lett 10:355–363
Stapley JS, Keogh JS (2005) Behavioral syndromes influence mating systems: Floater pairs of a lizard have heavier offspring. Behav Ecol 16:514–520
Stevenson-Hinde J, Stillwell-Barnes R, Zunz M (1980) Individual differences in young rhesus monkeys. Primates 19:473–482
Stevenson-Hinde J, Hinde CA (2011) Individual characteristics: Weaving psychological and ethological approaches. In: Weiss A, King JE, Murray L (eds) Personality and temperament in nonhuman primates. Springer, New York
Suomi JS (1987) Genetic and maternal contributions to individual differences in rhesus monkey biobehavioral development. In: Krasngor N (ed) Psychobiological aspects of behavioral development. Academic Press, New York
Suomi SJ, Novak MA, Well A (1996) Aging in rhesus monkeys: different windows on behavioral continuity and change. Dev Psychol 32:1116–1128
Suomi SJ (2004) How gene-environment interactions shape biobehavioral development. Lessons from studies with rhesus monkeys. Res Hum Dev 1:205–222
Thomas F, Adamo S, Moore J (2005) Parasitic manipulation: where are we and where should we go? Behav Process 68:185–199
Uher J (2011) Personality in nonhuman primates: What can we learn from personality psychology. In: Weiss A, King JE, Murray L (eds) Personality and temperament in nonhuman primates. Springer, New York
Veenema AH, Blume A, Niederle D et al. (2006) Effects of early life stress on adult male aggression and hypothalamic vasopressin and serotonin. Eur J Neurosci 24:1711–1720
Watters JV, Meehan CL (2007) Different strokes: Can managing behavioral types increase post-release success? Appl Anim Behav Sci 102:364–379
Weiss A, King JE, Perkins L (2006) Personality and subjective well-being in orangutans (*Pongo pygmaeus* and *Pongo abelii*). J Pers Soc Psychol 90:501–511
Werner EE, Anholt BR (1993) Ecological consequences of the trade-off between growth and mortality rates mediated by foraging activity. Am Nat 142:242–272
Wey T, Blumstein DT, Shen W et al. (2008) Social network analysis of animal behaviour: A promising tool for the study of sociality. Anim Behav 75:333–344
Whybrow PC (2005) American mania: When more is not enough. W.W. Norton & Company, New York
Wilson DS, Clark AB, Coleman K et al. (1994) Cautiousness and boldness in humans and other animals. Trends Ecol Evol 11:442–446
Wingfield JC, Hegner RE, Dufty AM et al. (1990) The challenge hypothesis – Theoretical implications for patterns of testosterone secretion, mating systems, and breeding strategies. Am Nat 136:829–846
Wolf M, van Doorn GS, Leimar O et al. (2007a) Life history tradeoffs favour the evolution of personality. Nature 447:581–585
Wolf M, van Doorn GS, Leimar O et al. (2007b) Wolf et al. reply. Nature 450:E5-E6
Wolf M, van Doorn GS, Weissing FJ (2008) Evolutionary emergence of responsive and unresponsive personalities. Proc Natl Acad Sci USA 105:15825–15830
Wright J (2007) Cooperation theory meets cooperative breeding: Exposing some ugly truths about social prestige, reciprocity and group augmentation. Behav Process 76:142–148
Yerkes RM (1939) The life history and personality of the chimpanzee. Am Nat 73:97–112

Index

A. Weiss et al. (eds.), *Personality and Temperament in Nonhuman Primates*, Developments in Primatology: Progress and Prospects, DOI 10.1007/978-1-4614-0176-6,

Printed by Books on Demand, Germany